OPEN TUBULAR COLUMN GAS CHROMATOGRAPHY

OPEN TUBULAR COLUMN GAS CHROMATOGRAPHY

Theory and Practice

MILTON L. LEE
Department of Chemistry
Brigham Young University
Provo, Utah

FRANK J. YANG
Varian Instruments
Walnut Creek, California

KEITH D. BARTLE
Department of Physical Chemistry
University of Leeds
Leeds, United Kingdom

A WILEY-INTERSCIENCE PUBLICATION

JOHN WILEY & SONS

New York   Chichester   Brisbane   Toronto   Singapore

Library of Congress Cataloging in Publication Data:

Lee, Milton L.
    Open tubular column gas chromatography.

    "A Wiley-Interscience publication."
    Includes index.
    1. Gas chromatography.   I. Yang, Frank J.   II. Bartle,
Keith D.   III. Title.

QD79.C45L43   1984   543'.0896   83-14780
ISBN 0-471-88024-8

Printed in the United States of America

10 9 8 7 6 5 4 3 2 1

PREFACE

Gas chromatography with open tubular columns has been used for over two decades, but it has generally been the province of the specialist chromatographer rather than that of the general analyst. Recent advances, however, have brought this technique into much more widespread use. In addition to improvements in sample introduction methods, oven temperature control, detector sensitivity, and electrometer stability and speed, the emergence of flexible fused-silica column technology has been the major factor that has led to its universal acceptance for both special and routine analyses in which high resolution, speed, and sensitivity are required.

For these reasons, it seemed timely to us to bring together in book form the relevant theory, the state-of-the-art instrumentation, the practice, and the application of open tubular column gas chromatography. It was felt that a comprehensive treatment such as this would complement the existing reviews and books already published in this area.

This book is intended to serve as a reference source for those interested in the theoretical and technical background of open tubular column gas chromatography, as well as for analysts concerned with details of the practice of the technique. Emphasis is placed on the reasons for selection of particular instrumentation and operating conditions for selected applications. Both the experienced chromatographer and the newcomer should find this book useful.

We are grateful to a number of friends without whom this book could not have been completed: Douglas W. Later and Bob W. Wright who searched and abstracted the vast literature on the subject and also contributed to the writings of Chapters 3 and 7; Kae Lin Doman who typed the manuscript and Rebecca W. Later who did the artwork; Harold G. Lee and Stephen T. Lee who assisted in proofreading and index compilation; Leslie S. Ettre who provided the stimulus for the initiation of this book; and finally to our families who encouraged us in its completion.

Milton L. Lee
Frank J. Yang
Keith D. Bartle

Provo, Utah
Walnut Creek, California
Leeds, United Kingdom
August 1983

FOREWORDS

When, in the Spring of 1958, I received the preprints of the papers to be presented at the forthcoming Second International Symposium on Gas Chromatography which I planned to attend, I glanced through them to note which presentations I should plan to attend and which, looking less exciting, would give me time to enjoy Amsterdam. The paper on the Theory of Chromatography in Open and Coated Tubular Columns With Round and Rectangular Cross-Sections by Marcel J. E. Golay seemed to belong to the second category. It had no chromatograms, but, instead, contained 93 equations in 15 pages. At that time, my activities mainly concerned practical quantitative determinations by gas chromatography, and thus it was easy to decide to skip that paper for a boat tour in the canals of Amsterdam. Little did I know then that Golay's paper would be the highlight of the Symposium, representing the basis of a fundamentally novel approach to gas chromatography. And I never dreamed that within a few months I would join in the United States the group in which Golay was working and spend the next 25 years trying to understand all of the intricacies of those 93 equations. Every time something comes up, I still find out that it has already been explained 25 years ago by Marcel in that wonderful paper which I missed.

Twenty five years have passed since the Amsterdam Symposium and the birth of open-tubular columns. Today, their application field is vast, encompassing every aspect of analytical chemistry, and it is still growing: It can be safely predicted that, in the not too distant future, most of gas chromatographic separations will be carried out on the columns originally invented by Golay.

In such a constantly changing field it is very important that up-to-date monographs be available for consultation to both the novices and the more experienced chromatographers when they are faced with a new problem. The present book by three very gifted young chromatographers fulfills this requirement. Besides presenting a detailed theoretical background, it deals with the questions related to both column preparation and evaluation and the system in which these columns are employed. Furthermore, it also systematically discusses their application in every important field.

I sincerely hope that the readers of this book find it useful in their day-to-day work and that it helps them in further mastering the use of Golay's columns, introduced just 25 years ago.

LESLIE S. ETTRE

Norwalk, Connecticut
July 1983

It is a pleasure to see a book on open tubular column gas chromatography written with such outstanding thoroughness and balance. Open tubular gas chromatography is growing by leaps and bounds; it is too important to be treated with any less excellence and completeness than provided here. With this book, those who wish to practice the art have at their disposal a thorough guide. However, I am glad to say, they do not have a cookbook.

Cookbooks are wonderful for specific problems but useless when treading on new ground. The scope of open tubular gas chromatography is so great that a cookbook recipe cannot be written for each possible application. The authors have wisely recognized that a theoretical understanding is necessary to cover the new ground.

The theoretical and conceptual foundations of the book are well developed. The chapter on theory provides a sound theoretical basis; it excels in showing how theory can be used to improve separation. The chapter on column technology shows how a column really works—or better, how it can be made to work toward specific objectives.

By the time one gets to the chapters on practical applications, the framework has been established for understanding the techniques described and any necessary variations.

This book represents a quantum jump in the literature of open tubular gas chromatography. I would like to commend its capable authors for their contribution to the field.

J. CALVIN GIDDINGS

Salt Lake City, Utah
September 1983

CONTENTS

ix

OPEN TUBULAR COLUMN GAS CHROMATOGRAPHY

ONE

INTRODUCTION

Because of its so far unsurpassed resolving power, speed, and small sample size requirement, gas chromatography (GC) is one of the most widely applied methods in analytical chemistry. Conventional GC columns up to 5 m long, packed with particles of a solid support which are coated with a film of stationary phase, allow resolution of mixtures containing tens of components. These are of little use, however, in applications where complex mixtures such as air pollution, tobacco smoke, body fluids, and so on, which may contain thousands of different compounds, are encountered. The low permeability of packed columns means that prohibitively high pressures are required to improve significantly the resolution, which depends on the square root of column length.

The proposal of Golay,[1] made in 1957, that the packed column be replaced by a narrow open tube with a thin film of stationary phase on the inner wall, proved to be truly significant. The low resistance to carrier gas flow means that very long columns can be employed, leading to a vast improvement in resolution.

By 1960, column efficiencies between 10^5 and 10^6 were reported for the speedy analyses of volatile and nonpolar mixtures such as light petroleum fractions.[2] Over the next decade, considerable progress was made toward the wider application of open tubular column GC, particularly by refinement of techniques in the areas of column technology (glass columns), derivatization of samples before chromatography to increase volatility, and detection—especially by mass spectrometry.[3]

Such research laid the foundation for the rapid increase in the use of open tubular columns in analyses since 1970. The number of papers devoted to or employing open tubular columns is now rising exponentially (Figure 1.1).

1

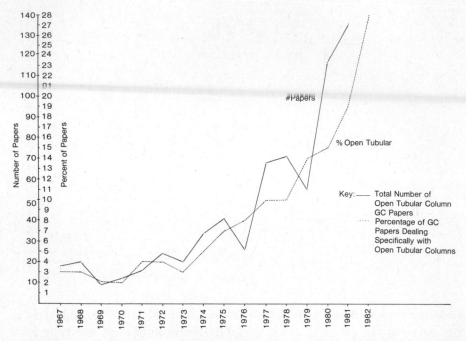

FIGURE 1.1   Number of open tubular column gas chromatographic publications as a function of year. Data compiled from *Chemical Abstracts*.

There is a concomitant increase in the proportion of all GC papers that describe the use of open tubular columns, so that the eventual displacement of the more traditional procedures now seems inevitable.

The virtually complete replacement of metal columns by those fabricated from glass during the 1970s, as the chemistry of the glass surface became better understood,[4] contributed particularly to the expansion of open tubular column GC. Moreover, the current requirements for the trace analysis of environmental and biomedical samples, and the resolution of mixtures such as are met in the chemistry of flavors and fossil fuels are matched by the new column technology. In particular, inert, thermostable, and efficient open tubular columns are available, along with associated selective detectors and injection methods, which allow on-column injection of liquid and thermally labile samples. The development of robust fused-silica columns which have performance generally similar or superior to that of glass columns, but are much more easily handled, is bringing open tubular column GC within the scope of every analytical laboratory.

Variants of conventional wall-coated open tubular columns are porous-layer open tubular columns (PLOT columns) and support-coated open tubular columns (SCOT columns).[5] Here, the stationary phase is coated on a porous layer (generally of particles) on the inner wall of the column. The available surface area is thus increased and the phase ratio is decreased since the total amount of stationary phase is greater. However, the thickness of the coating film in PLOT and SCOT columns remains similar to that of wall-coated columns, so that the advantage of ready diffusion of solute is retained. Since a greater volume of liquid phase has been introduced without increasing the film thickness, there is no extra resistance to mass transfer.

Three general methods are available for the preparation of PLOT and SCOT columns. In the first method, the inside wall is chemically treated, for example, by etching. In the second, a layer of particles may be deposited on the inside of the tube. This can be done by cementing a layer of particles of diatomaceous earth, for example, onto the inner wall with a fusible bonding agent during the capillary drawing process; the PLOT column is then coated with the stationary phase. In the third method, SCOT columns are conventionally prepared by lining the capillary with porous particles during coating, for example by passing through, or evaporating, a suspension of fine diatomaceous earth particles in a solution of the stationary phase. The latter method is probably the most convenient.

SCOT columns generally have properties intermediate between those of packed columns and wall-coated open tubular columns (Table 1.1 and Figure 1.2). Most important, the decreased phase ratio, β, results in (a) larger sample capacities and (b) greater retention of sample components.

The increased sample capacity of SCOT columns is useful in two main respects: (a) the limit of detectability of trace components is improved because of the increased sample size which can be injected; (b) higher column

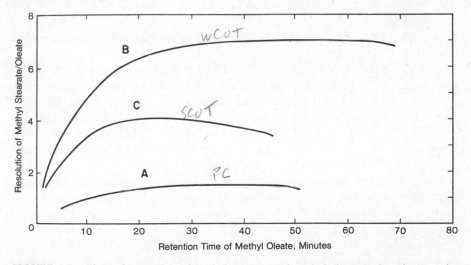

FIGURE 1.2. Plots of resolution of methyl stearate and methyl oleate against the retention time of methyl oleate on (A) a packed column, (B) a wall-coated open tubular column, and (C) a support-coated open tubular column. (Reproduced with permission from ref. 16. Copyright Elsevier Scientific Publishing Company).

TABLE 1.1.   Comparison of wall-coated open tubular, support-coated open tubular, and packed columns.

	Wall-coated open tubular	Support-coated open tubular	Packed
Length (m)	10–100	10–50	1–5
Internal diameter (mm)	0.1–0.8	0.5–0.8	2–4
Liquid film thickness (μm)	0.1–1	0.8–2	10
Capacity per peak (ng)	<100	50–300	10,000
Resolution	High	Moderate	Low

temperatures are necessary but are achievable because of extra stability of the stationary phase layer, so that narrow peaks with consequently greater intensity may be observed. Both of these factors led to a limit of detectability of only 0.36 ppm for an analysis of cyclohexane in high purity n-heptane.[5] The greater amounts of individual components eluted from the SCOT column result in considerable advantages if multiple or less-sensitive detectors are to be used.

The holding back of solutes by the SCOT column is especially useful for resolution of components in the early part of the chromatogram. Since they are eluted over a longer time, detection by a scanning method becomes more feasible. The higher temperatures necessary for SCOT column operation may also be advantageous for volatile compounds for which subambient operation may be required when using a wall-coated open tubular column. Therefore, a number of specialized areas exist in which PLOT and SCOT columns may be preferred over wall-coated open tubular columns, and there is some use in trace analysis and in combination with mass spectrometric and infrared detection.

The principal disadvantages of PLOT and SCOT columns are practical restrictions on length, coupled with the lower efficiencies per unit length which arise from both larger internal diameters and the β values generally used. The application of these columns in those areas where very high resolution is necessary is thus limited. In the following chapters, discussion is limited to wall-coated open tubular columns because of their wide usage and greater potential for resolving complex mixtures.

Of course, the enhanced resolution of open tubular column GC is manifest qualitatively in the narrowness of eluted peaks. For a given phase and temperature, the resolution of two peaks (measured by a resolution factor R) depends on not only the column efficiency and the relative retention but also on the capacity ratio (retention relative to an unretained peak). Baseline resolution is effected for $R = 1.5$, and if suitable selective (e.g., chiral or liquid crystalline) stationary phases are available, separation of isomers or other apparently hard-to-separate pairs may be perfectly possible on a packed column. The increase in the value of R by up to 10 times on an open tubular column, however, means that interposed peaks with retention intermediate between the two can now be resolved: this improvement is vital if complex mixtures are analyzed. For simpler "clean" mixtures, the higher efficiencies available for open tubular columns may also increase R to a value that allows complete separation of two or more peaks unresolved on a packed column.

Both advantages are illustrated in Figure 1.3, which compares the chromatograms of an essential oil on the same methylsilicone stationary phase, but from an open tubular column and a packed column.[6] A number of pairs that are incompletely resolved on the packed column, for example, 1-2, 3-4, and 5-6, are now completely separated on the open tubular column. The large number (approximately 20) of compounds resolved between 6 and 7 on the latter column are represented by two maxima and a shoulder in the packed-column chromatogram.

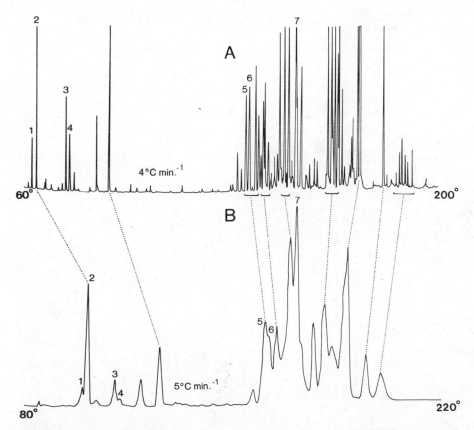

FIGURE 1.3.   Chromatograms of Calmus oil on (A) a 50 m × 0.3 mm i.d. OV-1 glass open tubular column and (B) a 4 m × 3 mm i.d. column packed with 5% OV-1 on 60/80 mesh Gaschrom Q. Both runs were independently optimized. (Reproduced with permission from ref. 6. Copyright Dr. Alfred Huethig Publishers.)

Results from what has been described[7] as the three generations of GC
are illustrated in Figure 1.4. The column material used in early open tubular
column GC work was almost exclusively stainless steel, which allowed dramatic improvement in resolution over that of packed columns. When the
problems associated with coating stationary phases onto glass were even-

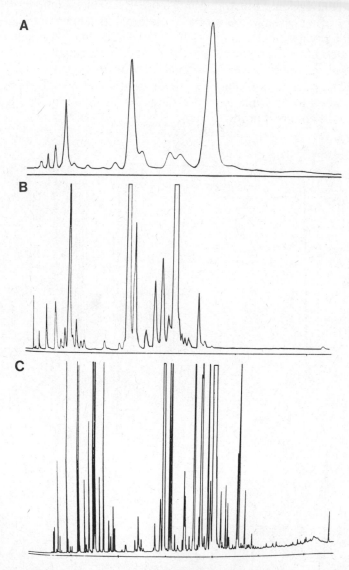

FIGURE 1.4.   Three generations in gas chromatography. Peppermint oil separated on (A) 6 ft
× ¼ in. i.d. packed column, (B) 500 ft × 0.03 in. i.d. stainless steel open tubular column, and
(C) 50 m × 0.25 mm i.d. glass open tubular column. All columns contained Carbowax 20M
stationary phase and were operated under optimized conditions. (Reproduced with permission
from ref. 7. Copyright Preston Publications, Inc.)

tually solved, the further improvement in resolution evident in Figure 1.4 was observed. Similar results are now routinely available with fused-silica columns with the added advantages of increased durability and ease of manipulation.

The evolution of the state-of-the-art open tubular GC method as regards resolution can be traced for a variety of sample types, but the separation of environmental polycyclic aromatic hydrocarbons (PAH) is particularly instructive.[8] The first open tubular column chromatogram of such a mixture was reported in 1964 (Figure 1.5A) on a 35 m glass column coated with a

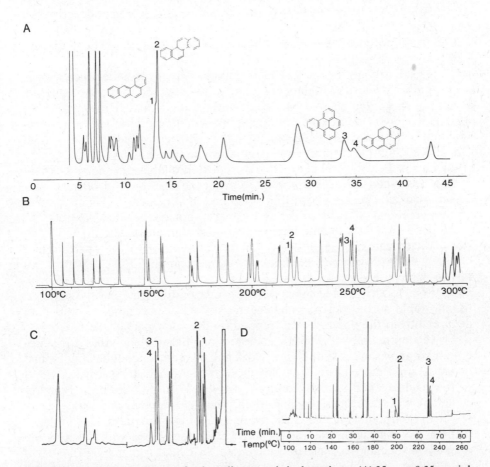

FIGURE 1.5. Chromatograms of polycyclic aromatic hydrocarbons. (A) 35 m × 0.35 mm i.d. glass column coated with SE-30, isothermal at 200°C. (Reproduced with permission from ref. 9. Copyright Elsevier Scientific Publishing Company.) (B) 65 m × 0.30 mm i.d. glass column coated with SE-52, temperature program from 100 to 300°C at 1.8°C min⁻¹. (Reproduced with permission from ref. 10. Copyright Preston Publications, Inc.) (C) 50 m × 0.5 mm i.d. stainless steel column coated with OV-17, isothermal at 270°C. (Reproduced with permission from ref. 11. Copyright Springer-Verlag.) (D) 22 m × 0.26 mm i.d. glass column coated with SE-52, temperature program from 100 to 260°C at 2°C min⁻¹. (Reproduced with permission from ref. 12. Copyright American Chemical Society.)

methylsilicone elastomer, SE-30.[9] The efficiency of separation, as witnessed by the incomplete resolution of the isomer pairs, benz[a]anthracene/chrysene and benzo[e]pyrene/benzo[a]pyrene, is poor. Improved resolution was obtained three years later (Figure 1.5B) on an SE-52 (methylphenylsilicone) coated glass column with temperature programming:[10] the above isomer pairs are somewhat better separated. While there was near baseline resolution of the test pairs on a stainless steel column (Figure 1.5C),[11] this was accompanied by peak tailing, loss of trace components by adsorption, and relatively broad peaks. Significant improvement was obtained during the mid 1970s, when acid leaching of Lewis acids from the glass before coating was shown to improve the efficiency and degree of deactivation of columns.[4] With these columns, extremely narrow and symmetrical peaks were produced for PAH (Figure 1.5D) with resolution of isomer pairs far superior to previous attempts.[12]

A special strength of open tubular column GC lies in its applications in trace analysis, since the limits of detection are extended to the subpicogram level: the sharp and narrow bands not only result in higher resolution, but also have increased intensity. Components present either at low concentration in mixtures or diluted in a sample matrix can be resolved and identified. On packed columns, trace components are either unresolved (especially from a major peak—see Figure 1.6) or are lost in the baseline noise.[13] There is also less bleed of stationary phase during the operation of open tubular column GC as a result of the thin films and lower operating temperatures so that greater detector stability and signal-to-noise levels are possible.

Modern splitless and on-column injection techniques allow the introduction of much greater quantities of dilute solutions onto the open tubular column. Refined ancillary methods for concentrating trace components of mixtures have also been developed. In particular, preconcentration of volatiles from food, polluted air and water, and body fluids and tissues onto thermostable porous polymers has been extremely successful. Mixtures are desorbed from the polymer and trapped on the early part of the column.

Trace compounds are often adsorbed on high-surface-area chromatographic supports in packed-column work and on metal open tubular columns, but are eluted more easily from lower-surface-area, well-deactivated glass or fused-silica open tubular columns. As discussed above, on a properly deactivated surface, it is possible to chromatograph polar and catalytically sensitive compounds at the trace level. Leaching of metal ions from glass surfaces coupled with reaction of surface silanol groups with deactivating reagents, especially silanes, polysiloxanes, or cyclic siloxanes, produces a suitable nonadsorptive open tubular column.[4] Although the surface of fused silica does not contain metal oxides, residual hydroxyl groups must still be deactivated.

Analysis of tobacco smoke is illustrative of progress in trace analysis by GC with open tubular columns.[14] In 1962, 70 peaks were resolved on a stainless steel column in an analysis of the dilute mixture that comprises the gas phase of cigarette smoke. By 1964, this number had increased to 150

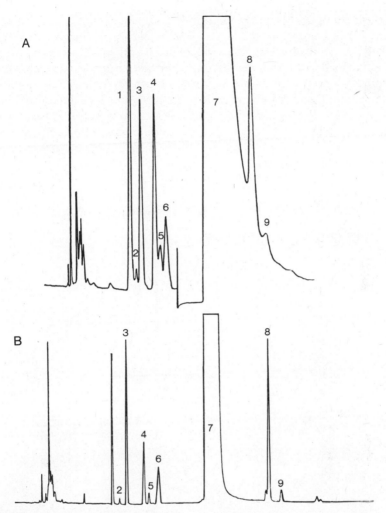

FIGURE 1.6. Chromatograms of styrene impurities (5–200 ppm) on (A) packed column, isothermal at 95°C, 34 min analysis time, and (B) 68 m × 0.28 mm i.d. open tubular column, isothermal at 80°C, 7 min analysis time. The stationary phase for both columns was 1,2,3-tris-(2-cyanoethoxy)propane. (Reproduced with permission from ref. 13. Copyright Dr. Alfred Huethig Publishers.)

peaks, but resolution of reactive trace components could not be increased because of adsorption on the metal surface. The application of glass columns, at first with nonpolar phases, allowed resolution of 300 peaks, many at trace levels, but acidic and basic substances were still strongly adsorbed. More thermostable nonadsorptive glass open tubular columns with polar stationary phases were then produced, which, with injection without stream splitting, yielded 800 peaks for the semivolatiles of cigarette smoke.

The power of the current open tubular column GC methods, with ancillary techniques in trace analysis, is exemplified by the identification of the red fox volatile scent compounds,[15] which have an olfactory communication function. Headspace samples of snow containing urine were concentrated on a porous polymer and then analyzed on a polar phase-coated glass open tubular column with both nitrogen- and sulfur-sensitive detection and combined GC-MS (Figure 1.7). Several compounds were identified as unique products of foxes (both sexes), but quinaldine was found only in male fox urine.

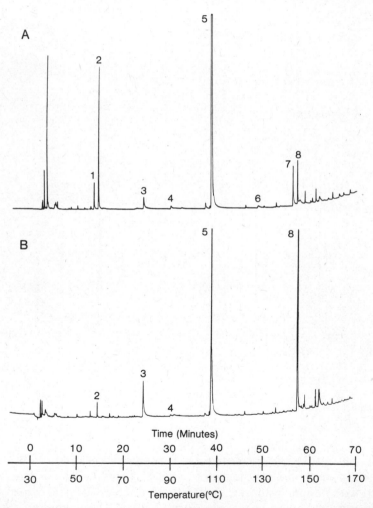

FIGURE 1.7. Chromatograms of urinary volatile compounds in the red fox (*Vulpes vulpes L.*) for (A) male and (B) female. Selected peak identifications: (2) Δ^3-isopentylmethylsulfide; (5) acetophenone; (7) quinaldine; and (8) geranyl acetone. (Reproduced with permission from ref. 15. Copyright American Association for the Advancement of Science.)

TABLE 1.2. Speed of analysis for packed and open tubular columns.[a]

		Packed column	Open tubular columns	
Length (m)		2.4	45.7	2.2
Comparison of retention time for given resolution[b]	Resolution of methyl stearate/ methyl oleate	1.54	7.06	1.54
	Retention time of methyl oleate (min)	35.6	51.7	2.5
Comparison of resolution in given time	Resolution of methyl stearate/ methyl oleate	1.22	5.85	
	Retention time of methyl oleate (min)	15.0	15.0	

[a] Diethylene glycol succinate stationary phase, column temperature 180°C. Data obtained from ref. 16.
[b] At optimum carrier gas velocity.

Because resistance to mass transfer in the liquid phase is much smaller in open tubular columns, the retention time for a given resolution (for samples with high capacity ratios) is always less than for a packed column, so there is a substantial advantage in analysis time. Ettre and March[16] have shown how column performance may be compared by calculating the retention time for a given resolution, by comparing resolution obtained in a given time, or by graphing resolution against time. Their data for the separation of methyl oleate and methyl stearate on diethylene glycol succinate are summarized in Table 1.2.

At the same temperature, base line resolution of these two compounds would take over 30 min on a packed column, whereas only 2.5 min are required on an open tubular column that is only 2.2 m long. If the columns are operated so that the retention times are the same, the test pair is almost five times as well resolved on the open tubular column. The best comparison of column performance is a graph of resolution against retention time (e.g., see Figure 1.2).[16] Again for the methyl stearate/methyl oleate pair, the best separation that can be achieved on a packed column is baseline resolution in about 30 min. For an open tubular column, however, far better resolution is achieved much more quickly. A short open tubular column is more efficient than a packed column and will also result in a shorter analysis time.

Lower column temperatures are possible for a given analysis on open tubular columns as compared to packed columns (e.g., see Figures 1.3 and

1.6), because of the higher phase ratio, β (ratio of volumes of gas and liquid phases), which results in a lower capacity ratio for a given mixture component. There are consequent advantages in ease of operation and stationary phase stability. The use of inert glass and fused-silica column materials has also allowed thinner films with less adsorption and still greater β values. High molecular mass compounds can be eluted from open tubular columns at readily accessible temperatures. For example, coronene (relative molecular mass = 300 daltons) can be chromatographed below 250°C on short (<20 m) glass open tubular columns, whereas over 300°C is required for elution from a packed column.[8]

The high resolution and reproducibility of sampling and retention data now routinely available in GC with open tubular columns mean that the chromatogram of a complex mixture contains detailed fine structure comprising a rich data set eminently suitable for use as a fingerprint. Increasing use is being made of open tubular column chromatography in a number of fingerprinting areas, especially in the analysis of environmental and forensic samples, flavors, and body fluids (see Chapter 7). Many polluting emissions and chemical spillages are complex mixtures, while single pollutants may degrade to many derivatives. Correlation of patterns with source, such as petroleum-derived mixtures in the environment, is common on the basis of open tubular column chromatograms, while this method is the standard method of identifying fire accelerants in arson cases. Food flavors can also be correlated with chromatographic profiles. Metabolic profiling—the diagnosis of disease from the detection of characteristic individual compounds or patterns of compounds in physiological fluids—is a particularly rapidly expanding area of application of open tubular column fingerprinting.

The simultaneous use of selective as well as universal detectors may provide additional fingerprints. For example, while homologous series of hydrocarbons dominate the flame ionization detector chromatogram of the nonpolar fraction of sewage effluent, electron capture detection reveals a second characteristic profile—this time of the chlorine containing compounds.[17] As many as four detectors may be simultaneously coupled to an open tubular column. Such advances, along with the increasing availability of computer capability for the storage and handling of large quantities of data, are allowing the combination of fingerprinting with pattern recognition techniques for the automatic screening and identification of profiles.

Several books have previously been published concerning various aspects of open tubular column GC.[18–22] The recent tremendous growth in areas such as fused-silica column technology, sample introduction techniques, multidimensional chromatography, and chromatographic applications has established a need for an integrated and in-depth compilation of the modern practice of open tubular column GC. In addition, a comprehensive treatment of the theory of this powerful technique and the relationship of theory to actual practice has long been needed. The following chapters in this book are the result of an attempt to bring together both the theory and practice of open tubular column GC.

REFERENCES

1. M. J. E. Golay, in *Gas Chromatography* (*1957 Lansing Symposium*), V. J. Coates, H. J. Noebels, and I. S. Fagerson, editors. Academic Press, New York, 1958, p. 1.
2. D. H. Desty, A. Goldup, and B. H. F. Whyman, *J. Inst. Petrol.* **45**, 287 (1959).
3. M. Novotny, *Anal. Chem.* **50**, 16A (1978).
4. M. L. Lee and B. W. Wright, *J. Chromatogr.* **184**, 234 (1980).
5. L. S. Ettre and J. E. Purcell, *Adv. Chromatogr.* **10**, 1 (1974).
6. K. Grob and G. Grob, *J. High Resoln. Chromatogr./Chromatogr. Commun.* **2**, 109 (1979).
7. W. Jennings, *J. Chromatogr. Sci.* **17**, 637 (1979).
8. M. L. Lee and B. W. Wright, *J. Chromatogr. Sci.* **18**, 345 (1980).
9. A. Liberti, G. P. Cartoni, and V. Cantuti, *J. Chromatogr.* **15**, 141 (1964).
10. N. Carugno and S. Rossi, *J. Gas Chromatogr.* **5**, 103 (1967).
11. G. Grimmer and H. Bohnke, *Z. Anal. Chem.* **261**, 310 (1972).
12. M. L. Lee, K. D. Bartle, and M. Novotny, *Anal. Chem.* **47**, 540 (1975).
13. G. Schomburg, *J. High Resoln. Chromatogr./Chromatogr. Commun.* **2**, 461 (1979).
14. K. Grob, *Chem. Ind.* (*London*), 248 (1973).
15. J. W. Jorgenson, M. Novotny, M. Carmack, G. B. Copland, S, R. Wilson, S. Katona, and W. K. Whitten, *Science* **199**, 796 (1978).
16. L. S. Ettre and E. W. March, *J. Chromatogr.* **91**, 5 (1974).
17. K. Grob, *Chromatographia* **8**, 423 (1975).
18. L. S. Ettre, *Open Tubular Columns in Gas Chromatography*. Plenum Press, New York, 1965.
19. W. Jennings, *Gas Chromatography with Glass Capillary Columns,* second edition. Academic Press, New York, 1980.
20. R. R. Freeman, editor, *High Resolution Gas Chromatography,* second edition. Hewlett-Packard, Avondale, Pennsylvania, 1981.
21. W. G. Jennings, *Comparisons of Fused Silica and Other Glass Columns in Gas Chromatography*. Huethig, Heidelberg, 1981.
22. W. G. Jennings, editor, *Applications of Glass Capillary Gas Chromatography*. Marcel Dekker, New York, 1981.

TWO

THEORY AND POTENTIAL

2.1 INTRODUCTION

The importance of chromatographic theory in the practice of open tubular column GC is (a) to state the principles of retention and resolution of sample components in the column, (b) to relate the fundamental thermodynamic parameters to the measurable chromatographic parameters, and (c) to examine the origins and implications of solute zone spreading so that chromatographic performance, instrumentation, and techniques can be optimized.

This chapter relates the basic chromatographic terms and definitions to the principles of peak retention and resolution. The interrelationships between a chromatographic peak and various column parameters are discussed. Rather than their mathematical importance, the practical usefulness of the chromatographic parameters and the implications of the zone spreading processes are stressed. The excellent works of Giddings,[1] Purnell,[2] and Golay[3] are recommended for those interested in pursuing the theoretical aspects in more depth.

The theoretical potential of open tubular columns is explored in terms of column efficiency, resolving power, speed of analysis, and peak capacity. The practical considerations of flow rate, inlet pressure, column sample capacity, sample volume, sampling time, sample detectability, sample concentration or mass range, and detector cell volume and time constant are also discussed.

2.2 PRINCIPLES OF PEAK RETENTION AND RESOLUTION

The retention and resolution of sample components (solutes) in open tubular column partition GC result from the differential distribution (partition) of

14

the solutes between the stationary liquid and the mobile gas phases. As a result of the solution–dissolution process of the solute molecules into and out of the stationary liquid phase, solute retention and resolution in the column are obtained. The magnitude of retention depends on the partition coefficient (K) which is defined as the ratio of the concentrations of solute in the stationary and mobile phases:

$$K = \frac{\text{solute concentration in the stationary phase}}{\text{solute concentration in the mobile phase}} \qquad (2.1)$$

The larger the value of the partition coefficient for a sample component, the higher the solubility and the longer the retention of that component in the stationary phase. An insoluble gas sample component (i.e., $K = 0$) migrates through the column with the velocity of the mobile phase and is eluted in one column gas phase volume (V_G). V_G can be measured either by effective column cross-sectional area multiplied by column length (i.e., $A_c \times L$) or by the corrected elution volume of a nonretained solute (V_M^o) which is given by

$$V_G = A_c L = V_M^o = F_c t_m j = V_M j \qquad (2.2)$$

where F_c is the mobile phase volumetric flow rate measured at the end of the column at a given chromatographc temperature; t_m is the elution time of a nonretained solute; V_M is the retention volume corresponding to the gas holdup time (t_m); and j is the James–Martin gas compressibility correction factor[4]

$$j = \frac{3}{2} \frac{P^2 - 1}{P^3 - 1} \qquad (2.3)$$

where P is the ratio of column inlet to outlet pressures, that is,

$$P = \frac{P_i}{P_o} \qquad (2.4)$$

A sample component that is soluble in the liquid stationary phase migrates through the column with an average zone velocity slower than the average velocity of the mobile phase and is eluted in more than one column gas phase volume, V_G. The retention volume (V_R) can be experimentally measured from the flow rate and the retention time (t_R):

$$V_R = F_c t_R \qquad (2.5)$$

The corrected retention volume (V_R^o) for a retained sample can then be calculated by using Equation (2.6):

$$V_R^o = V_R j \qquad (2.6)$$

The partition coefficient for a sample component can be related to V_R° and the volume of the stationary liquid phase volume in the column (V_L) by

$$K = \frac{V_R^\circ - V_G}{V_L} \tag{2.7}$$

The numerator in Equation (2.7) is defined as the adjusted or true retention volume (V_N) which is a measure of the net elution gas volume corresponding to the time (net retention time, t_N) the solute spends in solution in the stationary liquid phase in the column.

Equation (2.7) is the fundamental equation of gas–liquid partition column chromatography and is of significant theoretical importance because it relates the thermodynamic parameter K to the column retention parameters. Using Equation (2.7), we can also relate the partition coefficient to the column phase ratio (β) and partition (or capacity) ratio (k) by

$$K = \frac{V_G}{V_L} \left(\frac{V_R^\circ - V_G}{V_G} \right) = \beta k \tag{2.8}$$

where

$$\beta = \frac{V_G}{V_L} \tag{2.9}$$

and

$$k = \frac{V_R^\circ - V_G}{V_G} = \frac{V_R - V_M}{V_M} = \frac{V_N}{V_M} \tag{2.10}$$

In practice, K is constant for a given solute in a given stationary phase at a given temperature. Equation (2.8) indicates that a given column with a large β has a small k and vice versa. Generally, thick film columns have low phase ratios and long retention times for sample components. Low partition ratios and short retention times can be achieved by decreasing the stationary phase film thickness (i.e., increasing β) or by increasing column temperature (T_c) to decrease K according to Equation (2.11):

$$\ln K = - \frac{\Delta G^\circ}{\mathscr{R} T_c} \tag{2.11}$$

Here, ΔG° is the change in Gibbs free energy for the evaporation of solute from the stationary liquid phase. $\mathscr{R}$ is the gas constant.

The difference in the Gibbs free energies for the evaporation of solute components from the stationary liquid phase results in differential migration rates for individual solutes. The relative retention (α) for any two solute components can be expressed in terms of the ratio of either partition coefficient or partition ratio:

$$\alpha = \frac{K_2}{K_1} = \frac{k_2}{k_1} \tag{2.12}$$

Equation (2.12) gives the relative retention of two solute components with respect to their zone center of gravity. However, the effectiveness of chromatographic separation also depends on the control of solute zone spreading in order to avoid overlap. Detailed theoretical treatments of the dynamics of zone spreading have been given by Giddings[1] and Golay.[3] A brief discussion of the origin of zone spreading is now given.

2.3   ORIGIN OF ZONE SPREADING

Since the effectiveness of separation is determined not only by the relative retention of solute components but also by the control of zone spreading in order to minimize peak overlap, an understanding of the origin of zone spreading in an open tubular column is useful for the optimization of column technology, instrumentation, and chromatographic performance. The following discussion focuses on three major sources of solute zone spreading in an open tubular column; namely, (a) longitudinal molecular diffusional spreading, (b) resistance to mass transfer zone spreading in the mobile phase, and (c) resistance to mass transfer zone spreading in the stationary phase. The excellent monographs by Giddings[1] and Purnell[2] are recommended for more detailed treatments.

Longitudinal Molecular Diffusional Spreading (h_1)

Solute zone spreading due to longitudinal molecular diffusion occurs in both the mobile and stationary phases. The extent of diffusional spreading can be expressed in terms of peak variance, σ_l^2 (in length units), t_m, and t_N according to Einstein's equation[5]

$$\sigma_l^2 = 2D_G t_m \tag{2.13}$$

and

$$\sigma_l'^2 = 2D_L t_N \tag{2.14}$$

Here, D_G and D_L are solute molecular diffusion coefficients in the mobile and the stationary phases, respectively. If D_L is assumed to be 10^{-5} cm^2 s^{-1}, the variance contributed by molecular diffusional spreading in the stationary phase is normally negligible, except when t_N is so large that $\sigma_l'^2$ becomes important.

If the solute zone spreading per unit column length is defined as the height equivalent to an average theoretical plate (HETP or $\bar{h}$), the plate height (h_1) which originates from longitudinal molecular diffusional zone spreading in the mobile phase in an open tubular column can be written as

$$h_1 = \frac{\sigma_l^2}{L} = \frac{2D_G}{\bar{u}} \tag{2.15}$$

where the average mobile gas velocity ($\bar{u}$) equals Lt_m^{-1}.

According to Equation (2.15), h_1 is reduced as $\bar{u}$ is increased, or as D_G is decreased. This implies that the use of a low-diffusivity mobile phase (e.g., N_2) at high linear velocity could minimize the extent of zone spreading due to longitudinal molecular diffusion in the mobile phase. However, as will be seen shortly, high-viscosity mobile phases (i.e., low D_G) and high linear velocities may not be desirable if zone spreading due to mobile phase resistance to mass transfer dominates.

Zone Spreading Due to Resistance to Mass Transfer in the Mobile Phase (h_2)

Zone spreading also occurs while the sample migrates down the column due to the perturbation of solute molecular diffusion in the radial direction of the column by the nonuniform flow profile of the mobile phase. For a circular cross-sectional open tubular column, the mobile phase flow profile is parabolic under laminar flow conditions. Sample molecules diffusing into the center of the column cross-section are transported more rapidly down the column than those diffusing into the zone of slow local velocity near the column wall.

The extent of zone spreading in terms of HETP (h_2) due to the pertubation of nonuniform local mobile phase velocity in the lateral direction of the column is given[3] by Equation (2.16):

$$h_2 = \frac{r^2(1 + 6k + 11k^2)\bar{u}}{24D_G(1 + k)^2} \tag{2.16}$$

where r is the column radius. This indicates that h_2 is effectively independent of k, if k is greater than 10, since the function $(1 + 6k + 11k^2)/(1 + k)^2$ is essentially constant for large k values. At $k = 10$, Equation (2.16) can be rewritten as

$$h_2 = \frac{0.4r^2\bar{u}}{D_G} \tag{2.17}$$

Both Equations (2.16) and (2.17) indicate that h_2 is reduced by increasing D_G or by decreasing $\bar{u}$. The selection of the mobile phase and its velocity depends on the relative importance of h_1 and h_2. In general, when h_1 is the dominant zone spreading term, a low-diffusivity mobile phase and a high mobile phase linear velocity are preferred. For fast analysis at high $\bar{u}$, where h_2 becomes the major source of zone spreading, a gas with low viscosity and high diffusivity is desirable. Equation (2.16) or (2.17) also indicates that h_2 is reduced rapidly by decreasing the column radius. The theoretical potential and practical considerations of microbore open tubular columns are discussed in Sections 2.6 and 2.7.

Zone Spreading Due to Resistance to Mass Transfer in the Stationary Phase (h_3)

Since the open tubular column GC process discussed here is partition chromatography, diffusion-controlled mass transfer in the stationary liquid phase could become important if thick-film columns are used. Assuming that the stationary phase has a uniform film thickness (d_f), the transport of a solute into and out of the stationary liquid phase depends on D_L and K. Zone spreading in terms of h_3 is given[3] by Equation (2.18):

$$h_3 = \frac{k^3}{6K^2(1 + k)^2} \frac{r^2 \bar{u}}{D_L} \qquad (2.18)$$

In practice, $r \gg d_f$, and K can be written from Equation (2.8) as

$$K = \beta k = \frac{rk}{2d_f} \qquad (2.19)$$

By substituting Equation (2.19) into Equation (2.18), h_3 can be rewritten as

$$h_3 = \frac{2kd_f^2 \bar{u}}{3(1 + k)^2 D_L} \qquad (2.20)$$

Equation (2.20) indicates that zone spreading due to diffusion-controlled mass transfer in the stationary phase depends on d_f^2. By decreasing d_f, the contribution to zone spreading due to the resistance to mass transfer in the stationary liquid phase can be greatly reduced.

The relative importance of the stationary phase and the mobile phase mass transfer contributions to zone spreading is expressed in terms of the plate height ratio as

$$\frac{h_3}{h_2} = \frac{4D_G k}{\beta^2 D_L(1 + 6k + 11k^2)} \qquad (2.21)$$

As illustrated in Figure 2.1, if $D_G/D_L = 10^4$ and a maximum value for $k/(1 + 6k + 11k^2) = 0.079$ (at $k = 0.301$) is assumed, the ratio of h_3/h_2 is below 0.01 for $\beta \geq 550$. For $\beta \leq 100$, zone spreading due to h_3 can be as important as that contributed by h_2.

By increasing the partition ratio to 10, as is also shown in Figure 2.1, the ratio h_3/h_2 can be reduced because the value of $k/(1 + 6k + 11k^2)$ decreases. This implies that for a column of given β, the contribution to zone spreading of a well-retained solute ($k \gg 1$) from the resistance to mass transfer in the stationary phase (h_3) is less significant than that of h_2. In general, h_3 increases as β decreases. For low values of β and k, zone spreading due to h_3 predominates; for high values of β, h_2 dominates.

The sum of h_1, h_2, and h_3 is the average zone spreading or plate height for a solute peak in an open tubular chromatographic column with no gas compressibility effect ($\bar{h}$).

$$\bar{h} = h_1 + h_2 + h_3 = \frac{2D_G}{\bar{u}} + \frac{r^2(1 + 6k + 11k^2)\bar{u}}{24D_G(1 + k)^2} + \frac{2kd_f^2 \bar{u}}{3(1 + k)^2 D_L} \qquad (2.22)$$

A graphical representation of the effect of mobile phase linear velocity on plate heights, h_1, h_2, and h_3, is given in Figure 2.2. As indicated, h_1 decreases with increasing linear velocity since a shorter diffusion time is

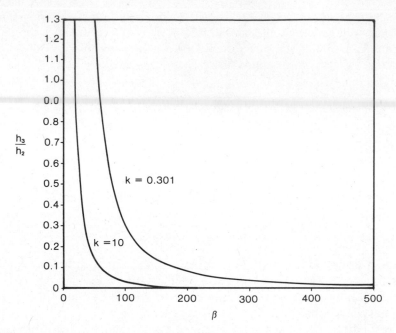

FIGURE 2.1. Variation of the ratio of the stationary phase to mobile phase mass transfer contribution to solute molecular zone spreading with β for solutes with capacity ratios of 0.301 and 10.

allowed. Figure 2.2 also shows that h_2 increases with respect to an increase in $\bar{u}$, because an increase in linear velocity increases the differences in flow stream velocities in the lateral direction of the column. h_3 is also increased as the mobile phase velocity increases. However, the slope of a plot of h_3 against $\bar{u}$ is normally small for high-efficiency thin-film columns. The sum of all three zone spreading contributions is also plotted against $\bar{u}$ in Figure 2.2. A minimum plate height ($\bar{h}_{\min}$) that corresponds to the optimum mobile phase average velocity ($\bar{u}_{\mathrm{opt}}$) and the maximum column efficiency is also indicated. In practice, mobile phase velocities higher than $\bar{u}_{\mathrm{opt}}$ are often used to reduce analysis time.

The above plate height equations are expressed in terms of D_G and $\bar{u}$. In practice, the solute–mobile phase binary molecular diffusion coefficient at the outlet pressure (normally at ambient pressure) of the column (D_G°) is easier to obtain. D_G and $\bar{u}$ can be related to values at the outlet of the column (D_G° and u_o) by

$$\bar{P}D_G = P_o D_G^{\circ} \qquad (2.23)$$

$$\bar{P}\bar{u} = P_o u_o \qquad (2.24)$$

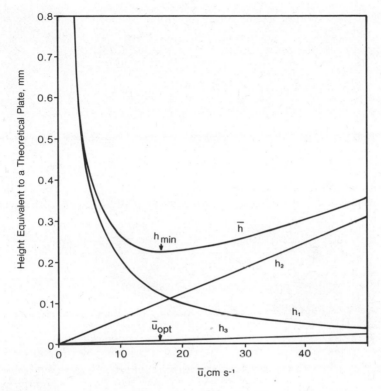

FIGURE 2.2. Graphical representation of the effect of mobile phase linear velocity on plate heights, h_1, h_2, h_3, and the total plate height, $\bar{h}$. $D_G = 0.1$ cm^2 s^{-1}, $D_L = 10^{-5}$ cm^2 s^{-1}, $d_f = 0.5$ μm, $k = 10$, and $r = 0.0125$ cm.

Here $\bar{P}$ and P_o are the average and the outlet mobile phase pressures for the column, respectively.

Substituting Equations (2.23) and (2.24) into (2.22), $\bar{h}$ can be expressed in terms of D_G° and u_o as

$$\bar{h} = \frac{2D_G^\circ}{u_o} + \frac{r^2(1 + 6k + 11k^2)u_o}{24D_G^\circ(1 + k)^2} + \frac{2kd_f^2\bar{u}}{3(1 + k^2)D_L} \qquad (2.25)$$

For columns with large internal diameters and short column lengths, for which the mobile phase pressure at the inlet is not much higher than the outlet, u_o can be approximated by $\bar{u}$. However, as column internal diameters are decreased (≤ 200 μm) and column lengths are increased, a moderate mobile phase pressure gradient across the column is required to obtain a desirable linear velocity. This high pressure at the column inlet compresses the mobile phase volume to a fraction of that present at the low-pressure column outlet. Since the same mass of gas must flow through each cross-section of the column in unit time, the linear velocity is therefore highest where the pressure is lowest (at the column outlet). $\bar{u}$ is related to the outlet velocity of the column by the factor j [Equation (2.3)]:

$$\bar{u} = ju_o \qquad (2.26)$$

$\bar{h}$, corrected for the velocity variation along the flow direction of the column due to the gas compressibility effect, can be expressed as

$$\bar{h} = \frac{2D_G^\circ j}{\bar{u}} + \frac{r^2(1 + 6k + 11k^2)\bar{u}}{24D_G^\circ(1 + k)^2 j} + \frac{2kd_f^2\bar{u}}{3(1 + k)^2 D_L} \qquad (2.27)$$

A more complete average plate height equation, which also accounts for zone spreading due to gas decompression as it migrates down the column, is given[6] in Equation (2.28):

$$\bar{h} = \frac{2D_G^\circ jf}{\bar{u}} + \frac{r^2(1 + 6k + 11k^2)\bar{u}f}{24D_G^\circ(1 + k)^2 j} + \frac{2kd_f^2\bar{u}}{3(1 + k)^2 D_L} \qquad (2.28)$$

Here, f is Giddings' plate height correction factor:

$$f = \frac{9}{8} \frac{(P^4 - 1)(P^2 - 1)}{(P^3 - 1)^2} \qquad (2.29)$$

Equation (2.28) gives more accurate theoretical plate height values for microbore open tubular columns than those calculated with Equations (2.22) and (2.25). It should be used for more precise theoretical evaluation of the performance of microbore (≤ 200 μm) open tubular columns, particularly, when f approaches 1.125 and $j \ll 1$.

If $\bar{h}$ is plotted against $\bar{u}$ using Equation (2.28), a hyperbolic curve such as that shown in Figure 2.2 can be obtained. $\bar{h}_{min}$, which corresponds to the maximum column efficiency can be written as

$$\bar{h}_{min} = 2 \left(\frac{f^2 r^2(1 + 6k + 11k^2)}{12(1 + k)^2} + \frac{4}{3} \frac{kd_f^2 jf D_G^\circ}{(1 + k)^2 D_L} \right)^{1/2} \qquad (2.30)$$

Equation (2.30) indicates that as d_f decreases and the pressure gradient across the column increases (i.e., j decreases), the second term, which represents the combined zone spreading effects of longitudinal molecular diffusion and the resistance to mass transfer in the stationary phase, becomes negligible due to the small values of d_f and j. This implies that $\bar{h}_{min}$ can be expressed by Equation (2.31) for thin-film microbore columns if $p \gg 1$

$$\bar{h}_{min} = 1.125r \left(\frac{1 + 6k + 11k^2}{3(1 + k)^2} \right)^{1/2} \qquad (2.31)$$

The optimum average velocity, $\bar{u}_{opt}$, corresponding to $\bar{h}_{min}$, is given by

$$\bar{u}_{opt} = 4D_G^\circ(1 + k)j \left(\frac{3fD_L}{r^2(1 + 6k + 11k^2)fD_L + 16kd_f^2 D_G^\circ j} \right)^{1/2} \qquad (2.32)$$

The term $16kd_f^2 D_G^\circ j$ approaches zero as d_f and j values are reduced for thin-film microbore columns. $\bar{u}_{opt}$ is thus written as

$$\bar{u}_{opt} = \frac{4D_G^\circ j(1 + k)}{r} \left(\frac{3}{1 + 6k + 11k^2}\right)^{1/2} \tag{2.33}$$

Equation (2.33) indicates that $\bar{u}_{opt}$ is proportional to the factor j and is inversely proportional to r. Since a small r gives a small value of j, $\bar{u}_{opt}$ for microbore columns is not increased as the column internal diameter is decreased (see Figure 2.14).

2.4 CHARACTERIZATION OF A CHROMATOGRAPHIC PEAK AND THE RELATED COLUMN PARAMETERS

Chromatographic peaks are normally characterized by their retention parameters and peak resolution. The two most frequently used retention parameters are V_R and t_R. V_R is a measure of the mobile phase volume flowing through the column from the time the sample is injected into the column inlet to the time the sample is detected at the column outlet. t_R is the sample elution time corresponding to the retention volume. In practice, t_R is more commonly used than V_R because of convenience in direct measurement and real-time chromatographic data acquisition with a computer data system. For a nonretained solute, the retention time corresponding to V_G is defined as t_m:

$$t_m = \frac{V_G}{\bar{F}_c} \tag{2.34}$$

Here, $\bar{F}_c$ is the average mobile phase flow rate which is related to F_c by

$$\bar{F}_c = F_c j \tag{2.35}$$

Equation (2.34) can also be written in terms of L and $\bar{u}$ as

$$t_m = \frac{L}{\bar{u}} \tag{2.36}$$

For a retained solute, the retention time can be related to t_m and k by substituting Equations (2.34), (2.35), and (2.10) into (2.5).

$$t_R = t_m(1 + k) \tag{2.37}$$

By rearranging Equation (2.37), k can then be measured from t_R as

$$k = \frac{t_R - t_m}{t_m} = \frac{t_N}{t_m} \tag{2.38}$$

By using Equation (2.38), α can be measured from retention time data according to Equation (2.39):

$$\alpha = \frac{t_{N_2}}{t_{N_1}} \tag{2.39}$$

A typical chromatogram defining retention time data and peak width parameters is shown in Figure 2.3.

As mentioned above, chromatographic peaks are also characterized by the separation of two adjacent peaks in terms of resolution (R),[7,8] which is defined as the ratio of the retention time difference (Δt) to the average peak base width for two adjacent peaks $(W_{b_1}$ and $W_{b_2})$.

$$R = \frac{\Delta t}{(W_{b_1} + W_{b_2})/2} \tag{2.40}$$

In practice, we may assume $W_{b_1} = W_{b_2}$ for two closely spaced peaks, and thus

$$R = \frac{\Delta t}{W_{b_2}} \tag{2.41}$$

For a Gaussian peak, W_{b_2} can be related to the width at half height (W_{h_2}) the standard deviation (σ_2) by

$$W_{b_2} = 1.70 W_{h_2} \tag{2.42}$$

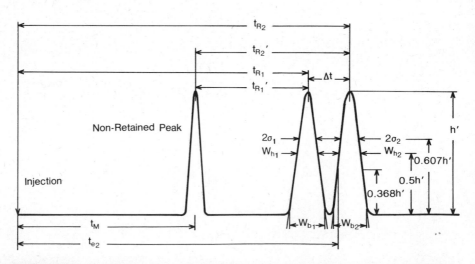

FIGURE 2.3. Chromatographic retention and zone broadening parameters—Gaussian distribution model.

and

$$W_{b_2} = 4\sigma_2 \tag{2.43}$$

R can therefore be expressed in terms of W_{h_2} and σ_2 by

$$R = \frac{\Delta t}{1.70 W_{h_2}} \tag{2.44}$$

$$R = \frac{\Delta t}{4\sigma_2} \tag{2.45}$$

Satisfactory resolution requires $R \geq 1$, that is, $\Delta t \geq 4\sigma_2$. Baseline resolution is obtained when $R \geq 1.5$ (i.e., $\Delta t \geq 6\sigma_2$).

Δt, as discussed in Section 2.2, is determined by the difference in the partition coefficients of two solutes in the stationary phase, and it increases in proportion to the distance migrated along the column. The width of the peak in terms of σ_2 increases with the square root of the distance migrated. As shown in Figure 2.4, for solute pairs that have more unequal solubilities in the stationary phase, Δt_1 rises rapidly and intercepts the 4σ curve after only a short migration along the column. Other pairs that have nearly equal solubilities have a more gradual increase in Δt_2, and, therefore, satisfactory resolution (i.e., $\Delta t_2 \geq 4\sigma$ or $R \geq 1$) is not achieved unless a long column (e.g., $L = 100$ m in this example) is used. R increases with $L^{1/2}$, and thus satisfactory resolution (i.e., $R \geq 1$) can always be reached if a sufficiently

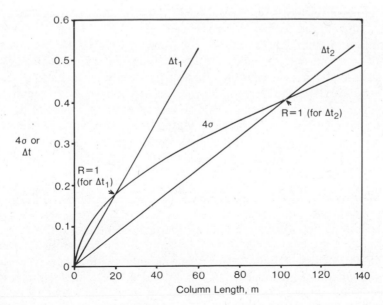

FIGURE 2.4. Graphical representation of chromatographic resolution with respect to column length. $D_G = 0.1$ cm^2 s^{-1} and $\bar{u} = 20$ cm s^{-1}.

long column can be used. However, a long column may pose difficulties in solute detection (too dilute) and may require long analysis times.

Using Equations (2.12), (2.37), (2.39), and (2.45), R can be expressed as

$$R = \frac{1}{4} \frac{t_{R_2}}{\sigma_L} \left(\frac{\alpha - 1}{\alpha} \right) \left(\frac{k}{1 + k} \right) \qquad (2.46)$$

The term, t_{R_2}/σ_2 equals the square root of the total column theoretical plate number (n) which is written as

$$n = \frac{L}{h} \qquad (2.47)$$

In terms of n, Equation (2.46) can be rewritten as

$$R = \frac{1}{4} n^{1/2} \left(\frac{\alpha - 1}{\alpha} \right) \left(\frac{k}{1 + k} \right) \qquad (2.48)$$

This equation is of considerable importance, since it relates resolution to the column efficiency, the fundamental distribution parameter, α, and the peak retention parameter, k. Column parameters, such as length and inner diameter, required for a given resolution of a solute pair can be calculated from Equation (2.48). Furthermore, the theoretical potential in terms of resolving power of an open tubular column can also be predicted (see Section 2.6).

In addition to the use of "resolution" (R) for characterizing column resolving power, the terms "effective peak number" (EPN),[9] "Trennzahl" (separation number, SN),[10] and "peak capacity" (N')[11] are also frequently used for column performance characterization and comparisons. The first two parameters measure the number of compounds that can be separated between two standard compounds such as two consecutive members of a homologous series. The peak capacity measures the number of peaks with unit resolution ($R = 1$) which can be eluted from a column between partition ratio $k = 0$ and a given k value. The peak capacity for an open tubular column of n theoretical plates can be written[11] as

$$N' = 1 + \frac{\ln(1 + k)}{\ln(n^{1/2}/2 + 1) - \ln(n^{1/2}/2 - 1)} \qquad (2.49)$$

This equation allows direct comparison of the potential of different columns for complex sample analysis. For example, under optimum chromatographic conditions, a 25 m $\times$ 250 μm i.d. open tubular column can resolve about 200 components eluted between $k = 0$ and $k = 10$ with unit resolution between each adjacent peak, in comparison to a peak capacity of 143 obtained with a 25 m $\times$ 500 μm i.d. open tubular column.

In the case of EPN, the number of peaks with baseline resolution ($R = $

1.5) between two standard compounds is calculated from Equation (2.50):

$$EPN = \frac{2(t_{R_2} - t_{R_1})}{W_{b_1} + W_{b_2}} - 1 \qquad (2.50)$$

SN requires $R = 1.177$ and is obtained from Equation (2.51):

$$SN = \frac{t_{R_2} - t_{R_1}}{W_{h_2} + W_{h_1}} - 1 \qquad (2.51)$$

In practice, both EPN and SN are useful for comparing the separation power of columns for the same pair of compounds eluted in the same retention range. The dependence of EPN and SN on the retention parameters has been recognized by many researchers.[12-14] The use of EPN and SN requires that the retention parameters, such as k and the retention index range, be indicated. Figure 2.5 shows a typical relationship between SN and k for the n-alkanes eluted from a nonpolar column (SE-30) at a given temperature (110°C) and mobile phase velocity (19.3 cm s^{-1}). The separation number increases rapidly as k increases from 0 to 5, and approaches a constant SN value when the partition ratio is above 10.

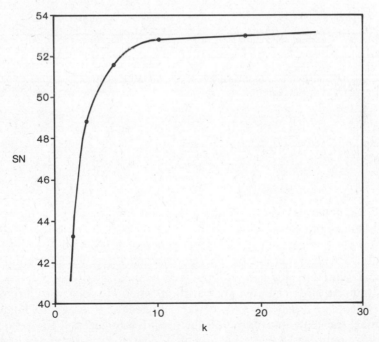

FIGURE 2.5. Graph of measured separation number against capacity ratio for n-alkanes eluted from a 50 m × 0.22 mm i.d. SE-30 fused-silica column. The indicated capacity ratios are for the first component of the pairs for the SN measurements. Conditions: He carrier gas at 19.3 cm s^{-1}, 110°C column temperature, 200:1 split sampling.

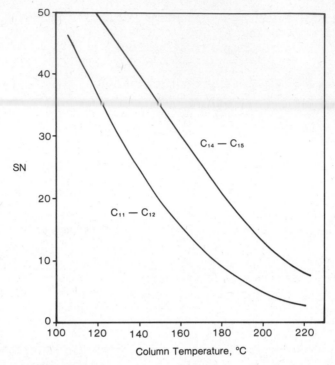

FIGURE 2.6. Graph of separation number vs. temperature for the pairs C_{11}–C_{12} and C_{14}–C_{15} *n*-alkanes eluted from a 50 m × 0.22 mm i.d. SE-30 fused-silica column. The range of partition ratios for the pair C_{11}–C_{12} was between 0.1 and 3.1, and for the pair C_{14}–C_{15} was between 0.25 and 18.3 for the temperature ranges measured. Conditions: He carrier gas, 30 psig column inlet pressure, 200:1 split sampling.

SN also depends[12] on the temperature, the column length, the nature of the stationary phase, and the mobile phase velocity. The effect of temperature on SN is the result of the dependence of K, N, and α on T_c. Figure 2.6 shows a typical relationship between SN and T_c for two adjacent *n*-alkane homologs eluted from a nonpolar column: a linear relationship exists for the lower temperature range, where partition ratios are relatively large and N does not change drastically between two adjacent homologs. At higher temperatures, the partition ratios for solutes are reduced and the relationship becomes nonlinear due to the fact that N is highly dependent on small changes in K. Figure 2.6 also shows that the separation numbers measured for the same pair of *n*-alkanes are reduced with respect to an increase in T_c. The higher the temperature and the smaller the k, the smaller is SN.

2.5 EXPERIMENTAL DETERMINATION OF THE NUMBER OF THEORETICAL PLATES AND HETP

For a Gaussian peak, the retention time, t_R, and the peak broadening parameter, σ, can be used[15] to calculate n:

$$n = \left(\frac{t_R}{\sigma}\right)^2 \tag{2.52}$$

In practice, W_h or W_b is more convenient to measure than σ. In terms of W_h and W_b, the number of theoretical plates can also be written as

$$n = 5.545 \left(\frac{t_R}{W_h}\right)^2 \tag{2.53}$$

and

$$n = 16 \left(\frac{t_R}{W_b}\right)^2 \tag{2.54}$$

For an asymmetric peak, the retention time (at the center of gravity for the peak) may deviate considerably from the peak apex. A more reliable expression for n as suggested by Glueckauf[16] and Bohemen and Purnell[2,17] is

$$n = \frac{t_R t_e}{\sigma^2} \tag{2.55}$$

Here, t_e is the retention time at $1/e$ of the peak height, that is, $0.368\ h'$ as shown in Figure 2.3.

There is little difference between using t_R^2 and $t_R t_e$ for determining n if n is large as in open tubular column GC. The percentage difference can be calculated from

$$\frac{\Delta n}{n}\ \% = \frac{t_R - t_e}{t_e}\ 100\% \tag{2.56}$$

Column efficiency is also given by

$$\bar{h} = \frac{L}{n} \tag{2.57}$$

and

$$\bar{h} = \frac{\sigma_l^2}{L} \tag{2.58}$$

where

$$\sigma_l = \frac{\sigma \bar{u}}{1 + k} \tag{2.59}$$

The smaller the $\bar{h}$ value, the higher the column efficiency for the same column length.

It is evident that the expressions given above for n and $\bar{h}$ do not represent the effective resolving power and separation efficiency of the column. This is because only the effective retention time (t_N) contributes to the column resolving power and separation efficiency. Therefore, the effective theoretical plate number (N) is also often used.[18,19]

$$N = \left(\frac{t_N}{\sigma}\right)^2 \qquad (2.60)$$

Since $t_N = t_m k$ and $t_R = t_m(1 + k)$, the relationship between N and n can be written as

$$N = n \left(\frac{k}{1 + k}\right)^2 \qquad (2.61)$$

Equation (2.61) indicates that N has the greatest deviation from n when measured for slightly retained components with low k values. An example in which the values of n and N are compared for a typical open tubular column is given in Figure 2.7. It can be seen that n and N approach each

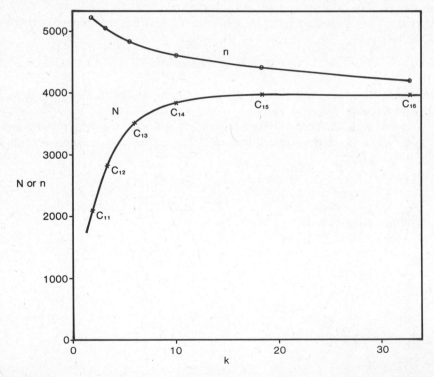

FIGURE 2.7. Graph of measured column theoretical plate number (n) and effective theoretical plate number (N) per meter against capacity ratio for n-alkanes from C_{11} to C_{16}. Conditions: 50 m $\times$ 0.22 mm i.d. SE-30 fused-silica column, He carrier gas at 19.3 cm s^{-1} linear velocity, 110°C column temperature, 200:1 split sampling.

other as k increases. Figure 2.7 clearly indicates that the effective separation efficiency and resolving power of an open tubular column can only be measured for well-retained compounds (i.e., high k values).

Since R is proportional to $n^{1/2}[k/(1 + k)]$, that is, $N^{1/2}$, from Equation (2.48), Figure 2.7 also indicates that the resolving power of open tubular columns is poor for slightly retained components. In general, open tubular column GC is most suitable for complex and well-retained compounds. For the analysis of gas mixtures, k values must be increased by increasing the stationary liquid phase loading or reducing the column temperature in order to increase retention and improve peak resolution.

2.6   THEORETICAL POTENTIAL

Open tubular column technology has advanced sufficiently to allow the preparation of columns with internal diameters as small as 30 μm.[18,20,21] With microbore open tubular columns, separation efficiency, resolving power, speed of analysis, and peak capacity can be greatly increased for the analysis of very complex and trace samples. The theoretical potential of open tubular columns is discussed below.

Separation Efficiency

As discussed previously, separation efficiency is measured by the magnitude of solute zone spreading and is expressed in terms of $\bar{h}$. According to Equation (2.28), $\bar{h}$ depends on the mobile phase velocity, column internal diameter, stationary phase film thickness, partition ratio, mobile phase viscosity, and the solute–mobile phase binary molecular diffusion coefficient. In order to minimize $\bar{h}$ for open tubular columns, the selection of (a) a thin stationary liquid film thickness and high phase ratio, (b) the optimum mobile phase velocity, (c) a low-viscosity mobile phase, and (d) microbore columns are all important.

In the case of thin stationary liquid film thickness and high phase ratio, for example, $\beta \geq 100$, zone spreading due to h_3 may be neglected. As illustrated in Figure 2.8, for columns with internal diameters less than 200 μm and phase ratios of 250, the h_3 values measured at u_{opt} for a solute with $t_R = 30$ min and $k = 10$ are less than 0.45% of that measured for h_2. The required d_f values for $\beta = 250$ with various column internal diameters are also plotted in Figure 2.8.

Equation (2.27) may thus be written as

$$\bar{h} = \frac{2D_G^\circ jf}{\bar{u}} + \frac{d_c^2(1 + 6k + 11k^2)\bar{u}f}{96D_G^\circ(1 + k)^2 j} \tag{2.62}$$

where d_c is the column internal diameter. From Equation (2.62), the average

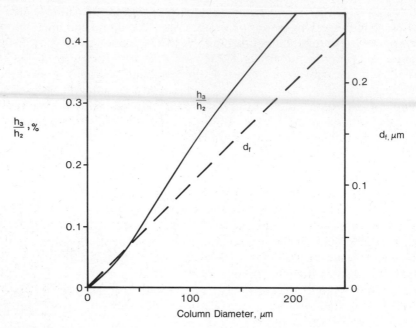

FIGURE 2.8. Graphs of h_3/h_2 percent ratio and film thickness against column internal diameter for columns with a phase ratio of 250. $t_r = 1800$ s, $k = 10$, and H_2 carrier gas at $\bar{u}_{opt}$.

minimum HETP ($\bar{h}_{min}$) can be expressed as

$$\bar{h}_{min} = \frac{d_c f}{2} \left(\frac{1 + 6k + 11k^2}{3(1 + k)^2} \right)^{1/2} \tag{2.63}$$

and the corresponding mobile phase optimum average velocity ($\bar{u}_{opt}$) is

$$\bar{u}_{opt} = \frac{2D_G^\circ j}{d_c} \left(\frac{3 + (1 + k)^2}{1 + 6k + 11k^2} \right)^{1/2} \tag{2.64}$$

Equation (2.63) shows that $\bar{h}_{min}$ is directly proportional to d_c and f for a given k. f is in the range 1–1.125 for increasing column resistance to mobile phase flow. The capacity ratio term $\{(1 + 6k + 11k^2)/[3(1 + k)^2]\}^{1/2}$ is in the range 0.58–1.9 for k ranging from 0 to 100. $\bar{h}_{min}$ is therefore between $0.29d_c$ and $1.07d_c$ for a given column and solutes retained with $k \leq 100$. Figure 2.9 is a plot of $\bar{h}_{min}$ vs. d_c for partition ratios of 1 and 10. A solute retention time $t_R = 30$ min in helium carrier gas is assumed.

As shown in Figure 2.9, $\bar{h}_{min}$ is of the order of 0.8–1 column internal diameter for $k = 10$. For a solute with $k = 1$, $\bar{h}_{min}$ is of the order of 0.61–0.69 column internal diameter. This implies that one million theoretical plates can be realized in an analysis time of 30 min using a 20 m open tubular column with an internal diameter of 20–25 μm.

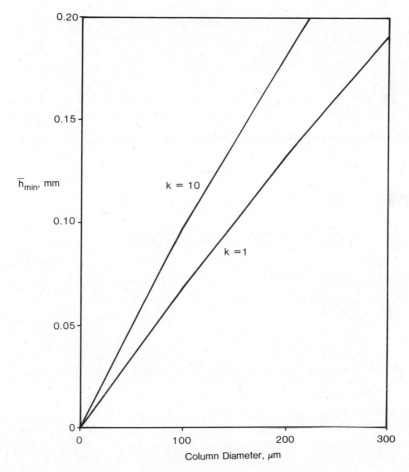

FIGURE 2.9.   Graph of average plate height against column internal diameter. He carrier gas at $\bar{u}_{opt}$ and a solute retention time $t_r = 30$ min.

Resolving Power

As discussed previously, sample components are separated from each other because of the differences in the migration rate of each individual solute. The ratio of the migration rates of any two sample compounds can be measured in terms of α. Since the migration rate of each solute depends on the change in the Gibbs free energy resulting from solute–stationary liquid phase interaction, the relative retention is therefore a measure of the difference in $\Delta G°$ values $[\Delta(\Delta G°)]$ for a pair of compounds eluted under given chromatographic conditions. The potential of a column for the separation of a pair of compounds with a given $\Delta(\Delta G°)$ value can be written as $T \ln \alpha = \Delta(\Delta G°)$ and can thus be evaluated in terms of α. As shown in Figure 2.10,

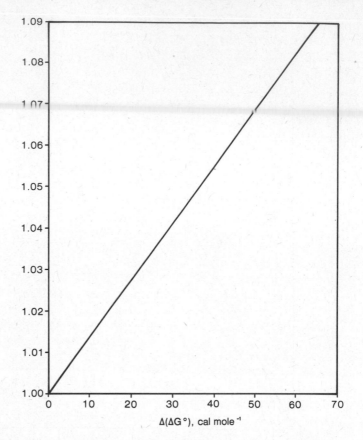

FIGURE 2.10. Variation of relative retention, α, for the separation of two adjacent components with the difference in their Gibbs free energy, $\Delta(\Delta G°)$. 100°C column temperature.

the smaller the values of α, the greater the resolving power of the column for compounds with small differences in their free energies of interaction with molecules of the stationary liquid phase.

α can also be related to column internal diameter, mobile phase velocity, and analysis time by using Equations (2.48), (2.57), (2.62), and (2.64).

$$\alpha = \left[1 - \left(\frac{(1 + k)(1 + 6k + 11k^2)(1 - \chi^2)R^2 d_c^2 f}{6t_R D_G° k^2 \chi^2 j} \right)^{1/2} \right]^{-1} \quad (2.65)$$

Here $\chi = \bar{u}/\bar{u}_{opt}$, so that the mobile phase velocity can be expressed in terms of $\bar{u}_{opt}$.

Equation (2.65) allows the resolving power of an open tubular column to be examined with respect to the column internal diameter, mobile phase solute binary molecular diffusivity, mobile phase velocity and compressibility, and allowed analysis time. As shown in Figure 2.11, assuming $k = 10$, $D_G^o = 0.1$ cm^2 s^{-1}, $R = 1$, $\chi = 1$ and 50, $t_R = 1800$ s, and helium and hydrogen mobile phases, the smaller the column internal diameter, the better the resolving power. There is a near linear relationship between column internal diameter and resolving power for columns with internal diameter larger than 30 μm. Below a column internal diameter of 30 μm, the resolving power increases dramatically in a nonlinear fashion. From Equation (2.65), it can be seen that when all other variables are held constant, α depends on column internal diameter and the mobile phase compressibility correction factors j and f. In the case of wide-bore open tubular columns, where the pressure gradient across the column is small, the column resolving power increases linearly with decreasing column internal diameter. For microbore columns, where j and f factors are not equal to 1, a remarkable increase in column resolving power is available due to the nonlinear behavior predicted by Equation (2.65).

Equation (2.65) also indicates that α decreases rapidly as carrier gas velocity increases. As shown in Figure 2.11, with a velocity 50 times higher than $\bar{u}_{opt}$, the column resolving power is greatly reduced. Figure 2.11 also

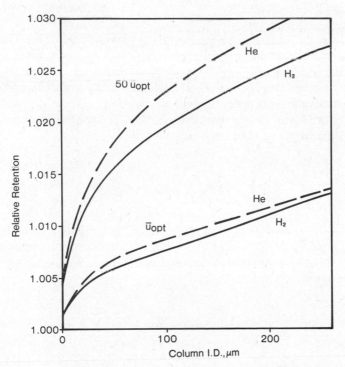

FIGURE 2.11.   Graph of relative retention against column internal diameter: $t_R = 1800$ s, $k = 10$, $D_G^o = 0.1$ cm^2 s^{-1}, $R = 1$.

shows that H_2 is preferred as a carrier gas to He at higher flow rates due to a larger j factor and better resolving power in terms of α.

Speed of Analysis

The theoretical potential of open tubular column GC can also be examined from the point of view of the separation efficiency in terms of speed of analysis or rate of generation of theoretical plates. This parameter may be expressed[1] as

$$\frac{dn}{dt} = \frac{\bar{u}}{\bar{h}(1 + k)} \qquad (2.66)$$

From Equation (2.66), the number of plates generated in a given time increases with increasing $\bar{u}$ and with decreasing h and k. The rate of generation of theoretical plates at $\bar{u}_{opt}$ is expressed by

$$\left(\frac{dn}{dt}\right)_{\bar{u}_{opt}} = \frac{48D_G^\circ(1 + k)}{d_c^2(1 + 6k + 11k^2)} \frac{j}{f} \qquad (2.67)$$

Equation (2.67) indicates that the speed of analysis increases rapidly with decreasing column internal diameter due to the inverse relation between $(dn/dt)_{\bar{u}_{opt}}$ and the square of column internal diameter. However, the effect of decreasing the column internal diameter is reduced by the decrease in j values as the resistance to mobile phase flow increases. Figure 2.12 shows the relationship of $(dn/dt)_{\bar{u}_{opt}}$ to d_c. It indicates that a gradual improvement in the speed of analysis is observed with decreasing column internal diameter down to about 30 μm. Below that diameter, the generation of more than 500 theoretical plates per second becomes possible.

Peak Capacity

Equation (2.49) allows direct comparison of the potential of columns for complex sample analysis. From Equation (2.49), the maximum peak capacity for a column can be expressed in terms of column internal diameter, analysis time, and j and f.

$$N'_{max} = 1 + \frac{\ln(1 + k)}{\ln\left[\left(\dfrac{12D_G^\circ(1 + k)t_R j}{(1 + 6k + 11k^2)d_c^2 f}\right)^{1/2} + 1\right] - \ln\left[\left(\dfrac{12D_G^\circ(1 + k)t_R j}{(1 + 6k + 11k^2)d_c^2 f}\right)^{1/2} - 1\right]} \qquad (2.68)$$

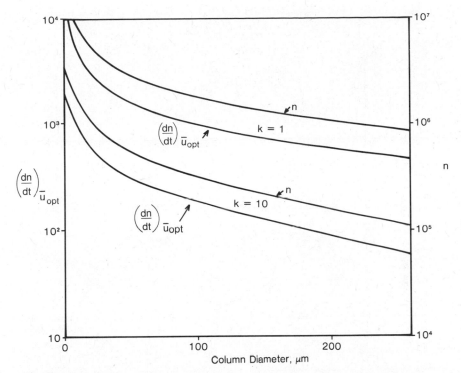

FIGURE 2.12.   Graphs of the rate of generation of theoretical plates (dn/dt) and total column plate number (n) generated in 30 min against column internal diameter. Hydrogen carrier gas at $\bar{u}_{opt}$.

A plot of N'_{max} against d_c using $t_R = 1800$ s, $k = 1$ and 10, and hydrogen at optimum flow velocity is shown in Figure 2.13. The peak capacity increases gradually with decreasing column internal diameter. A more dramatic increase in peak capacity to more than 500 solute components eluted between $k = 0$ and $k = 10$ is also shown for a column internal diameter below 30 μm.

Peak capacity depends also on the range of partition ratios according to Equation (2.68). By decreasing the partition ratio from 10 to 1 (i.e., solute components eluted between $k = 0$ and $k = 1$) and holding all other parameters constant, the peak capacity is reduced by a factor of 2 (Figure 2.13). This implies that microbore columns with lower β ratios are desirable for complex sample analysis because a higher k and N' are obtained. However, at lower phase ratios, the contribution to solute zone spreading due to h_3 may become significant in determining the total solute zone spreading. Phase ratios between 100 and 500 are preferred for high-resolution open tubular column GC.

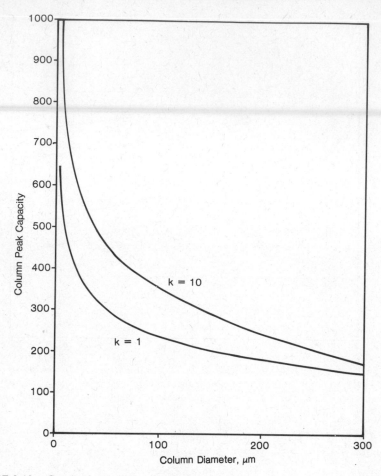

FIGURE 2.13. Graph of column peak capacity as a function of column internal diameter. t_R = 1800 s, β = 250, and hydrogen carrier gas at $\bar{u}_{opt}$.

2.7 PRACTICAL CONSIDERATIONS IN MICROBORE OPEN TUBULAR COLUMN GC

From Section 2.6, it is clear that microbore open tubular columns have great potential for improving column performance in terms of column efficiency, resolving power, speed of analysis, and peak capacity. Tremendous gains in theoretical performance are predicted for columns with internal diameters $\leq$ 30 μm, because of the nonlinear relationship between performance and column internal diameter. However, to use the available potential of microbore columns, the gas chromatograph must have a matching level of performance. Practical considerations of flow rate, inlet pressure, sample capacity, detectivity, sample volume, sampling time, and detector cell volume and time constant are discussed next.

Flow Rate and Inlet Pressure

Using Equation (2.64), plots of $\bar{u}_{opt}$ and the corresponding required inlet pressures against column internal diameters are shown in Figure 2.14. It is clear that $\bar{u}_{opt}$ for microbore columns is greatly affected by j values. The latter are small because of the required high pressure gradient across the microbore columns. Optimum average mobile phase velocities for 5–200 μm i.d. columns are between 10 and 20 cm s^{-1}. This linear velocity range corresponds to volumetric flow rates of between 10^{-4} and 0.4 cm^3 min^{-1}. These small flow rates cannot be accommodated by modern flow controllers, and pressure-controlled pneumatics are required. In addition, such small flow rates will grossly affect the operation of commonly used inlet systems, such as splitters, in terms of split linearity and injector dead volume effects.

An additional difficulty in using microbore columns is the high inlet pressure required. The use of columns with internal diameters under 50 μm requires inlet pressures much higher than are available on modern chromatographs. Inlet pressures of several hundred bars are required to operate narrow-bore columns with internal diameters ≤50 μm at velocities several times $\bar{u}_{opt}$.

Sample Capacity and Detectability

Sample capacity is a measure of the maximum sample mass that can be used without overloading the column. The sample overloading effect occurs when

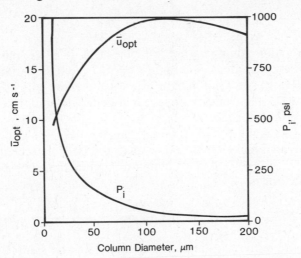

FIGURE 2.14.   Graphs of optimum carrier gas velocity and column inlet pressure against column internal diameter: t_R = 1800 s, k = 10, and H$_2$ carrier gas.

the solute distribution isotherm between the mobile and the stationary phase is nonlinear, and is manifested by distorted peak shape and loss of resolution.

The total sample mass of a Gaussian peak is expressed by

$$q = \frac{C_m V_R (2\pi)^{1/2}}{n^{1/2}} \qquad (2.69)$$

where C_m is the solute concentration at the peak apex. The retention volume can be written in terms of column dimensions as

$$V_R = \frac{\pi}{4} d_c^2 L (1 + k) \qquad (2.70)$$

and n can be expressed in terms of column internal diameter as

$$n = \frac{L}{\hat{h} d_c} \qquad (2.71)$$

where $\hat{h}$ is the reduced plate height and is defined as

$$\hat{h} = \frac{h}{d_c} \qquad (2.72)$$

Substitution of Equations (2.70) and (2.71) into (2.69) gives

$$q = C_m d_c^{5/2} (1 + k)(L\hat{h})^{1/2} \qquad (2.73)$$

Thus, the sample capacity of an open tubular column is strongly dependent on column internal diameter. If two peaks with the same peak heights (i.e., same C_m), partition ratios, and reduced plate heights are eluted from two columns, a comparison of their sample capacities can be written as

$$\frac{q_1}{q_2} = \left[\left(\frac{d_{c_1}}{d_{c_2}} \right)^5 \frac{L_1}{L_2} \right]^{1/2} \qquad (2.74)$$

An open tubular column with $d_c = 250$ μm, $L = 25$ m, and $\beta = 250$ has a sample capacity of about 100 ng at $k = 10$. From Equation (2.74), a 30 μm i.d. column of the same length has a sample capacity of only 0.5 ng. This very low sample capacity poses great demands on detector sensitivity and low noise detector electronics.

Furthermore, as the column internal diameter decreases, the range of the sample component mass which can be analyzed is also affected. The overall mass range for an open tubular column in conjunction with a gas chromatographic detector is

$$M = \frac{q}{Q_0} \qquad (2.75)$$

where Q_0 is the minimum detectable quantity for the sample with a given detector. Q_0 measured for narrow-bore columns may be better than for wide-bore columns. The enhanced detection factor, E, for open tubular columns and a mass flow-rate-dependent detector may be written[22]

$$E = \frac{t_{R_1}}{t_{R_2}} \left(\frac{n_2}{n_1} \right)^{1/2} \tag{2.76}$$

For example, a 30 μm i.d. column generates about five times more theoretical plates than a 200 μm i.d. column in the same analysis time (see Figure 2.12) and gives an E 2.2 times greater than that obtained using the 200 μm i.d. column.

M for open tubular columns with a flame ionization detector (FID) as a function of column diameter is given in Figure 2.15. This indicates that as

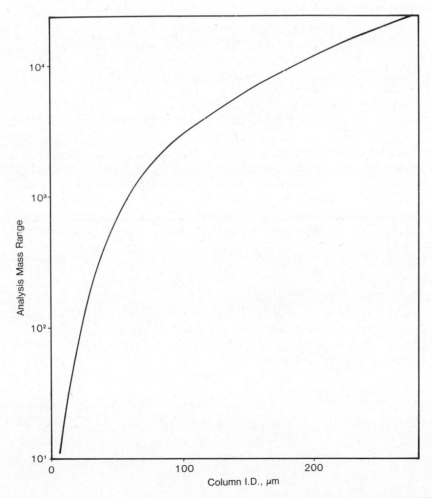

FIGURE 2.15. Graph of analysis mass range against column internal diameter with an FID: t_R = 1800 s, β = 250, k = 10, D_G^o = 0.1 cm^2 s^{-1}, D_L = 10^{-5} cm^2 s^{-1}, and H$_2$ carrier gas at $\bar{u}_{opt}$.

column internal diameter decreases M decreases quite rapidly, becoming limited to about 10 at 8 μm i.d.

In the case of detection with a concentration-dependent detector, such as an electron capture detector (ECD), a thermal conductivity detector (TCD), or a photoionization detector (PID), Q_0 may be greatly reduced due to the enhancement factor.[22]

$$E = \frac{L_1}{L_2} \left(\frac{d_{c_1}}{d_{c_2}}\right)^2 \left(\frac{n_2}{n_1}\right)^{1/2} \tag{2.77}$$

For example, if a Q_0 of 10 fg (10^{-14} g) of sample on a 200 μm i.d. column is assumed, then a Q_0 of 80 ag (8×10^{-17} g) is expected for a 30 μm i.d. column. The mass range of analysis according to Equation (2.75) for a 30 μm i.d. open tubular column with an electron capture detector has a lower mass limit of 80 ag and an upper limit of 8 fg with a 10^4 linear range. Thus, microbore open tubular columns with electron capture detection can afford significant advantages in the analysis of trace samples.

Sample Volume and Sampling Time

In addition to considerations of sample capacity in terms of the sample mass that can be injected into a microbore open tubular column, it is also important to reduce the sample volume. Serious consequences, such as peak tailing, stationary phase stripping, retention time shift, and nonlinear splitting (discrimination), could occur due to too large a sample volume.

The maximum sample vapor volume which will not increase peak variance by more than a fraction, θ, can be expressed by Equation (2.78) [from the rearrangement of Equation (2.73) and assuming a plug injection profile].

$$V_{max} = 2.72\theta(1 + k)(L\hat{h}d_c^5)^{1/2} \tag{2.78}$$

As shown in Figure 2.16, for a 1% (i.e., $\theta = 0.01$) increase in peak variance, a 1 μL volume is allowed for compounds with $k = 1$ and 10 and $t_R = 1800$ s eluted from a 200 μm i.d. column. As the column internal diameter decreases, the maximum sample volume also drops, becoming less than 10 nL for a 30 μm i.d. column. This small volume presents no problem if an inlet splitter is used. However, at the required split ratio of the order of 1000:1 to 40,000:1 for a 1 μL liquid sample, splitter linearity and quantitative accuracy may be affected.

The sampling time depends on the elution time for the compound injected. For a well-retained compound, the sampling time is normally not significant enough to affect the peak resolution. However, as the column efficiency increases and the retention time decreases, rapid sampling is required. If the peak variance increases by θ, the sampling time can be calculated according to Equation (2.79):

$$t_s = \frac{0.0346\theta t_R}{n^{1/2}} \tag{2.79}$$

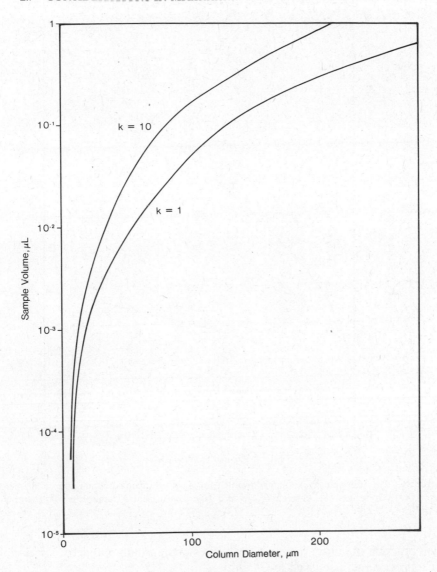

FIGURE 2.16. Graph of maximum allowable sample vapor volume against column internal diameter: $t_R = 1800$ s, H_2 carrier gas at $\bar{u}_{opt}$. Plug injection profile is assumed along with a 1% increase in peak variance due to the sample volume initial bandwidth contribution.

As shown in Figure 2.17, a sampling time of the order of 0.1 s for a compound with $k = 10$ and $t_R = 1800$ s is required for columns with internal diameter ≤ 100 μm. A sampling time of the order of 1 ms can easily be achieved with a splitter injector or a rapid sampling valve.

The above discussion for small sample volumes and fast sampling times assumes that no solvent effect or solute focusing techniques are applied.

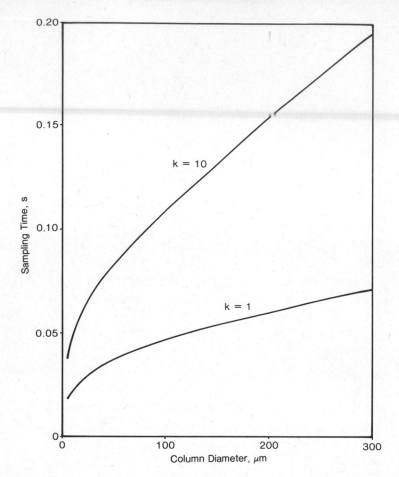

FIGURE 2.17. Graph of maximum allowable sampling time against column internal diameter: $t_R = 1800$ s, hydrogen carrier gas at $\bar{u}_{opt}$, and 1% increase in peak variance due to the sampling time.

The practice of solute focusing for large sample volume injectors at a slow injection rate is discussed in Chapter 5.

Detector Cell Volume and Time Constant

On the assumption that the detector cell operates as a finite volume mixing chamber, its contribution to an increase in peak variance may be written[23]

$$V_d = \frac{\theta t_R F_c}{n^{1/2}} \tag{2.80}$$

where F_c is the volumetric flow rate. If the detector cell is assumed to have the same internal diameter as the column, Equation (2.80) can be rewritten in terms of column internal diameter and reduced plate height as

$$V_d = \frac{\pi}{4} \theta[t_R u' \hat{h}(1 + k)d_c^5]^{1/2} \qquad (2.81)$$

where u' is the gas velocity in the detector. For the case where $u' = \bar{u}$, the required detector cell volume falls off rapidly with decreasing column internal diameter. As shown in Figure 2.18, for $\theta = 0.01$ and $k = 10$, the detector cell volume falls below 10^{-5} cm^3 for a 50 μm i.d. column. A flame-

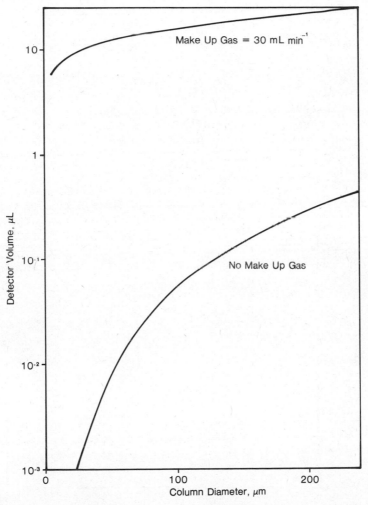

FIGURE 2.18. Graph of detector cell volume against column internal diameter required for contribution of no more than 1% increase in peak variance due to mixing chamber effects: t_R = 1800 s, β = 250, D_G^o = 0.1 cm^2 s^{-1}, D_L = 10^{-5} cm^2 s^{-1}, k = 10, and H$_2$ carrier gas at $\bar{u}_{opt}$.

based detector such as an FID or an on-column detector could be used. Other detectors, such as the ECD, PID, and TCD, however, are unsuitable for use without make-up gas. As is also shown in Figure 2.18, when a 30 cm^3 min^{-1} makeup gas flow (carrier gas or fuel gas) is introduced into the detector, the cell volume needs to be no smaller than 10^{-2} cm^3 for a 30 μm i.d. column, and up to 2.5×10^{-2} cm^3 for a 250 μm i.d. column, due to the increase in mobile phase linear velocity by a factor of about 5000.

The detector response time constant (τ_d) also can make a significant extra-column contribution to peak broadening. The required time constant for a θ fractional increase in peak variance depends on the retention time and column efficiency as

$$\tau_d = \frac{\theta t_R}{n^{1/2}} \tag{2.82}$$

As illustrated in Figure 2.19, a time constant of 30–50 ms is required for a contribution of no more than 1% (i.e., θ = 0.01) increase in peak variance for an eluted solute with $t_R = 1800$ s and $k = 10$ from columns of 50–250 μm i.d. For columns with internal diameter ≤30 μm, however, the required time constant decreases rapidly due to the rapid decrease in peak widths obtained with microbore columns.

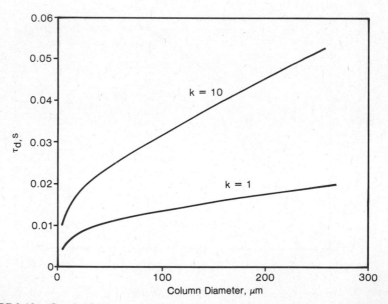

FIGURE 2.19. Graph of detector time constant (τ_d) against column internal diameter required for contribution of no more than 1% loss in peak resolution: $t_R = 1800$ s, $\beta = 250$, $k = 10$, $D_G^0 = 0.1$ cm^2 s^{-1}, $D_L = 10^{-5}$ cm^2 s^{-1}, and H$_2$ carrier gas at $\bar{u}_{opt}$.

SYMBOLS USED

A_c	Column cross-sectional area
C_m	Solute concentration at the elution peak apex
d_c	Column internal diameter
d_f	Stationary phase film thickness
D_g	Average binary molecular diffusion coefficient between solute and mobile phase molecules in the column
D_G°	Solute–mobile phase binary molecular diffusion coefficient at the outlet pressure of the column
D_L	Solute–stationary phase binary molecular diffusion coefficient
E	Detection enhancement factor
EPN	Effective peak number
F_c	Mobile phase volumetric flow rate measured at the end of the column
$\overline{F}_c$	Average mobile phase volumetric flow rate
f	Giddings' plate height correction factor
ΔG°	Change in the Gibbs free energy for the dissolution of solute into the stationary liquid phase
$\overline{h}$ or HETP	Height equivalent to an average theoretical plate
h_1	HETP originated from longitudinal molecular diffusional zone spreading
h_2	HETP originated from resistance to mass transfer zone spreading in the mobile phase
h_3	HETP originated from resistance to mass transfer zone spreading in the stationary phase
$\overline{h}_{\min}$	Minimum HETP corresponding to optimum carrier gas velocity
$\hat{h}$	Reduced plate height
h'	Peak height
j	James–Martin gas compressibility correction factor
K	Partition coefficient or distribution ratio
k	Partition ratio or capacity ratio
L	Column length
M	Sample mass range
N	Total column effective plate number
n	Total column plate number
N'	Peak capacity
P_o	Mobile phase pressure at the outlet of the column
P_i	Mobile phase pressure at the inlet of the column

P	Ratio of P_i to P_o
$\bar{P}$	Average mobile phase pressure in the column
q	Sample mass or sample capacity
Q_0	Minimum detectable quantity
R	Resolution between two adjacent peaks
r	Column radius
SN	Separation number
T_c	Column temperature
t_e	Retention time at $1/e$ of peak height
t_m	Column void time or the elution time of a nonretained solute
t_n	Adjusted or net retention time for a retained solute
t_R	Retention time
t_s	Sampling time
τ_d	Detector time constant
Δt	Retention time difference between two adjacent peaks
$\bar{u}$	Average mobile phase linear velocity in the column
u_o	Mobile phase linear velocity measured at the outlet of the column
$\bar{u}_{opt}$	Optimum mobile phase average velocity
u'	Gas velocity in the detector
V_d	Detector cell volume
V_G	Column gas phase volume
V_L	Stationary liquid phase volume
V_M	Retention volume corresponding to the gas hold-up time
V_M°	Corrected elution volume of a nonretained peak
V_{max}	Maximum sample volume
V_N	Adjusted or true retention volume for a retained peak
V_R°	Corrected retention volume
V_R	Retention volume for a retained peak
W_b	Base width of a peak
W_h	Width of a peak at half height
α	Relative retention for a pair of peaks
β	Column phase ratio
σ^2	Peak variance in time
σ_l^2	Peak variance in length
$\mathscr{R}$	Gas constant
χ	Ratio of mobile phase velocity to optimum mobile phase velocity

REFERENCES

1. J. C. Giddings, *Dynamics of Chromatography*. Marcel Dekker, New York, 1965.
2. J. H. Purnell, *Gas Chromatography*. John Wiley & Sons, New York, 1967.
3. M. J. E. Golay, in *Gas Chromatography*, D. H. Desty, editor. Butterworths, London, 1958.
4. A. T. James and A. T. P. Martin, *J. Biochem.* **50**, 679 (1952).
5. A. Einstein, *Ann. Phys.* **17**, 549 (1905).
6. J. C. Giddings, S. L. Seager, L. R. Stucki, and G. H. Stewart, *Anal. Chem.* **32**, 867 (1960).
7. T. Ellerington, in *Gas Chromatography*. D. H. Desty, editor. Butterworths, London, 1958.
8. W. L. Jones and R. Kieselback, *Anal. Chem.* **30**, 1590 (1958).
9. R. A. Hurell and S. G. Perry, *Nature* **196**, 571 (1962).
10. R. E. Kaiser, *Anal. Chem.* **189**, 1 (1962).
11. J. C. Giddings, *Anal. Chem.* **39**, 1027 (1967).
12. L. S. Ettre, *Chromatographia* **8**, 291 (1975).
13. R. E. Kaiser and R. Rieder, *Chromatographia* **8**, 491 (1975).
14. S. P. Cram, F. J. Yang, and A. C. Brown III, *Chromatographia* **10**, 397 (1977).
15. J. J. Van Deemter, F. J. Zuiderweg, and A. Klinkenberg, *Chem. Eng. Sci.* **5**, 271 (1956).
16. E. Glueckauf, *Trans. Faraday Soc.* **51**, 34 (1955).
17. J. Bohemen and J. H. Purnell, in *Gas Chromatography*, D. H. Desty, editor. Butterworths, London, 1958.
18. D. H. Desty, A. Goldup, and W. T. Swanton, in *Gas Chromatography*, N. Brenner, J. E. Callen, and M. D. Weiss, editors. Academic Press, New York, 1962.
19. J. H. Purnell, *J. Chem. Soc. (London)*, 1268 (1960).
20. F. J. Yang and S. P. Cram, 31st Pittsburgh Conference on Analytical Chemistry and Applied Spectroscopy, paper No. 115, Atlantic City, New Jersey, March, 1981.
21. C. P. M. Schutjes, E. A. Vermeer, J. A. Rijks, and C. A. Cramers, *Proceedings of the Fourth International Symposium on Capillary Chromatography*, p. 687. Hindelang, Germany, May 1981.
22. F. J. Yang and S. P. Cram, *J. High Resoln. Chromatogr./Chromatogr. Commun.* **2**, 487 (1979).
23. J. G. Sternberg, in *Advances in Chromatography*, Vol. 2, J. C. Giddings and R. A. Keller, editors. Marcel Dekker, New York, 1966, p. 205.

THREE

COLUMN TECHNOLOGY

3.1 INTRODUCTION

Although efficient sample introduction devices, sensitive detectors, sophisticated electronically controlled ovens, high-speed recorders and integrators, and other devices are essential components in modern high-resolution GC systems, the column remains the heart of the analytical instrument. The ultimate quality of any separation cannot be any better than the column itself. The growth and widespread use of open tubular column GC has paralleled the development of column technology. As the use of open tubular columns becomes more widespread, it would seem logical that even greater demands will be placed on their ability to perform more complex and difficult separations. To realize the full potential available for the resolution of complex mixtures into individual components, the preparation of highly efficient, well-deactivated, and thermally stable columns is imperative.

Since the inception of open tubular column GC in 1958, when Golay[1] presented its theoretical basis, a mass of information and experience has been accumulated on the parameters involved in preparing high-quality columns. The early developments of open tubular column GC have been reviewed by Ettre.[2] Originally, materials such as plastic, stainless steel, copper, and other metals were used for column fabrication. However, with the development of the glass drawing machine by Desty,[3] glass became the material of choice. Individual investigators were then able to draw columns to the length and dimensions of choice as well as having a more inert and easily modifiable material. Although the glass columns were somewhat fragile, they essentially replaced the earlier materials and have been used extensively. Most of the existing column technology was originally developed for glass columns and has been reviewed by several authors.[4-8] In 1979, with the introduction of flexible fused-silica columns by Dandeneau and co-workers,[9,10] the beginning of another new era in open tubular column GC was initiated. Although there has been some controversy over the utility of this material, fused silica is finding extensive use and is replacing glass as the material of choice. This is a natural result of its stronger mechanical properties and its higher intrinsic inertness. Unfortunately, the technology involved in drawing fused silica is more complicated than drawing glass, and it is generally not possible for chromatographers to fabricate their own column material.

In this chapter, the state-of-the-art column technology for glass and fused-silica columns will be discussed. Because of the expanding interest in fused silica, the major emphasis will be centered around this material. A comprehensive review of glass capillary column technology is contained elsewhere.[6,8] Included in this chapter will be a fundamental discussion of silica surface chemistry and its implications in the chromatographic process. Column preparation methods will be discussed, including deactivation, coating, and cross-linking. Also, a discussion of stationary phases currently being used and special stabilization parameters will be given. Finally, a rigorous scheme for column evaluation will be presented.

3.2 SILICA SURFACE CHEMISTRY

Composition and Structure

Fused silica has been described as the most simple glass. The bulk material consists entirely of silica tetrahedra. Such tetrahedra can be attached to none, one, two, three, or four other tetrahedra by silicon–oxygen bonds at their corners to form a three-dimensional network. In amorphous (noncrystalline) silica structures, the tetrahedra are linked together in a totally random fashion. One model proposes that there are regions of order alternating with connecting regions of disorder. A more favored model postulates a random network that lacks any periodicity of symmetry.[11,12] A two-dimensional schematic representation depicting a random silica network is shown in Figure 3.1. Because these types of silica possess a highly cross-linked structure, their melting points are extremely high ($\sim 2000°C$) which makes fabrication of these materials difficult. This same structure, however, imparts a high tensile strength that permits a very thin-wall capillary of high flexibility to be fabricated.

In the manufacture of glass, various metal oxides are added to the silica to modify its chemical properties. The addition of Na_2O, CaO, MgO, B_2O_3, Al_2O_3, and other materials generally softens and lowers the melting point of the original silica by disrupting the Si—O—Si bonds. Characteristics such as viscosity, thermal expansion, solubility, and durability are also controlled

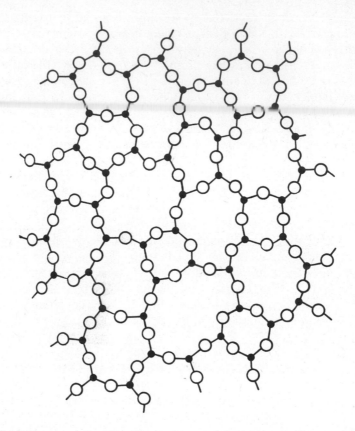

FIGURE 3.1. Representation of a silica surface, assuming a random network of SiO_4 tetrahedra.

by the addition of these modifiers. Soft glass and Pyrex glass are commonly used for column preparation. Their bulk compositions are listed in Table 3.1.

Both natural quartz and synthetic fused-silica materials are used in capillary column manufacture. Natural quartz from rock crystals is melted or fused at extremely high temperatures under vacuum by flame or electric fusion to produce fused quartz. The purity of this material depends on the purity of the original quartz and on any contamination arising from the fusion process. Generally, the metallic impurities are less than 100 ppm. In purified fused quartz, the level of contaminants is reduced to 10–50 ppm. Fused silica, however, is prepared synthetically from the flame hydrolysis of high-purity silicon tetrachloride. The $SiCl_4$ undergoes reaction with the water vapor in the combustion process and SiO_2 is formed.[13] The SiO_2 is condensed on a substrate and built up to the desired thickness. The final product contains less than 1 ppm of metallic impurities. The distribution of metallic impurities in the above materials is listed in Table 3.2.[14]

Although glass is generally considered to be inert with regard to adsorption and catalytic behavior, it can manifest undesirable activity in capillary column applications. To a lesser extent, untreated fused silica and quartz also display undesirable surface activity. Column wall activity can be attributed to the silica surface structure and to impurities found in the surface monolayers of the glass matrix. The various metallic oxides, added during the manufacture of glass, that are present on or near the surface of the glass can act as Lewis acid sites.[15,16] These sites are considered to be cationic in which the positive charge is concentrated on a cation of small radius while the negative charge is distributed over the internal bonds of the incomplete silica tetrahedra.[17] Lewis acids function as adsorption sites for lone-pair donor molecules such as ketones and amines. Molecules containing π bonds, such as aromatic compounds and olefins, also interact with Lewis acid sites. It has also been firmly established that boron impurities in silica provide surface Lewis acid sites that are capable of chemisorbing electron-donating molecules.[18] Since the bulk composition of fused-silica materials is essentially free of metallic impurities, it is unlikely that the surface composition would contain any Lewis acid sites. It is the absence of these adsorptive sites that gives fused silica its higher degree of intrinsic inertness.

Undoubtedly, the single most important structural detail of the silica surface is the hydroxyl groups that are attached to the surface silicon atoms. These silicon atoms are presumably tetrahedrally coordinated to three other oxygen atoms and, thus, to the bulk silica. This infers that at low temperatures the surface silicon atoms prefer to complete their coordination requirements by attachment to monovalent hydroxyl groups rather than by formation of strained siloxane bridges or charged species.[19] Several types of hydroxyl groups have been identified on high-surface-area porous silica. Presumably, these same structures exist on glass and fused-silica surfaces and exhibit similar behavior. Hydroxyl groups which are attached to adjacent silicon atoms are termed *vicinal,* and when two hydroxyl groups are attached

TABLE 3.1. Typical bulk glass compositions.

Glass type	Component							
	SiO_2	Na_2O	CaO	Al_2O_3	B_2O_3	MgO	BaO	K_2O
Kimble R6 flint[a] soda-lime soft glass	67.7	15.6	5.7	2.8	—	3.9	0.8	0.6
Pyrex 7740[b]	81.0	4.0	0.5	2.0	13.0	—	—	—

[a] Kimble Gass Works, Toledo, Ohio.
[b] Corning Glass Works, Corning, New York.

TABLE 3.2. Metal impurities in various types of fused silica.

Types	ppm										
	Al	Ca	Cu	Fe	Mg	Mn	Ti	Na	K	B	P
Natural quartz	30–50	1.0	0.8	2.0–3.3	1.0	0.03	3	2.0	2.0	0.3	0.5
Purified natural quartz	1–10	0.1	—	0.5	0.3	0.01	—	<0.5	1.0	—	—
Synthetic fused silica	0.1	0.1	0.004	0.2	0.1	0.01	—	0.4	0.001	0.01	0.1
High-purity synthetic fused silica	0.03	—	0.01	0.03	—	—	—	0.004	0.005	—	—

54

to the same silicon atom the term *geminal group* has been applied. In addition to surface hydroxyl groups, there are also internal hydroxyl groups within the silica structure which are usually termed *intraglobular hydroxyls*.[20]

Many of the surface hydroxyls on porous silica are hydrogen bonded to one another and are described as "bound." Those that are not perturbed or involved in any interactions are described as "free." Whether two adjacent hydroxyl groups are bound or free is determined by the distance of one hydroxyl group from the oxygen atom of the adjacent hydroxyl group. Hydroxyls separated from adjacent oxygen atoms by more than 3.1 Å appear to be incapable of hydrogen bonding.[21] Since vicinal hydroxyl groups are separated by at least 3.1 Å, it is unlikely that they are hydrogen bonded to one another. Geminal hydroxyl groups also are probably not bonded to their partners because a five- or six-membered ring is normally needed for intramolecular hydrogen bonding. It is thought that approximately 50% of the surface hydroxyl groups are hydrogen bonded to one another.[22]

Under normal atmospheric conditions, water is adsorbed to the hydrogen bonded surface hydroxyls.[23] Heat treatments can remove the physically adsorbed water, leaving only the surface and intraglobular hydroxyls. Prolonged and more intense heating actually dehydrates the silica surface by the condensation of neighboring hydroxyl groups in which siloxane bridges are formed. From room temperature to approximately 165°C, only physically adsorbed water is removed from the surface of the silica. Between 165 and about 400°C, hydroxyl groups are thermally removed from the surface. Upon cooling and re-exposure to water, these sites hydrate, re-forming the original hydroxyl groups. Above 400°C, hydroxyl groups continue to be removed from the surface as the temperature is increased. However, as the treatment temperature increases, a decreasing number of hydroxyl groups can be re-formed on the surface. After heat treatment at about 800°C, the rehydration of the surface does not occur spontaneously. Between 165 and 400°C, the hydroxyl groups removed from the surface are those that are hydrogen bonded to one another: the most strongly hydrogen bonded groups disappear first, and the number of free groups remains almost unchanged.[24,25]

Even for porous silica, the concentration of surface hydroxyl groups has been the subject of much controversy. It is now generally agreed[26] that on an annealed silica surface that is fully hydroxylated, there is one hydroxyl group per silicon atom at a concentration of 4.6 OH nm^{-2}. Unfortunately, the methods commonly used for determining surface hydroxyl concentration on porous silica, such as IR spectroscopy, NMR spectroscopy, thermogravimetry, reaction titrations, deuterium exchange, and so on, are not applicable to low-surface-area materials such as glass and fused silica. However, in a recent experiment involving gaseous tritium exchange of the surface hydroxyl protons, it was possible to quantify the surface hydroxyl groups on actual glass and fused-silica capillary column surfaces.[27] These results are shown in Table 3.3. The concentration found on hydroxylated glass was somewhat less than the value of similar porous silica (2.8 vs. 4.6). The value

TABLE 3.3. **Surface hydroxyl concentration on fused silica and glass as determined by tritium exchange.**

Column type	Pretreatments	Surface hydroxyl concentration $(OH\ nm^{-2})$
Hydroxylated glass	Acid leached, water rinsed, heat treated at 150°C for 12 h	2.04 ± 0.09
Dehydroxylated glass	Heat treated at 600°C for 8 h	0.36 ± 0.02
Fused silica	Heat treated at 150°C for 4 h	0.21 ± 0.04

for the dehydroxylated glass was also less than the 1.6 groups found on similarly treated porous silica.[28] It appears that the hydroxyl groups on glass are fewer in number and more easily removed than those on porous silica. The extremely low concentration of 0.2 groups on the fused silica is consistent with its higher drawing temperature. It is also consistent with its lower chromatographic adsorptivity as compared to leached and dehydrated glass.

The hydrogens of the surface hydroxyl groups (also named *silanol groups*) are partially acidic, owing to d-electron cloud vacancies in the silicon atoms.[29] Consequently, the silanol groups are available as proton donors for hydrogen bonding sites. Hence, molecules containing high peripheral electron densities are adsorbed on a hydroxylated surface. A peripheral concentration of negative charge density arises from the π-electron bonds in unsaturated or aromatic hydrocarbons, and also from the free electron pairs of the oxygen and nitrogen atoms in hydroxyl, ether, carbonyl, and amino groups. Although the mechanism of adsorption is more complicated than an electrostatic hydrogen bond interaction, the type of adsorption involved closely resembles hydrogen bonding.[30] The heat of adsorption of polar molecules is highly dependent on the silica surface, and increases with greater surface hydroxylation.[31] Although mutually hydrogen bonded surface hydroxyl groups interact only slightly with lone-pair adsorbates, water interacts strongly with these groups and can form several molecular layers. The adsorbed water can then act as specific adsorption sites for molecules containing high electron densities in much the same manner as free surface hydroxyls.[23]

During heat treatment, surface hydroxyls condense to form water and siloxane bridges.[32] Such a bridge could have longer than optimal bond lengths between the oxygen and silicon atoms, thus decreasing its stability and increasing its reactivity. Such a strained structure could possess a high degree of ionic character.[33] Kunawicz et al.[34] have reported that the siloxane bridges resulting from the high-temperature (700°C) evacuation of silica are

more reactive than the remaining free hydroxyls. However, extremely high
temperatures (>900°C) are believed to release the strain from these bonds
and decrease their reactivity.[32] The siloxane bridge can function as a proton
acceptor in hydrogen bonding interactions. It has been shown that alcohol
molecules can interact significantly with the siloxane surface by van der
Waals interactions.[35] These interactions become more marked as the chain
length of the adsorbate alcohol increases. In a chromatographic study relating retention to adsorption, it was concluded that the siloxane bridge was
an active site.[36] Studies have also shown that strained siloxane bridges can
lead to the chemisorption of amines and alcohols.[37] The higher fabrication
temperature for fused silica should minimize the number of strained siloxane
bridges formed. However, it has been shown by Raman spectroscopy that
thermally induced Si^+ ^-O—Si structures remain in the material when rapid
cooling occurs,[38] as is probably the case with fused silica.

Surface Wettability

To achieve a high separation efficiency with an open tubular column, a
uniform and homogeneous film of stationary phase must be applied to the
inner wall of the tube. Furthermore, this thin film must maintain its integrity
and not rearrange to form droplets as the temperature is varied.

When a liquid droplet is placed on a solid surface, it may spread to cover
the surface or it may remain as a stable drop. The angle formed by the tangent
to the liquid drop from the edge at which it contacts the solid surface is
defined as the contact angle (θ). When θ is equal to zero, the liquid spreads
freely over the surface. As θ increases, the tendency for a liquid to spread
decreases. Therefore, the contact angle is a useful inverse measure, and cos
θ is a direct measure, of wettability.

The wettability is a thermodynamic function of the equilibrium between
the cohesion forces inside the liquid and the energy of the solid surface. The
cohesion forces inside the liquid are characterized by the surface tension,
and the energetics of the solid surface by the surface free energy. The contact
angle depends on the specific surface free energy of the solid and of the
liquid. Spreading generally occurs when the specific surface free energy of
the liquid is less than that of the solid. Compositions of both the solid surface
and the spreading liquid are of primary importance since the surface atoms
of both phases are attracted to each other by London dispersion forces.

Zisman[39] defined the critical surface tension (CST, γ_C) as that value of
the liquid surface tension above which liquids show a finite contact angle
on a given surface. By graphing the cosine of the contact angle versus the
surface tension for a series of liquids, a straight line or a narrow rectilinear
band is generally obtained (*Zisman plot*). The intercept on the surface tension axis where cos θ = 1 is designated as the critical surface tension. This
has significance in that liquids with a surface tension less than the surface's

TABLE 3.4. Surface tensions of selected stationary phases at 25°C.

Stationary phase	Chemical description	Surface tension (dyne cm^{-1})	Polarity $\Sigma\Delta I^a$
OV-101	Methylpolysiloxane	20.4	229
OV-3	10% Phenyl methylpolysiloxane	21.8	423
OV-210	50% Trifluoropropyl methylpolysiloxane	22.8	1520
Squalane	C$_{30}$ Alkane	30.0	0
OV-17	50% Phenyl methylpolysiloxane	31.4	884
UCON 50-HB-2000	Polyethylene-propylene glycol	35.7	1582
Carbowax 400	Polyethylene glycol	44.2	2587
1,2,3-Tris(cyanoethoxy)propane		49.2	4145

$^a \Sigma\Delta I = \Delta I_{benzene} + \Delta I_{1\text{-butanol}} + \Delta I_{2\text{-pentanone}} + \Delta I_{pyridine} + \Delta I_{nitropropane}$.

critical surface tension completely wet the surface, while those with a surface tension greater than the critical surface tension wet the surface incompletely. Furthermore, the slope of this line is indicative of the concentrations of wettable and nonwettable groups on the surface.[40] Typical stationary phases have surface tensions ranging between 19 and 50 dyne cm^{-1}. A representative list is given in Table 3.4.[41] An in-depth discussion describing the mathematical approach to surface wettability and its applicability to open tubular chromatographic columns is contained elsewhere.[41]

Glass is usually described as a high-energy surface[42] and is assumed to have an energy on the order of a few thousand erg cm^{-2}.[43] Such surfaces can easily undergo adsorption and hydration which changes their properties to low-energy surfaces[44] that are nonwettable by many organic liquids. Because of this tendency, there has been some confusion in the past about the exact nature of capillary column surface wettability. Also, since the intro-

TABLE 3.5. Critical surface tensions of various surfaces.

Surface	Treatment	Critical surface tension (dyne cm^{-1})
Pyrex glass	None (freshly drawn)	>72
Pyrex glass	Oxygen purge during drawing	>72
Fused silica (supplier A)	As received	>72
Fused silica (supplier B)	As received	>50
Fused silica (supplier A)	Equilibrated with water vapor	>50
Fused quartz	As received	>60

duction of fused-silica and quartz columns, there has been speculation that these materials had intrinsically low-energy surfaces.[14] However, recent studies[41] have verified that clean glass, fused silica, and fused quartz all have high-energy surfaces. These results are in accord with nonchromatographic wettability measurements.[45] Critical surface tensions for glass, fused silica, and quartz are given in Table 3.5.[41]

3.3 COLUMN PREPARATION

Column Drawing

During the cooling process in the manufacture of glass tubing, the product is exposed to oils and greases. In addition, organic vapors may adsorb on the glass surface during storage. Consequently, a number of different procedures have been used to clean the inner walls of glass tubing prior to drawing. These include dilute acids[46] and bases, and organic solvents such as acetone, diethylether, methanol, and methylene chloride.

The fused-silica or quartz preform tubing as obtained from the manufacturers invariably contains small scratches and pits on the inner bore and outer surfaces. To obtain strong, reproducible, and reliable capillary tubing, it is necessary to remove these imperfections. This is usually done by treating the preform with a dilute (~5%) hydrofluoric acid solution[14] followed by careful rinsing with distilled water. Additionally, the preform is usually annealed and carefully fire polished in a glass turning lathe at temperatures just below its softening point.

The first glass open tubular column drawing machine was designed and built by Desty and co-workers.[3] Two pairs of rollers, one before an electrically heated furnace and one after, control the draw ratio, and hence, the length and diameter of the column. As the column is drawn from the molten glass in the furnace, it is forced through a heated coiling tube to form a helix. Typical operating temperatures vary between 600 and 750°C depending on the type of glass. A more detailed description of the drawing process is contained elsewhere.[4]

Recently, the fabrication of thin-wall flexible glass columns using a modified commercial drawing machine was described.[47] These thin-wall columns were drawn to dimensions similar to fused silica (described below) and then coated with a polyimide outer coating to impart strength. Although it was claimed that these columns were as strong as fused-silica columns, it is questionable whether or not they will prove as durable and stable. Furthermore, glass demonstrates more surface activity than fused silica or quartz, and more extensive deactivation procedures are required. In some cases, the thin walls prevent the use of certain surface roughening and deactivation procedures.

In 1975, Desty[48] modified his original glass drawing machine to attain the necessary drawing temperature for quartz. The electrically heated oven was replaced with a special propane/oxygen burner. The inability to build suitable coiling tubes was the limiting factor in the use of this prototype machine, and only preliminary work was successfully accomplished. The discovery by Dandeneau and co-workers[9,10] that thin-wall open tubular columns of high flexibility could be drawn straight and then coiled to normal column dimensions by merely bending them into the desired shape has proven to be a significant advancement in capillary column technology. Generally, the wall thickness of a flexible fused-silica open tubular column is less than 25 μm.

Fused silica is drawn at high temperatures (1800–2200°C, depending on the bore) and is generally produced using advanced fiber optics technology and a clean-room atmosphere. The capillary can be used as a light conduit (fiber optic) so that infrared laser radiation can be beamed continuously through the capillary, and when coupled to microprocessor-controlled feed-back control circuitry, it provides a method of obtaining extremely uniform and precisely drawn capillary tubing. Induction furnaces, graphite furnaces, and hydrogen/oxygen burners have been used in fused-silica drawing machines to attain the necessary temperatures. A schematic diagram of a simple fused-silica drawing machine is shown in Figure 3.2.

Although thin-wall fused-silica columns have a high tensile strength (i.e., 1.38×10^5 kPa for a 0.3 mm o.d. column coiled in a 150 mm diameter), this strength is greatly reduced (factor of 10^4)[10] by slight surface imperfections caused by handling. Consequently, a polymeric sheath is applied directly to the capillary tubing as it emerges from the drawing furnace. This material needs to be mechanically durable and thermally stable to high temperatures to prevent contributions to column bleed when operating at the upper temperature limit. Initially, silicone rubber and other low-temperature polymers were used. Since then, polyamide and polyimide polymers have been used. They are stable to temperatures of 350°C, and to over 400°C for short time periods. Metal-covered open tubular columns may soon be available which may prove advantageous. Vitreous carbon has also been used as an exterior covering for fused silica.[49] This is formed by passing the column through a butane atmosphere immediately following the drawing stage.

Due to the sophisticated technology involved in drawing fused-silica open tubular columns, most chromatographers and suppliers rely on the fiber optics industry actually to fabricate the capillary tubing. Currently, there are several companies that offer a range of sizes, outer coatings, and glass compositions.

Surface Treatments

Modifications of the column walls prior to stationary phase coating have two primary purposes: (a) deactivation of the active surface sites and (b) chemical and/or physical modification to enhance surface wettability. Unfortunately, it is difficult to accomplish completely both of these goals si-

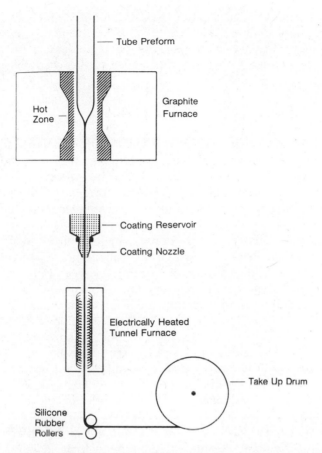

Tube Preform

Hot Zone

Graphite Furnace

Coating Reservoir

Coating Nozzle

Electrically Heated Tunnel Furnace

Take Up Drum

Silicone Rubber Rollers

FIGURE 3.2. Schematic diagram of a fused-silica drawing machine. (Reproduced with permission from ref. 14. Copyright Preston Publications, Inc.)

multaneously. When surfaces are deactivated to render them nonadsorptive and chemically inert for sample solutes, the original high-energy surface is converted to a lower-energy surface. Consequently, only stationary phases of the appropriate surface tensions spread evenly on the deactivated surface. As stationary phase polarity increases, so does its surface tension, making it more difficult to coat evenly. This does not present a problem for most nonpolar phases since they are usually stable on deactivated surfaces. Also, deactivation is not usually necessary for extremely polar phases since they tend to form their own deactivation layers and block the active sites. However, for mid-range polarity stationary phases, serious problems can arise in obtaining both a deactivated surface and one that still possesses sufficient surface energy to remain wettable. With these considerations in mind, it is necessary to employ different preparation procedures for different polarity stationary phases. Current state-of-the-art column technology is more advanced for nonpolar columns, but significant improvements have been made in understanding the parameters and devising treatments for more polar phases. In this section, treatments to enhance surface wettability and deactivation methods will be discussed.

Surface Roughening

One method of increasing surface wettability to compensate for losses in surface energy is the physical roughening of the surface. Generally, a liquid will spread better on a rough surface than on a smooth one because the surface covered by the liquid drop releases more energy due to interfacial forces. For a rough surface, there is more area under the liquid drop, and therefore, more energy is released. The influence of surface roughness becomes apparent by a decrease in the contact angle. Wenzel[50] defined a roughness factor that equates decreased contact angle directly to the increased surface area. The validity of this relationship has been fully proven.[51]

Physical surface roughening of glass open tubular columns is one of the most widely used modification techniques. Roughening can be accomplished by some form of surface corrosion or by the deposition of microparticulate matter on the column surface. Methods employing corrosion are generally not utilized in the preparation of fused-silica columns since the thin-wall columns become brittle and excessively fragile.[52] Surface roughening by aqueous surface corrosion, induced crystal growth with HCl or HF gas, and deposition of barium carbonate, sodium chloride, or silica will be briefly discussed below.

The roughening of glass open tubular column inner surfaces has been reported using solutions of ammonia,[53] sodium hydroxide,[54] hydrofluoric acid,[55] hydrochloric acid,[56] potassium hydrogen difluoride,[55] and successive treatments with several of these solutions.[57] Most of these aqueous methods deeply attack the glass surface and produce strongly adsorbing columns that are useful only in some special applications in gas–solid chromatography.

The roughening of soft glass columns can be accomplished by the formation of NaCl crystals induced by the reaction of HCl gas at elevated temperatures.[58-62] Pyrex columns of low alkali content show little reaction with HCl gas and remain transparent, while soft glass is very reactive and turns an opaque white after treatment. The crystal growth and distribution are dependent on the time and temperature of the reaction. Temperatures ranging from 300 to 400°C for times of 1–24 h have been suggested. The formation of NaCl crystals on the glass surface has been discussed in detail by Franken et al.[61]

Surface roughening can also be accomplished by reaction of the glass surface with hydrogen fluoride gas.[63-69] By using high enough concentrations of hydrogen fluoride gas, silica whiskers can be formed.[64,65] The length and shape of these whiskers are highly dependent on the exact reaction conditions. Whiskers ranging from fine and filamentary, resembling glass wool, to short and wide can be formed.[67] The glass surface area can be increased up to 1000-fold.[68] This procedure works for both soft and Pyrex glass columns, but is usually used for Pyrex glass. When this procedure was applied to fused silica, holes rather that outgrowths were formed.[70] Reagents that have been used include HF gas,[67] a fluoroether that cleaves off HF gas on heating,[63–65] and ammonium hydrogen difluoride which dissociates to produce HF gas on heating.[68] The latter appears to be the reagent of choice since it is easy to handle, forms no troublesome carbon deposits (as do fluoroethers), and reacts to form maximum whisker growth and symmetry after only 3 h.[68]

Sandra et al.[69] have summarized the advantages and disadvantages of whisker surfaces. An obvious advantage is that with the increase in surface area a larger sample capacity is obtained. Furthermore, silica whiskers stabilize all stationary phases, and droplet formation is seldom observed. The main disadvantage is that whisker surfaces are extremely active and most deactivating methods are inadequate.[71,72] In addition, the high degree of roughening decreases the separation efficiency.

A procedure based on the production of a layer of barium carbonate crystals grown from nuclei on the glass was reported by Grob et al.[73–75] Barium carbonate crystals were formed on the surface by reacting carbon dioxide gas with a liquid coating of barium hydroxide. It was found that the structure of barium carbonate layers produced on the glass surfaces is influenced by a large number of experimental variables including glass surface structure, crystallization temperature, and surfactants.[75] Although this procedure successfully roughens the surface, it also produces an active surface that is difficult to deactivate.[71] The barium carbonate layers also lead to phase degradation and increased bleeding rates.[76]

Surface roughening can also be done by the direct deposition of NaCl crystals on the column surface from an aqueous sodium chloride solution.[77–79] Although these procedures have only been reported for glass columns, they could be applied to fused-silica surfaces as well. The degree of crystal coverage is dependent on the concentration of the sodium chloride solution, with a 10% solution giving approximately a 20% surface coverage. The amount of roughness imparted by this procedure was not adequate to stabilize and maintain thermostable columns of the more polar stationary phases.[78]

German and Horning[80] developed a method of surface roughening by suspending fine particles of silanized silicic acid in the stationary phase as it is coated on the inner wall of the column. Since these particles do not adhere to the glass surface or self-aggregate, they must be coated with a binder such as the stationary phase itself and, in some instances, a surfactant like BTPPC (benzyltriphenylphosphonium chloride). Experience has shown that when an adequate amount of stationary phase, solvent, and Silanox are suspended together, the solution is too viscous to coat properly. Therefore a two-step coating procedure is employed. In the first step, a dilute solution of stationary phase is suspended together with the Silanox to act as a binder and is passed slowly through the column leaving behind a film of silica particles bound to the glass wall by the stationary phase. In the second step, a more concentrated stationary phase solution is passed through the column, thus increasing the total amount of stationary phase on the column wall.[81] Since two steps are involved, it closely resembles SCOT (support-coated open tubular) column preparation procedures. A variety of polar columns have been successfully produced by a modification of the two-step dynamic method,[82,83] including the use of polar surfactants,[82] one-step dynamic procedures,[84] and the use of unsilylated fumed silica.[85,86] The deposition of a thin film of colloidal silicic acid on the inner wall of borosilicate glass has also been described.[87] Another method[88] involves depositing a minute amount of silica on the inner wall of the capillary column.

Surface Deactivation and Chemical Modification

Another method that has been used to enhance surface wettability is chemical modification of the surface. This is particularly desirable since excessive surface roughening degrades column efficiency, especially in smaller bore columns. By this approach, attempts are made to bond or incorporate groups onto the glass or silica surface that possess sufficient surface energy to facilitate coating of the desired stationary phase. Chemical modifications are also used to deactivate the active surface sites. In glass columns that contain Lewis acid sites, more extensive deactivation procedures are necessary. In fact, untreated fused silica is sufficiently inert that reasonably good quality columns can be prepared without any surface treatments. In this section, the various approaches to glass and fused-silica deactivation are discussed. In some cases these deactivation procedures also function to enhance surface wettability.

One of the earliest forms of deactivation was by the use of surface-active agents, which adhere to the inner wall of the open tubular column, forming an oriented monomolecular layer. These materials usually have a polar end that is adsorbed on the column wall and deactivates the active sites. The other end of the molecule presents a surface that decreases the surface tension of the liquid stationary phase. Several workers have used BTPPC[66,88,89] or sodium tetraphenylborate (Kalignost)[88] for deactivation of glass open tubular columns. Other surface-active agents such as polydentate phosphonium salts,[90] amines,[66,69] or ammonium salts[66] have been used. Also, thin coatings of basic salts such as K_2CO_3, Na_3PO_4,[91,92] and KF[93] have been successfully used in the preparation of polar phase glass columns.

The main drawbacks of these approaches are the ready displacement of the monolayers by other substances, their limited thermal stabilities, and their inherent activity toward many sensitive compounds. It is also possible that these additives may affect the retention characteristics of the stationary phase.

Various procedures[73,94] have been described in which a nonextractable layer of Carbowax can be formed on the column surface. Thin films of Carbowax are coated on the column wall and, after a heat treatment of about 280°C, are exhaustively extracted with solvent.

Surface wettability measurements made on both glass and fused-silica columns treated by this procedure indicate that a surface having a critical surface tension of 45 dyne cm^{-1} is formed.[41] Consequently, stationary phases such as Carbowax and those less polar, including apolar phases,[95] are easily coated on column surfaces treated in this manner. The first fused-silica columns prepared by Dandeneau and Zerenner[9] were deactivated with Carbowax 20M using the above procedure.

The exact bonding mechanism of Carbowax to the glass surface is not known. It is believed by some[96] that pyrolysis of Carbowax produces organic species, such as oxirane, which are chemically bound to the surface by reaction with the surface hydroxyl groups. The actual formation of chemical bonds between the thin polymer layer and the silica surface has been verified by diffuse reflectance infrared spectroscopy.[97]

In addition to Carbowax 20M, a number of other organic polymers have been used for deactivation. Some of these include Carbowax 1000,[74] Carbowax 400,[59] Superox-4,[98] ethoxycarbonylpolyphenylene,[99] alkylpolysiloxanes,[72] poly-N-β-hydroxyethyl aziridine,[66] N-isopropyl-3-azatidinol,[61] and N-cyclohexyl-3-azetidinol;[66] the latter two, although not polymeric, can be heat polymerized at relatively low temperatures.

The thermal stabilities of Carbowax deactivated columns are a matter of controversy. Although claims have been made of temperature stabilities above 300°C, studies have shown[72] that these columns are not stable for extended use beyond 250°C. In addition to thermal instability, polar polymer deactivation layers can also influence the polarity of the stationary phase. This is especially noticeable for columns coated with thin films of apolar phases. Unfortunately, this effect is pronounced enough on fused-silica columns that the Carbowax method is no longer considered a viable form of deactivation.[14]

Leaching of open tubular columns serves two primary functions. For glass columns, leaching is used to remove the metallic cations from the glass matrix forming a silica-rich layer. This process also increases the number of surface hydroxyl groups. Since fused-silica columns are free of surface impurities, the only effect of leaching is to increase the hydroxyl density[52] which may be desirable for subsequent treatments. In glass columns, the removal of the metallic Lewis acid sites results in improved chromatographic performance and lends a higher degree of column reproducibility by minimizing the effects of glass variety.

A tremendous amount of work has been done studying the leaching process and in developing optimized procedures for its successful use. Dealkalinization of HCl gas treated surfaces with acidic solutions were first used.[99,100] In other procedures, concentrated (20%) HCl was used directly to leach the metallic cations from the glass matrix. Two widely used procedures are a static method[101-103] in which the column is partially filled with leaching solution and sealed, and a dynamic method[104,105] in which a continual flow of fresh leaching solution is flushed through the column.

The success of leaching for the formation of a silica-rich surface layer has been fully validated from surface analysis studies of leached open tubular columns.[104,105] These studies [105] have also shown that ion migration back to the surface occurs, but can be minimized with appropriate heat treatments and additional leaching. Other studies have also been conducted on leaching[106] and the formation of silica-rich surface layers.[107] Water vapor has been used to leach fused-silica columns.[70]

One of the most practical methods of eliminating the undesirable activity of the surface hydroxyl groups is to replace them chemically with inert groups. A similar method involves covering them with an inert substance that reacts with the surface hydroxyl groups. Furthermore, silica surface wettability can be enhanced by choosing modifying groups that are compatible with the desired stationary phase.

One of the most widely applied methods of chemical surface modification in glass open tubular column preparation is silylation or silanization, the terms being synonomous. Using this method, the surface hydroxyl groups on the silica are replaced with silyl-ether groups. Modifications produced in this manner are extremely stable, owing to the strength of the Si—O—Si linkage. The polarity and chemical characteristics of the modified layer that is formed by silylation can be controlled by the choice of constituents of the silylation reagent.

Most column silylation procedures have utilized trimethylchlorosilane, hexamethyldisilazane, or mixtures of the two to form a trimethylsilyl surface layer.[63,88,102,105,108-113] Other silylation procedures utilizing reagents with more polar groups have been described by Bartle and Novotny,[114] Grob and Grob,[115] and Welsch et al.[116] Early experience with low-temperature silylation was disappointing because columns deteriorated rapidly.[88] Evidence indicates that reaction temperatures of 300–400°C are necessary to ensure complete available reaction.[28,117,118] When silylation procedures were performed at these higher temperatures,[105,111,112] highly deactivated (for nonpolar phases) and very stable deactivation layers were formed. The critical surface tension of leached and trimethylsilylated glass and fused silica was found to be about 21 dyne cm^{-1}.[41] It is not surprising that the critical surface tension of such a surface is similar to the surface tension of a methylpolysiloxane stationary phase, since the modified surface should have a similar composition. From the same study,[41] the increased effectiveness of silylation at 400°C over lower temperatures was also demonstrated. Zisman plot data[41] for several variously treated column surfaces are listed in Table 3.6. The less negative slopes of the Zisman plots correspond to a decreasing con-

TABLE 3.6. Surface wettability data derived from Zisman plots for variously treated capillary column surfaces.

		Zisman plot data	
Surface	Chemical treatment	Critical surface tension (dyne cm^{-1})	Slope (cm dyne^{-1} × 10^{-3})
Pyrex glass	HCl leach, 400°C silylation	21	−34
Pyrex glass	150°C silylation	21	−18
Fused silica	400°C silylation	21	−24
Fused silica	NaOH and HCl leach, 400°C silylation	21	−29
Pyrex glass	HCl leach, methylpolysiloxane	21	−37
Pyrex glass	HCl leach, D$_4$ treatment	21	−31
Fused silica	D$_4$ treatment	21	−34

centration of trimethylsilyl groups. Clearly, the degree of surface coverage by silylation is related to the number of hydroxyl groups on the surface (at least to a certain limiting concentration). Grob et al.[110] have reported that a maximum density of silanol groups increased the degree of deactivation and thermostability obtainable by silylation. Data from surface wettability measurements[41] also show that hydroxylated fused silica gives rise to a higher surface coverage by silylation (see Table 3.6).

Schomburg et al.[119] have described a deactivation procedure in which a nonextractable layer of methylpolysiloxane phase is produced on the glass surface (PSD = polysiloxane degradation). With this technique, a polysiloxane phase such as OV-101, OV-1, SE-54, OV-17, and so on, is pyrolyzed on the column surface at temperatures of approximately 450°C. During heat treatment, the polysiloxane partially decomposes with probable bonding taking place between the decomposition products and the surface silanol groups. Wettability measurements[95] indicate that this deactivation method using OV-101 forms a very dense surface coverage with the same critical surface tension as is produced by silylation (Table 3.6). After heat treatment, the column is rinsed with solvent to remove the nonbonded stationary phase. The column is then recoated with the same stationary phase to produce a film of defined thickness and polarity. Low-polarity methylpolysiloxane phases have been successfully used for the deactivation of fused-silica columns. Phases such as OV-17, however, are not amenable for the deactivation of fused-silica surfaces.[70]

Recently, a renewed effort has been made in producing deactivation layers on glass and fused silica that are chemically compatible with the stationary

phase. This is a natural result of attempts to prepare inert columns coated
with polar stationary phases. Compatability between surface and phase can
be obtained by surface modification at high temperature with cyclic siloxanes
with the same functional sidegroups as the silicone stationary phase. Meth-
ylcyclosiloxanes were used by Stark et al.[120] to deactivate fused-silica surfaces for subsequent coating of nonpolar stationary phases. This same procedure has been applied by Verzele et al.[70] Wettability measurements[41] made
on fused silica and glass for D_4 (octamethylcyclotetrasiloxane) treatments
indicate that surfaces similar to those produced by silylation are formed (see
Table 3.6). An example of the effectiveness of the D_4 treatment is shown in
Figure 3.3 in which chromatograms are shown of acidic and basic test mixtures on a fused-silica column coated with SE-54. The deactivation consisted
of rinsing the column with methanol, followed by D_4 treatment. Both highly
acidic components and very basic components were eluted at low levels with
nearly perfect peak shape.[121]

Blomberg et al.[122,123] have described similar deactivations in which cyclic
trifluoropropyl methylsiloxanes were used to obtain deactivated glass surfaces suitable for coating with OV-215 (50% trifluoropropyl methylsilicone
gum). Modifications were also described using cyanopropylsiloxanes[123] to

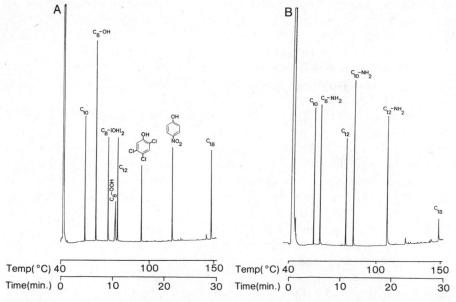

FIGURE 3.3. Chromatograms of (A) an acidic test mixture and (B) a basic test mixture obtained on the same column, illustrating the high degree of inertness possible on a deactivated
fused-silica column. Note: 11 m SE-54 fused-silica column (rinsed with methanol and D_4 deactivated prior to coating with 0.25 μm film thickness), temperature program from 40 at 4°C min^{-1}
after a 2 min isothermal period, H_2 carrier gas at 45 cm s^{-1}, sensitivity set for full-scale response
for 1 ng of dodecane. (Reproduced with permission from ref. 121. Copyright Elsevier Scientific
Publishing Company.)

deactivate glass surfaces suitable for coating with very polar phases such as Silar 10C. Similar work[124] in which the surface was alkylated with pentafluorobenzyl bromide at 60°C was reported to produce a glass surface compatible with OV-225 (75% phenyl, 25% cyanopropyl methylsilicone).

Two novel methods of deactivating fused silica have been described by Pretorius et al.[125,126] One method[125] involved forming a thin layer of pyrocarbon on the inner surface of the capillary. This was done by the high-temperature pyrolysis (1200°C) of hydrocarbons in the gas phase. Pyrocarbon is an allotropic form of carbon having essentially an inert surface and a critical surface tension of 34 dyne cm^{-1} which makes it wettable by low- and medium-polarity phases. In the study in which these results were reported, thick-wall fused-silica tubing which did not require a polymer outer coating was filled with propane and heated at 1200°C. It would be possible to form the same type of surface in thin-wall fused-silica columns during the drawing procedure by purging the column with the appropriate hydrocarbon mixture.

In the second method,[126] a thin film of elemental silicon was deposited on the inner surfaces of glass open tubular columns. Elemental silicon, if deposited in a very thin film, forms a smooth and inert surface of high surface energy. Such films were produced from the static pyrolysis of monosilane at temperatures between 250 and 500°C.

Column Coating

The main objective in coating the open tubular column with the stationary phase is to provide a uniform film, usually 0.1–1.5 μm thick, throughout the length of the column. This is necessary in order to obtain the highest possible separation efficiency and resolution. Coating can be accomplished by either the dynamic or static method. The static coating method has generally been considered to be superior to the dynamic method,[60,88] producing columns that are more efficient. This procedure is more difficult and time consuming, however. Both methods are discussed below.

The dynamic method was first described by Dijkstra and de Goey[127] and generally consists of filling 2–15 coils of the column with a solution of the stationary phase, followed by forcing this volume through the column at a velocity of approximately 1–2 cm s^{-1} with helium pressure. A thin film of this solution is left behind on the column wall. Continual flushing with helium after coating evaporates the remaining solvent and leaves a thin coating of stationary phase.

Oftentimes, nonuniform films are obtained by this method. The reasons for this nonuniformity are: (a) As the coating solution is discharged from the end of the column, the coating velocity increases sharply, which results in a thicker film at that end of the column. A buffer column is often attached to the end of the column to avoid this problem. (b) The consumption of some of the coating solution during the coating process results in a faster linear velocity of the solution plug as coating proceeds, and therefore, an increasing film thickness. (c) When a fairly large amount of solution is deposited on the column wall, the liquid may drain from the walls and collect in the lowest parts of the coils. Proper orientation of the column in a horizontal direction around a cylinder can help reduce this effect.[4,128] (d) The solvent evaporation step involves transport of some of the stationary phase toward the end of the column, resulting in increasing film thickness along the column.[129,130] (e) Small temperature differences or fluctuations along the column cause the solvent to distill from the warmer parts of the column and condense at the cooler parts. This results in droplet formation and film nonuniformity.[131]

The formation of droplets or plugs (assuming the glass surface is wettable) can be largely avoided, or at least greatly reduced, by controlling the coating speed,[131] temperature,[131] concentration of the coating solution, solvent volatility,[132] and rate of solvent evaporation.[130] Methylene chloride has been a popular solvent for the dynamic coating procedure because of its low boiling point (41°C), good solvent properties, and nonflammability. However, the high vapor pressure of methylene chloride increases the probability of droplet and lens formation during coating. Temperature programming the capillary column in a water bath from 22 to 32°C during coating, and up to 42°C during the solvent evaporation step, has been found to be effective in eliminating this problem by ensuring that the temperature never falls below the dew point of the solvent vapor.[132] Maintaining gas flow through the column for several hours after coating helps evaporate the solvent and reduce the formation of droplets. Blomberg[130] found that at low solvent evaporation rates ($\sim$0.25 cm^3 min^{-1}) some of the stationary phase is transported from the column as hundreds of moving droplets of solution, leaving a very thin film of stationary phase uniformly distributed throughout the column. With rapid evaporation ($\sim$4.00 cm^3 min^{-1}), only lenses of short duration are formed and, as a consequence, there is no phase transport except at the very beginning of the column. This results in thicker, uniform films. Intermediate flow rates have resulted in nonuniform films.

Several different approaches to improving the dynamic coating method have been proposed. Levy et al.[133] have described the use of a flow restriction device, placed downstream from the column to be coated, which stabilizes the flow rate in the column during the entire coating process. Van Dalen[134] and McConnell and Novotny[135] connected the open tubular column to a syringe pump, filled the entire length of tubing with the solution of stationary phase, applied gas pressure at the other end of the column, and withdrew the solution by operating the syringe in the withdrawal mode. The syringe pump serves as a brake, preventing a change in coating speed as the plug length diminishes. The variation in film thickness obtained by this method was within 2%.

Probably the most significant development in the dynamic coating method was the introduction of the "mercury plug" method of Schomburg and co-workers.[136,137] This method involves adding a mercury plug between the solvent plug and the driving gas which, because of its high surface tension, wipes most of the coating solution off the surface as the plug moves through the column. More concentrated solutions are used in this procedure, resulting in the formation of films that resist drainage during the drying step.

Predictions of the film thickness (d_f) in dynamically coated open tubular columns have been made using several mathematical relationships. From an evaluation and comparison of these equations with experimental results, Bartle[129] concluded that the Fairbrother–Stubbs equation[138] suggested by Novotny and co-workers[109,139] was the most useful. In this relationship, d_f depends directly on r (column radius) and $v^{1/2}$ (coating velocity) as

$$d_f = \frac{rc}{200} \left(\frac{v\eta}{\gamma} \right)^{1/2} \tag{3.1}$$

where η and γ are the viscosity and surface tension of the solution, respectively.

The value of d_f has also been calculated from the difference in volume of the coating solution before and after coating,[2,131] from the weight of stationary phase that can be rinsed out of the column,[140,141] and from the shortening of a plug of stationary phase solution over a given column length.[131]

The static coating method was first developed by Golay[1] and later described for glass open tubular columns by Bouche and Verzele.[142] The general procedure involves filling the column with a dilute solution of the stationary phase, sealing the column at one end, and evaporating the solvent from the other end under vacuum. This leaves a thin film of stationary phase, the thickness of which can be easily calculated from

$$d_f = \frac{r}{2\beta} \tag{3.2}$$

where r is the radius of the column and β is the phase ratio. An important advantage of the static method is that the phase ratio is known accurately and, therefore, the film thickness can be accurately determined. To calculate the phase ratio from a weighed amount of stationary phase, the density of the stationary phase must be known. Rutten and Rijks[143] have recently tabulated a number of stationary phase densities which are listed in Table 3.7.

A number of papers[142–146] reported the details of various static coating procedures. In practice, it is important that the coating solution be dust-free and degassed to eliminate bumping during the solvent evaporation step, and that no air or vapor bubbles exist in the column, especially at the sealed end.[142] The column must be kept at a constant temperature to prevent non-uniform film deposition. A thermostated water bath is not adequate because temperature fluctuations, however small, produce sufficient contraction and expansion to cause the deposition of the stationary phase in the form of bands.[143] An air or water bath of constant temperature is usually sufficient. In the case where a higher than ambient temperature is needed, a simple

TABLE 3.7. Specific weights of some stationary phases.

Phase	Specific weight	Phase	Specific weight
AN-600	1.08	OV-101	0.96
DC-200	0.97	OV-105	0.99
DC-510	1.00	OV-210	1.32
DC-550	1.07	OV-225	1.09
DC-710	1.10	OV-275	1.16
DEGS	1.26	PEG 400	1.13
OS-124	1.21	QF-1	1.32
OV-1	0.98	SE-30	0.96
OV-3	1.00	SE-54	0.98
OV-7	1.02	SF-96	0.97
OV-11	1.06	Silar 5CP	1.13
OV-17	1.09	Silar 10C	1.12
OV-22	1.13	SP-2401	1.30
OV-25	1.15	Squalane	0.83
OV-61	1.09	XE-60	1.08

and inexpensive solution is to place the column into a water bath that is within another water bath.[131,145] Although the outer container is controlled by a thermostat, the heat transfer from the outer to the inner container is very gradual.

Although methylene chloride was originally used as a solvent for static coating, pentane (when the stationary phase solubility allows) is now preferred, since evaporation can be accomplished in approximately half the time necessary for methylene chloride solutions. Average coating times at room temperature range from approximately 15 h for methylene chloride solutions to about 8 h for pentane for a 20 m × 0.3 mm i.d. column.[144] Goodwin[146] claimed that many solvents, including methylene chloride, deposit the phase as droplets of about 0.01–0.05 mm diameter. Diethylether, however, was found to deposit the phase in a uniform film.

Evidence indicates that columns coated with stock solutions of SE-30 that have aged for several weeks produce more active and less thermostable columns.[147] It was suggested that trace amounts of HCl, present in the coating solution of methylene chloride, was responsible for some stationary phase degradation. Venema et al.[148] have verified that solutions of SE-30 and SE-54 stationary phases which have been slightly acidified degrade rapidly, as evidenced by a decrease in molecular weight. Consequently, it is necessary to use fresh stationary phase solutions prepared from pure solvents for maximum column performance and stability.

Although a sodium silicate solution (water glass) was originally used to seal the ends of columns,[142–144] several other methods of making seals have been suggested. These include cements,[59] silicone gums,[140] adhesives,[145,149,150] Apiezon N,[151] Vaseline hardened by addition of a small amount of histological wax,[146] and paraffin wax.[152] Mechanical attachment of a plug with shrinkable Teflon tubing has also been advocated.[150] Static coating and mechanical closure techniques have been reviewed by Grob and Grob.[153]

In Golay's original work,[1] open tubular columns were coated by filling them with a dilute solution of liquid phase, sealing one end, and drawing the column, open end first, through an oven. Ilkova and Mistryukov[91,154,155] and later Jennings[156] revised this method for glass columns by introducing the open end of the coil into a high-temperature oven and rotating the column around its coiling axis into the oven. As the column is screwed into the oven, the solvent evaporates and escapes through the open end leaving the stationary phase on the column wall. A recent[157] modification of this procedure involves introducing the column into a heated liquid bath instead of an air bath. This procedure reportedly provides more homogeneous coating.

3.4   STATIONARY PHASES

Numerous stationary phases have been used in GC, although only a few have really been used successfully in open tubular column GC. The performance requirements for high efficiency and detectability are so demanding that most commercial phases do not possess the necessary properties that lead to efficient, inert, and thermally stable open tubular columns. A list of commercially available phases that have been used in open tubular column applications is given in Table 3.8. It is interesting to note that the polysiloxanes and polyglycols account for the bulk of the phases used.

Stationary Phase Stability

For an open tubular column to retain its efficiency at its upper operating temperature and with prolonged use, it is necessary that the stationary phase remain stable as a thin, uniformly deposited film. Although it is possible (and has been for several years) to prepare very stable columns with apolar stationary phases, the same success has not been achieved for polar phases. With the introduction of fused silica, the awareness of this problem was intensified since many of the moderately polar phases that coat on glass produced columns of very low efficiency and low thermal stability when coated on fused silica. At first, it was thought that the surface energy of the fused silica was insufficient for the more polar phases, but experiments[41] have shown that clean fused silica is a high-energy surface. Capillary rise experiments[41] have shown that untreated glass, fused silica, and quartz surfaces all have critical surface tensions greater than 50 dyne cm^{-1}. This value is higher than the surface tension of one of the most polar stationary phases, 1,2,3-tris[cyanoethoxy)propane. By surface roughening (glass columns) it is

TABLE 3.8. Commonly used commercial stationary phases.

Name	Chemical nature	Temperature range (°C)	Polarity
Squalane	C_{30} Alkane	−50–100	0
SF-96, OV-101	Methylsilicone oil	0–200,260	205,229
SE-30, OV-1	Methylsilicone gum	20–350	216,217
SE-52	5% Phenyl methylsilicone gum	20–350	334
SE-54	1% Vinyl, 5% phenyl methylsilicone gum	20–350	337
UCON LB 550	Polyethylene polypropylene glycol, ca. 10/90	−20–160	496
OV-7	20% Phenyl methylsilicone oil	20–300	592
OV-1701	7% Phenyl, 7% cyanopropyl methysilicone gum	40–300	819
OV-17	50% Phenyl methylsilicone oil	20–250	884
OV-17 (gum)	50% Phenyl, 2% vinyl methylsilicone gum	20–300	—
OV-25	75% Phenyl methylsilicone oil	20–300	1175
OV-210	50% Trifluoropropyl methylsilicone oil	20–200	1520
OV-215	50% trifluoropropyl, 2% vinyl methylsilicone gum	20–200	1545
UCON HB 5100	Polyethylene propylene glycol, ca. 50/50	20–200	1706
OV-225	25% cyanopropyl, 25% phenyl methylsilicone oil (or gum)	50–220	1813
Superox-4	Polyethylene glycol gum	50–300	2238
Superox-0.1	Polyethylene glycol gum	50–280	2301
Carbowax 20M	Polyethylene glycol gum	80–250	2308
Superox 20M	Polyethylene glycol gum	50–300	2309
SP2300	36% Cyanopropyl methylsilicone oil	50–240	2424
Silar 5CP	50% Cyanopropyl, 50% phenyl silicone oil	50–240	2428
SP2340	75% Cyanopropyl methylsilicone oil	100–240	3678
Poly-S 179	Polyphenylether sulfone	200–400	—
Silar 10CP	100% Cyanopropyl silicone oil	100–240	3682
OV-275	100% Cyanoethyl silicone oil	100–240	4938
Chirasil-Val	Chiral polysiloxane	50–240	—

possible to stabilize sufficiently polar phases so that successful columns can be prepared.

In comparing glass and fused silica, fewer stationary phases are available on fused-silica columns. This is principally a result of the fact that the glass surface can be easily roughened, increasing both its wettability and film stability for a wider range of stationary phase polarities and viscosities. The thin walls inherent to fused-silica columns largely prevent such surface modification, and other solutions have been sought.

An important factor concerning the stability of stationary phases on both glass and fused-silica surfaces is the change in wettability parameters as the temperature increases. Both the surface tension of the stationary phase and the critical surface tension of the capillary surface change with temperature. In a series of experiments[41] at temperatures up to 250°C for selected stationary phases and surfaces, these changes parallelled one another, and no effective change in wettability occurred. However, it is possible that for some systems this situation would not occur, and conditions of nonwettability would be established at some elevated temperature.

Another important factor in the stability of coated stationary phases is the viscosity of the deposited film. A recent study[158] has shown that the coating efficiencies and stabilities of coated phases correlate with their viscosities. This agrees well with the long-term observation that viscous gum phases coat more efficiently and are stable to higher temperatures than less viscous phases. Nonchromatographic studies[159,160] have shown that the rate of film disruptions is inversely proportional to the viscosity of the film. Nonpolar polysiloxane phases exhibit nearly a constant viscosity as a function of temperature.[161] However, due to their structure, the viscosities of phenyl-containing and other polar phases drop off rapidly at elevated temperatures[161] (see Figure 3.4), resulting in lower film stabilities.

Recently, advances have been made in overcoming these problems. They include the *in situ* free radical cross-linking of coated stationary phases and the development of polar gum phases of high viscosity that are capable of being cross-linked. Both of these developments are described below.

Cross-Linkable Stationary Phases

The silicones possess a number of desirable qualities that make them the most frequently used stationary phases in GC. Principal among these is their excellent thermal stabilities over relatively wide temperature ranges. In addition, these polymers demonstrate exceptionally high permeability to solute vapors due to a high degree of chain mobility that provides openings permitting diffusion. This explains the high efficiencies often observed with columns coated with these phases. Blomberg[162] has thoroughly reviewed the fundamental reasoning and practical aspects relating to contemporary stationary phases in GC.

The excellent qualities of the stationary phases that have a polysiloxane backbone are a result of the chemical structure of the polysiloxane mole-

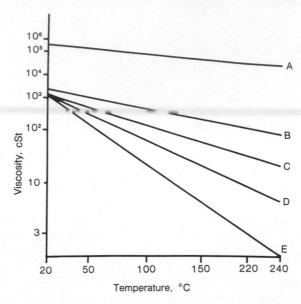

FIGURE 3.4. Viscosity dependence on temperature for several polysiloxanes and a mineral oil: (A) high-viscosity methylpolysiloxane similar to OV-1, (B) medium-viscosity methylpolysiloxane similar to OV-101, (C) medium-viscosity and low phenyl content methylphenylpolysiloxane, (D) medium-viscosity and high phenyl content methylphenylpolysiloxane similar to OV-17, and (E) medium-viscosity mineral oil. (Reproduced with permission from ref. 160. Copyright Springer-Verlag.)

cule.[163] It is believed that the methylpolysiloxane molecule has a helical conformation. A rise in temperature has two effects on the polysiloxane chain: it will increase the mean intermolecular distance while at the same time, it will expand the helices and thus diminish this distance. The expansion of the helices opposes the effect of the increase of intermolecular distance, so that the net intermolecular distance, and therefore the viscosity, appears to be only slightly affected by the temperature.[158]

The substitution of bulkier groups, such as phenyl and cyanopropyl, in the polysiloxane chain distorts the regular helical conformation of the polymer. This reduces the compensating effects of bond distance and helix expansion, with the net result being a greater change in viscosity with temperature (Figure 3.4). Nevertheless, these polysiloxanes retain their viscosities better than other stationary phase materials. The change in viscosity with temperature for a medium viscosity mineral oil is shown in Figure 3.4 for comparison.

The importance of viscosity on stationary phase stability has intensified efforts to synthesize tailor-made polysiloxane gums that could be cross-linked by free radical initiation. Cross-linking greatly controls the tendency to lose viscosity with increasing temperature during temperature programming, and it produces a stationary phase film that is resistant to wash-out by organic solvents. This is especially important for the moderately polar to polar stationary phases.

There have been two basic approaches used in the production of nonextractable cross-linked stationary phases. The first approach involves the formation of Si—O—Si bonds, while the second approach results in carbon–carbon bonds, usually between methyl groups attached to silicone atoms (Si—C—C—Si). The advantages of the enhanced film stability obtained by cross-linking has been discussed by Grob and Grob.[164] Large volumes of solvent can be introduced into the column during injection or for rinsing out nonvolatile materials with minimal phase stripping, and changes in viscosity with temperature are minimized.[158]

Probably the first report of the preparation of immobilized, thermally stable stationary phases was by Sinclair et al.[165] who described the oxidation of SE-52 on Gas Chrom Q. Grob[166] first attempted surface bonding on open tubular columns using organolithium compounds and *in situ* polymerization of polyolefins. Madani and co-workers[167–171] prepared methyl- and methylphenylpolysiloxane prepolymers by hydrolysis of dimethyl- and diphenylchlorosilanes. After dynamically coating the polymeric mixture on the capillary wall, the column was filled with ammonia gas, sealed, and heated to a high temperature for 24 h. Thus, chemical bonding was accomplished under base catalyzed conditions. Blomberg et al.[172–176] reported the *in situ* synthesis of methyl and phenylpolysiloxanes by dynamically coating the column with silicon tetrachloride. The column was then sealed and heated to 320°C for over 20 h. Using either method, the condensation of hydroxy and alkoxy groups to split out water, alcohols, or ethers left Si—O—Si cross-links. Problems associated with this approach included the observations of higher column activity and lower column efficiency than were obtained from columns merely coated with commercial phases. These problems were caused by (a) residual silanol or alkoxy groups left in the phase after cross-linking which cannot be chemically deactivated or thermally removed because of steric problems, and (b) sample molecules being less soluble in the stationary phase as a result of the cross-linking levels (10–50% depending on prepolymer chain length) required by this approach. The thermal stabilities of these phases, however, were extremely good owing to the stability of the Si—O—Si bond. It would appear that if longer polymer chains were used and less cross-linking was needed, this would be a more attractive approach.

The free radical cross-linking of silicone polymers through carbon–carbon bonds to form insoluble rubbers is well documented and has been performed industrially for years.[177,178] Unfortunately, this procedure has not been used for the preparation of nonextractable stationary phases for open tubular column chromatography until recently. Free radical cross-linking of polysiloxane stationary phases with peroxides to form insoluble rubbers has been reported by several workers.[164,179–186] In addition to peroxides, certain azo compounds[121,187,188] and gamma radiation[189,190] have also been used as free radical generators. Handling difficulties as well as the large cost associated with a radiation source make it impractical for many laboratories to use radiation for cross-linking. Furthermore, gamma radiation is less effective in cross-linking phases more polar than the methylpolysiloxanes, and it leads to deterioration of the polyimide outer coating.

The majority of studies to date on the immobilization of stationary phases by free radical cross-linking have been done with nonpolar methylsilicone stationary phases.[164,179–183] This involves the production of free radicals by proton abstraction and the subsequent combination of free radicals on adjoining molecules to form Si—C—C—Si bonds. The free radical generators that have been used are: t-butyl peroxide (TBP), azo-t-butane (ATB), dicumyl peroxide (DCP), benzoyl peroxide (BP), 2,4-dichlorobenzoyl peroxide (DCBP), azo-t-octane (ATO), and azo-t-dodecane (ATD).

The coating and cross-linking procedures differ depending on the type of free radical initiator used.[121] For those that are solids at room temperature (DCP, BP, DCBP, and ATD), the peroxide is doped directly into the stationary phase solution, and the column is coated in the normal fashion. However, for the free radical initiators that are liquids at room temperature (TBP, ATB, and ATO), the columns are usually coated first and then saturated with the vapors of the free radical initiator. This is accomplished by bubbling nitrogen gas through the liquid free radical initiator and purging the coated columns for approximately 2 h. Because of its low vapor pressure, ATO can also be doped directly in the stationary phase solution and coated in the same manner as the solid free radical initiators.

Cross-linking is accomplished by sealing both ends of the column and heating from 40°C to the curing temperature (~120–200°C, depending on the free radical initiator) at 4°C min^{-1} and holding for a specified time period.

The aroyl peroxides (DCBP and BP) are the most reactive, and cross-linking can be done at lower temperatures and with lower levels. However, these peroxides also form the most polar decomposition products (i.e., carboxylic acids), and they have limited solubilities in methylpolysiloxane stationary phases. This low solubility leads to irregularities in the film after cross-linking. The decomposition products are incorporated in the stationary phase to some extent and change its polarity.[121] For these reasons, the aroyl peroxides are not recommended.

The peroxide that has the least deleterious effect on the activity and polarity of the cross-linked phase is DCP. This peroxide has been the most popular free radical generator for open tubular column preparation. However, oxidation of polysiloxane polymers can occur during cross-linking which increases the polarity and decreases the stability of the phases.[187]

Azo compounds are fast becoming the free radical generators of choice. Recent work has shown that there are no changes in polarity or increases in column activity caused by the decomposition products of these azo compounds.[121,187,188] Oxidation does not occur, and the entire range of stationary phases from nonpolar to polar can be adequately cross-linked. Figure 3.5 shows chromatograms of acidic and basic test mixtures obtained on an SE-

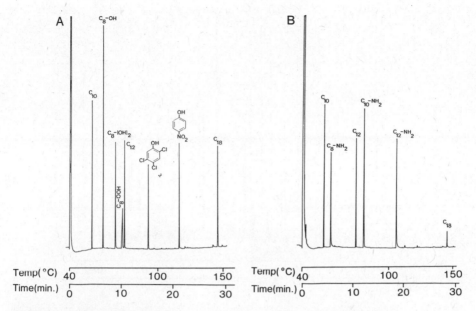

FIGURE 3.5. Chromatograms of (A) an acidic test mixture and (B) a basic test mixture obtained on a column containing a cross-linked (azo-*t*-butane) SE-54 stationary phase. Note: Chromatographic conditions the same as in Figure 3.3. (Reproduced with permission from ref. 121. Copyright Elsevier Scientific Publishing Company.)

54 column that was cross-linked with azo-*t*-butane.[121] These chromatograms are essentially identical to those obtained on a non-cross-linked column (see Figure 3.3).

Initial efforts to stabilize phenyl-containing phases by free radical cross-linking were not completely successful.[164,179] With fewer methyl groups present (being replaced by phenyl groups), there are fewer chances for methyl radicals to form and combine. Therefore, it is necessary to incorporate functional groups in the polymer during synthesis which cross-link more easily than methyl groups. Vinyl groups have generally been incorporated because lower levels of free radical initiators are necessary to achieve similar or higher levels of cross-linking as compared to methyl groups alone.[191] Peaden et al.[184] reported the synthesis of 50% and 70% phenyl polysiloxane gum phases that were more viscous than commercially available phases and that contained 1–4% vinyl groups for cross-linking. Figure 3.6 shows a chromatogram of a coal tar sample obtained on a cross-linked 50% phenyl polysiloxane stationary phase.[184] Blomberg and co-workers have also used vinyl groups in methylphenyl- and cyanopropylpolysiloxane stationary phases.[185,186]

Tolyl groups have also been used in place of vinyl groups. Blomberg and co-workers reported the synthesis of a 50% tolyl polysiloxane polymer and various cyanopropylsilicone gums containing tolyl groups.[185,186] Richter et

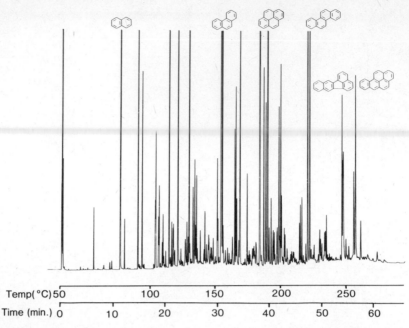

FIGURE 3.6. Chromatogram of a coal tar obtained on a cross-linked 50% phenyl, 1% vinyl polysiloxane stationary phase. Note: 12 m fused-silica column, temperature program from 40 to 250°C at 4°C min^{-1} after a 2 min isothermal period, H$_2$ carrier gas at 75 cm s^{-1}. (Reproduced with permission from ref. 184. Copyright Friedr. Vieweg and Sohn.)

TABLE 3.9. Recently synthesized gum phases.

Chemical nature	Temperature range (°C)	Reference
50% Phenyl methylsilicone gum	60–350	193
50% Phenyl, 1% vinyl methylsilicone gum	40–350	184
70% Phenyl, 4% vinyl methylsilicone gum	80–400	184
50% Phenyl, 3% vinyl methylsilicone gum	60–350	185
50% Tolyl methylsilicone gum	70–300	185
33% Cyanopropyl, 33% tolyl methylsilicone gum	40–300	186
50% Cyanopropyl, 25% tolyl methylsilicone gum	40–300	186
~95% Cyanopropyl vinylsilicone gum	40–300	186
70% Tolyl methylsilicone gum	80–300	187
90% Cyanopropyl, 10% tolyl silicone gum	50–300	187

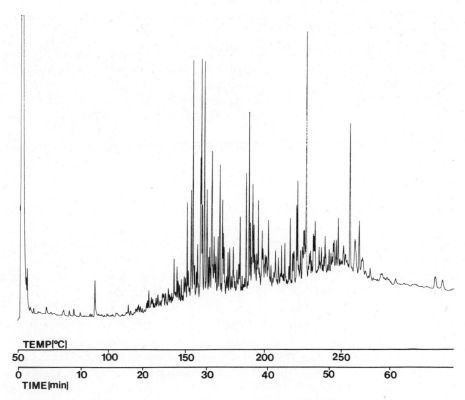

TEMP[°C]

50	100	150	200	250

0	10	20	30	40	50	60

TIME [min]

FIGURE 3.7. Chromatogram of an amino polycyclic aromatic hydrocarbon fraction of a coal liquid obtained on a cross-linked (azo-*t*-butane) 90% cyanopropyl, 10% tolyl polysiloxane stationary phase. Note: 12 m fused-silica column, temperature program from 50 to 250°C at 4°C min[−1] after a 2 min isothermal period, H_2 carrier gas at 75 cm s[−1].

al.[187] reported the synthesis of a 70% tolyl polysiloxane stationary phase as well as a 90% cyanopropyl polysiloxane polymer containing tolyl groups. A chromatogram obtained on the latter phase is shown in Figure 3.7.[192] The use of tolyl groups permits cross-linking to occur with lower levels of free radical initiators. However, because of their high reactivity, tolyl groups are easily oxidized by peroxides during cross-linking, and other free radical initiators are essential. A summary of the gum phases that have been recently synthesized and described is given in Table 3.9.

In addition to specially synthesized phases, several authors have also investigated the use of commercially available phases of moderate polarity to produce cross-linked phases. Grob and Grob,[181] Schomburg et al.,[189] Buijten et al.,[185,193] and Sandra et al.[194] have reported successful cross-linking of OV-1701 (6.8% phenyl, 6.8% cyanopropyl, 86.4% methylpolysiloxane). Sandra et al.[194] also reported the use of RSL-310, a polar liquid phase of unspecified structure, but related to the polyols (Ucons and Pluronics). This phase can also be stabilized by free radical cross-linking.

3.5 COLUMN EVALUATION

There are three parameters that are of primary importance in the evaluation of a gas chromatographic open tubular column. They are separation efficiency, surface inertness, and thermal stability. Although these parameters are generally evaluated independently, they are interrelated. For instance, surface activity manifested as peak tailing reduces chromatographic efficiency, lowers thermal stability, and increases column bleed. Likewise, low thermal stability can lead to stationary phase degradation with an accompanying loss of efficiency and an increase in surface activity. Consequently, it is important to consider carefully the interactions giving rise to a specific chromatographic behavior to evaluate a column accurately.

The general approach to column evaluation involves separating various test substances under controlled conditions and monitoring the peak shape and symmetry to obtain diagnostic data about separation efficiency and surface inertness. Thermal stability is usually characterized by bleed-rate experiments followed by efficiency and inertness evaluations. The main difficulties in chromatographic evaluation lie in choosing the appropriate test compounds, defining standardized chromatographic parameters, and in making meaningful measurements. Although various approaches have been described, no single method is currently considered to be universal or adequate to fulfill the needs of individual situations. Different polarity mixtures have been described by Grob et al.,[141,195] Schomburg et al.,[72,136] Alexander et al.,[60] Hartigan and Ettre,[196] Sandra and Verzele,[66,197] Welsch et al.,[111] Cram et al.,[198] deNijs et al.,[79] Dandeneau et al.,[9,10] Wright et al.,[105,121] and others. These mixtures contain test compounds with a variety of functional groups, imparting a wide range of polarity and acid–base properties. In addition to obtaining information on the quality of a column, it is possible to infer details about the surface chemistry from the specific behavior of the test compounds. Alkanes, aromatic compounds, alcohols, diols, aldehydes, ketones, phenols, free fatty acids, esters, and amines have been used. A chromatogram of the Grob polarity mixture is shown in Figure 3.8.[195]

As column technology has improved and more efficient and inert columns produced, it has been necessary to utilize more difficult-to-chromatograph compounds as test components. For instance, the classical acid–base test mixture of dimethylphenol and dimethylaniline is nearly quantitatively eluted with perfect peak symmetry on nondeactivated SE-54 fused-silica columns.[199] To ascertain the acidity or basicity of such columns requires the use of more acidic and basic test components such as chloro- or nitro-substituted phenols and primary alkyl amines, respectively. The choice of chromatographic test parameters also greatly influences the outcome of an evaluation. Oftentimes, the conditions of a test are chosen to optimize the performance of a column rather than providing a rigorous test of its properties. For example, injections made with the column at high isothermal temperatures (with respect to typical elution temperatures for the test com-

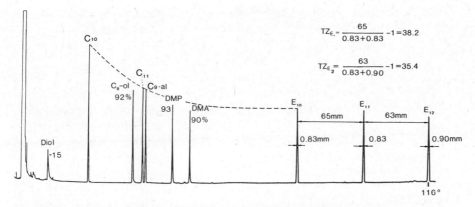

FIGURE 3.8. Chromatogram of the Grob polarity mixture on an SE-52 column. Peak assignments: Diol = 2,3-butanediol; C_{10} = n-decane; C_8—OH = octanol; C_{11} = n-undecane; C_9—al = nonanal; DMP = 2,6-dimethylphenol; DMA = 2,6-dimethylaniline; E_{10} = methyldecanoate; E_{11} = methylundecanoate; E_{12} = methyldodecanoate. (Reproduced with permission from ref. 195. Copyright Elsevier Scientific Publishing Company.)

pounds) allow the test solutes to elute rapidly with little chance for adsorption. Also, the injection of large quantities of a component saturates active sites and prevents observation of the peak tailing that would occur for small injections. The types of chromatographic measurements selected depend on the amount of information desired. In some cases, a casual qualitative examination of a test chromatogram is sufficient; oftentimes, a more rigorous and quantitative approach may be desired.

In this section the chromatographic parameters of separation efficiency, column surface inertness, and thermal stability, and the quantitative approaches available for their evaluation, will be discussed. Also, many of the problems commonly encountered in performance testing, as briefly outlined above, will be discussed more fully, with the goal of establishing the relevant criteria for a rigorous evaluation.

Separation Efficiency

The performance of open tubular columns has been reviewed in detail by Ettre[2] and Ettre and Purcell.[200] Practically speaking, the separation efficiency of a column is a measure of its ability to resolve the components of a mixture into individual compounds. A highly efficient column is characterized by sharp narrow peaks. There are several quantitative approaches to the evaluation of column efficiency. These include variations of the theoretical plate concept, regression and graphical interpretations of peak distribution, coating efficiency calculations, and separation number computations.

The most commonly used expressions utilizing the theoretical plate concept are described as follows: The number of theoretical plates (n) is given by

$$n = 16 \left(\frac{t_R}{w_b}\right)^2 \tag{3.3}$$

The value of t_R is determined by the retention time of the compound of interest, and w_b is the width of the peak at its base. Oftentimes, the width at half height (w_h) is used instead of w_b. This requires a constant equal to 8 ln 2 (or 5.545) instead of 16 to give

$$n = 5.545 \left(\frac{t_R}{w_h}\right)^2 \tag{3.4}$$

The number of effective theoretical plates (N), proposed by Desty et al.,[201] can be calculated from

$$N = 5.545 \left(\frac{t_N}{w_h}\right)^2 \tag{3.5}$$

The term t_N is the adjusted retention time and is calculated from $t_N = t_R - t_m$, where t_m is the elution time of an unretained compound. The effective plate number can be calculated from n by[202]

$$N = \left(\frac{k}{k + 1}\right)^2 n \tag{3.6}$$

where the capacity ratio (k) can be determined from

$$k = \frac{t_N}{t_m} \tag{3.7}$$

Since open tubular columns are usually operated at low capacity ratios, their efficiencies are lower than indicated by the theoretical plate number. It is important to emphasize that both n and N are strongly dependent on the k value of the evaluated peak,[137,203] and the required pressure drop for the optimum carrier gas flow has an influence on the separation efficiency per meter. With small k values, very large plate numbers can be generated. When making efficiency evaluations, the composition of the test mixture should be appropriate for the liquid phase so that the isothermal operating temperature can be kept as low as possible and a $k > 2$ achieved for the peak(s) from which the measurements are being made.[196]

Other formulations utilizing the theoretical plate concept have also been described. These include the number of real theoretical plates (N_{real}) proposed by Kaiser[204] and Said,[205] the geometric mean of n and N proposed by Golay,[206] the reduced plate number proposed by Giddings[207] and by Horne et al.,[208] and the mean specific plate number (N_{ms}) proposed by Brown.[209] The mean specific plate number has the advantage over most parameters in that it has only a small dependence on the partition ratio and allows for the column diameter.

The number of theoretical plates is uniquely defined only when flow velocity, the nature of the carrier gas, inlet and outlet pressures, column dimensions, stationary phase type and distribution, and the mass distribution coefficients are specified. The measurement also depends on the diffusion of the solute in the stationary and mobile phase.[210] Since the outcome of efficiency evaluations are based on so many hard-to-define parameters, alternative approaches using regression analysis of chromatographic data have been suggested.[210,211] This approach is usually recognized as the so-called A-B-T concept. The variance of the solute band is defined from chromatographic data such as retention and peak broadness. Although this approach is somewhat empirical, it has promise of providing an unambiguous evaluation of column efficiency.

Another approach to describing capillary column quality is the comparison of experimentally obtained plate numbers with theoretically predicted values. The coating efficiency (CE) has been defined as the ratio of theoretical to experimental plate height at optimum conditions,[4,142]

$$CE = \left(\frac{h_{theor}}{h_{exp}} \right)_{min} \tag{3.8}$$

The theoretical plate height (h_{theor}) is usually represented by the simplified Golay–Giddings equation[212] as

$$h_{theor} = r \left(\frac{11k^2 + 6k + 1}{3(1 + k)^2} \right)^{1/2} \tag{3.9}$$

h_{exp} or the experimental HETP is defined as the column length (L) divided by the number of theoretical plates (n). Cramers et al.[213] discussed a more general treatment of the coating efficiency, which includes the effects of resistance to mass transfer in the liquid phase and the pressure drop. Provided the diffusion coefficients of the solute in the stationary phase and carrier gas are known, a more accurate determination of the coating efficiency can be made.

The separation value concept and its meaning and relationship to existing column efficiency terms and chromatographic parameters has been discussed in detail by Ettre.[214] The Trennzahl or separation number (TZ) as described by Kaiser and Rieder[215] is calculated from

$$TZ = \frac{t_{R2} - t_{R1}}{w_{h2} + w_{h1}} - 1 \tag{3.10}$$

where the subscripts 1 and 2 refer to two specified components in a chromatogram, usually in a homologous series such as the n-alkanes. The calculated separation number is defined, therefore, as the number of peaks separated by approximately twice the width at half height that can be fitted between the two standards. More general and detailed approaches have been described.[205,216]

One problem in using n-alkanes for TZ measurements is their very different retentions on different stationary phases. Consequently, it is necessary to use different pairs of compounds for different phases to do evaluations at similar capacity ratios.[195] By using fatty acid methyl esters as standards, Grob et al.[195] reported similar capacities on different phases. It was also found that there was no significant difference between isothermal and temperature-programmed TZ values.[195]

Recently, several authors[217-219] have criticized this method of column evaluation. They found that the magnitude of the separation number varied inversely with column temperature and directly with partition ratios. Consequently, the magnitude of this test value is highly dependent on the exact chromatographic conditions and can be manipulated to give very high results if desired.[218] However, Grob and Grob[220] argued that column resolution, which is a practical measure of column efficiency, is also temperature dependent and that it is directly related to the TZ measurements.

A parameter that is often overlooked in open tubular column evaluation, and has an effect on efficiency, is the sample capacity. An "overloaded" peak causes tailing on the leading edge of the peak and reduces the efficiency. Keulemans[221] and Klinkenberg[222] defined the maximum permissible sample size as the maximum amount that can be injected into the column without more than 10% loss in efficiency. Grob and Grob[223] have defined this to be when the recorder pen takes more than twice the time to rise from baseline to peak maximum than to return to baseline.

Unfortunately, any type of efficiency evaluation can be manipulated to give artifically high values. For example, compounds eluted at very low capacity ($1 < K < 1.5$) always exhibit high efficiencies. The real value of efficiency measurements comes from direct comparisons of columns tested in exactly the same manner.

Column Deactivation

Probably the most important parameter in the assessment of quality in an open tubular column is the inertness or degree of deactivation. Column inertness influences the efficiency and thermal stability of the finished column. At least three different types of influences contribute to the activity of the column. They are reversible adsorption, irreversible adsorption, and catalytic degradation. The most obvious form of column activity is reversible adsorption, which is characterized by skewed and tailing peaks. Irreversible adsorption leads to the removal of sample solute and is characterized by lowered peak response or, in severe circumstances, by the complete disappearance of a sample solute. Catalytic degradation of labile compounds by column active sites also occurs and is evidenced by tailing peaks and reduced peak response. An inert and noncatalytic column would elute polar and sensitive compounds with perfect peak symmetry and quantitative peak response. A number of test procedures have been devised to characterize the influences just described. They are discussed below.

Reversible adsorption is most easily characterized. Methods of evaluating this parameter rely on the shape or symmetry of an eluted peak. Goretti and Liberti[224] defined an asymetry factor (A_s) as

$$A_s = \frac{a + b}{(a + b) - (a - b)} \tag{3.11}$$

where a and b are the baseline half widths measured from the perpendicular drawn through the peak maximum. The asymmetry factor measures the deviation from a Gaussian distribution and is an indicator of the interactions occurring in the elution process. Schieke and Pretorius[225] defined a tailing factor (TF) as a percentage according to

$$TF = \left(\frac{a}{b}\right) 100 \tag{3.12}$$

where a and b are defined as above, except that they are measured at 10% of the peak height above baseline. Cram et al.[198] have described the use of numerical methods to characterize open tubular peak shapes. If a column is not sufficiently deactivated, it will contribute to the retention of sample solutes and, thus, relative retention and retention indices of components containing different functional groups will change. Consequently, retention reproducibility is also a useful tool for evaluating column deactivation.[226,227]

To obtain diagnostic chromatographic peak shapes, it is necessary to employ some test mixture under a specified set of chromatographic conditions. One of the most common test mixtures has been described by Grob et al.[195,227] A chromatogram of this mixture is shown in Figure 3.8. The Grob test[195,227] also includes optimization of carrier gas flow rate and temperature-programming rate.

Hydrocarbons are included in test mixtures to ascertain the integrity of the chromatographic instrumentation. A lack of symmetry of these peaks indicates instrumental problems rather than column adsorption. Alcohols are usually more sensitive to adsorption than most other functional groups. A number of studies have been conducted using only two components in the test mixture: an alcohol and an alkane.[72,224,225] 1-Octanol is a sensitive indicator of adsorption because it is well retained. The adsorption of the alcohol can be caused by hydrogen bonding to the silanol groups or siloxane bridges on the silica surface and by interaction with Lewis acid components of the glass. A dihydroxy compound, such as 2,3-butanediol, is included as a more rigorous test. The acid–base properties of the column can be tested by the addition of compounds such as 2,6-dimethylaniline and 2,6-dimethylphenol to the test mixture. The addition of 2-ethylhexanoic acid and dicyclohexylamine can be made for a more rigorous acid–base test. It is obvious that all mixtures do not put the same requirements on column performance.[228] Particularly difficult to chromatograph are the McReynolds standards, pyridine and nitropropane, and bifunctional compounds such as vanillin, methylsalicylate, nitrophenols, and nitroanilines. However, with the introduction of fused-silica columns, compounds that were traditionally very difficult to chromatograph elute very easily with minimum peak tailing. For instance, alcohols, free fatty acids, acidic phenols, and so on, elute from untreated fused-silica columns quantitatively with nearly perfect peak symmetry. The acid–base test with dimethylphenol and dimethylaniline indicates an almost neutral surface. Consequently, it was necessary to develop more difficult chromatographic tests. Polarity mixtures containing highly acidic and basic test components have been adopted which contain short alkyl-chain free fatty acids, diols, nitro and chlorophenols, and primary alkyl amines.[121] Such test mixtures are shown in Figures 3.3 and 3.5. Several problems may be encountered in using a polarity test mixture for open tubular column evaluation. Components of the mixture may react with each other or decompose with time. Thus, care must be taken to include appropriate test compounds and to ensure that they are fresh. During chromatography, strong polar compounds may deactivate the surface for the subsequently eluted species.[229]

Testing the behavior of the column surface before coating has been helpful for the evaluation of different treatments during column preparations.[72,230,231] Schomburg et al.[72] have described a method for testing both coated and uncoated columns. The chromatographic system consisted of an open tubular column with a polar stationary phase such as Carbowax 20M on which peaks of a polarity test mixture are eluted with perfect symmetry. A piece of the column to be tested is connected to this column by means of shrinkable Teflon tubing. The change in peak shape after passing through the test column indicates the extent of surface activity. This procedure eliminates the problem of unknown adsorption in the injection port and indicates only column activity.

One of the most overlooked parameters in evaluations of peak shapes is the choice of chromatographic conditions. Many tests are conducted at high isothermal temperatures so that compound retention is low. Besides allowing a shorter residence time for adsorption influences to be observable, adsorption forces are less at higher temperatures, and peak tailing is minimized. For instance, Grob et al.[195] have shown that a very active column which produced a malformed octanol peak at 50°C produced an acceptable looking octanol peak at 105°C. For a fair and rigorous evaluation of column inertness, it is imperative that temperatures as low as possible be used. This would suggest that temperature-programmed runs at low programming rates would be more rigorous.

A more difficult component of column evaluation is irreversible adsorption. With this form of adsorption, it is possible for a portion of the solute to be eluted with proper peak shape, while the rest is irreversibly lost. Such adsorption is detected by quantifying the response of the peak under question. This is usually done by ratio testing[9,229,232] in which the response of the compound under question is compared to the response of a supposedly inert compound such as a hydrocarbon. This is usually done at decreasing levels of concentration in order to eliminate the effect of saturating the active sites. If the ratio of a particular compound remains constant over the entire range, the column would appear to be inert for that compound. An example of test results using this method is shown in Figure 3.9.[232] The particular column evaluated showed irreversible adsorption for the alcohol and the alkylamine. These types of tests emphasize the necessity of injecting small sample amounts during performance evaluation. At high injection levels, the adsorption losses would not be discernible. High injection levels can also improve the apparent performance of a column with regard to reversible adsorption. Peak tails appear less noticeable when large quantities are chromatographed. Higher levels mask the active sites and, compared to the total amount, display an insignificant amount of adsorption. For accurate testing, sample loads should be less than 1 ng per component. In fact, instrument sensitivity should be high enough that 1-ng loads give full-scale or near full-scale peak heights.

The most elusive form of inertness defect is catalytic degradation. In addition to active column sites, temperature can also lead to catalytic deg-

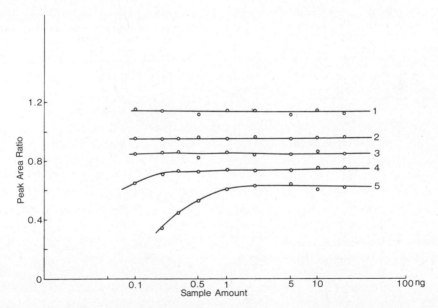

FIGURE 3.9. Plot showing quantitative recovery of polar compounds, illustrating irreversible adsorption at trace levels of certain components. Note: 25 m untreated fused-silica column coated with OV-101 and operated isothermally at 120°C. (1) dicyclohexylamine/n-C_{11}; (2) n-C_{12}/n-C_{11}; (3) 2,6-dimethylphenol/n-C_{11}; (4) octanol/n-C_{11}; (5) laurylamine/n-C_{11}. (Reproduced with permission from ref. 232. Copyright Elsevier Scientific Publishing Company.)

radation. For catalytic degradation of a sample, reduced peak response is also evident. Catalytic influences of a column can be differentiated from adsorptive influences by chromatographing several solutes of different functionality at different temperatures. Adsorptive losses vary inversely and catalytic losses vary directly with temperature.[233] Thus, as the column temperature increases, those solutes whose relative peak areas increase are prone to adsorption interactions, and those solutes whose relative peak areas decrease are prone to catalytic degradation on the column.

In summary, there are essentially three elements necessary for a high-quality evaluation of column inertness: (a) low test temperatures to ensure that adsorption effects are not masked and that solutes have adequate residence time to be fully acted upon by active sites; (b) low injection levels to prevent saturation of active sites; and (c) sensitive test components must be utilized which are in reality susceptible to acid–base interactions, hydrogen bonding attractions, and catalytic influences.

Thermostability

Since most chromatographic separations on open tubular columns are performed at elevated temperatures or employ temperature programming, the temperature stability of a coated column is of the utmost importance. For a column to perform satisfactorily at higher temperatures, the coated stationary phase must remain stable. That is, its decomposition must be minimal and it must remain as a thin, uniform surface film, and not break up into droplets. In addition, the surface beneath the stationary phase, that is, the deactivation layer, must remain stable. In fact, the temperature stability of the deactivation layer is as important as the stability of the stationary phase itself.

The amount of volatile decomposition products formed in the column per unit time depends on the stationary phase, temperature, surface properties of the support, surface area covered by the stationary phase, and the film thickness. The mass flow of bleeding products at the column outlet depends on the column length, carrier gas flow rate, and the rate of formation of bleed products. Assessment of the temperature stability of columns is difficult and depends on the type of analytical application involved and on the required standard of performance and reliability of the analyses. As suggested by Schomburg et al.,[119,136,234] the following properties should not deteriorate or change substantially during extended high temperature operation: (a) The capacity ratio (k) should not decrease significantly (indicating that the stationary phase is not decomposing to any great extent). (b) The separation efficiency in terms of theoretical plates per meter should remain essentially constant (indicating that the quality and integrity of the stationary phase film is not being disrupted). (c) The background signal arising from bleed products should remain constant and not exceed a defined limit (also indicating that the stationary phase and/or deactivating precoating layers are not decomposing significantly). (d) The polarity in terms of retention indices of standard compounds should remain constant (indicating that the chemical nature of the stationary phase and/or the deactivation is not changing). (e) The extent of surface deactivation for strongly polar solutes when using apolar stationary phases should remain stable. That is, the tailing behavior and adsorption of polar solutes at trace concentration levels should not become more pronounced after high-temperature operation of the column.

Deterioration of the stationary phase results from two primary mechanisms. The first, and probably least serious, is temperature oriented and is related to the chemical composition of the stationary liquid phase. At elevated temperatures, a point is reached where the stationary liquid exhibits a significant vapor pressure, and loss of the phase by evaporation becomes substantial. Further increases in temperature can result in chemical decomposition of the phase by pyrolysis. The maximum allowable operating temperature (MAOT) is usually specified by the manufacturer of the liquid phase. In practice, however, the specified MAOT is not usually obtainable in open tubular column chromatography. This is partially due to surface catalytic effects (the second mechanism) which induces stationary phase decomposition. This mechanism of decomposition is of major importance since steps can be taken to alleviate or eliminate the undesirable consequences. Surface effects are particularly noticeable in open tubular columns where the ratio of surface area to the amount of stationary phase is high.

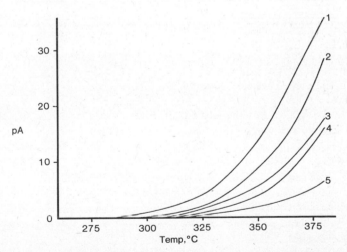

FIGURE 3.10. Plots showing results of bleeding rate experiments obtained from 20 m Duran glass open tubular columns. Note: temperature program at 5°C min^{-1} to 380°C, hydrogen carrier gas at 50 cm s^{-1} (measured at 250°C). (1) SE-30, d_f = 0.30 μm; (2) bonded methylsilicone rubber, d_f = 0.72 μm; (3) bonded methylsilicone rubber, d_f = 0.35 μm; (4) SE-30, d_f = 0.12 μm; (5) bonded methylsilicone rubber cross-linked with trichlorosilanes, d_f = 0.35 μm. (Reproduced with permission from ref. 176. Copyright Elsevier Scientific Publishing Company.)

Removal of alkali from glass surfaces increases the stability of methyl-siloxanes considerably. It has been known for years that alkali tends to depolymerize polysiloxanes[235] and, in chromatographic applications, it has been shown that both alkaline and acidic surfaces cause decomposition of silicone oil phases.[235] This is consistent with observations made by Schom-burg et al.[234] that columns prepared from borosilicate glasses (lower alkali content) exhibited much better thermostability. Fused-silica columns also exhibit better thermostability than glass columns.

In studies[236,237] of the chemical resistance of various stationary phases to specific compounds present on glass surfaces or compounds used for roughening techniques, it was found that several lead to stationary phase degradation. $BaCO_3$, metal chlorides, and aluminum oxide showed the strongest effect, while sodium chloride and several metal oxides showed little effect. Such studies indicate that surface impurities can lead to stationary phase instability.

The thermostability of open tubular columns is commonly evaluated by bleed-rate experiments. In this type of test, columns of different specifications are heated and the amount of column bleed plotted as a function of temperature. An example of such an evaluation is shown in Figure 3.10.[176] Of course, these experiments are only valid when the thermal history of the evaluated columns is similar. Another common method of evaluating thermal stability is to compare efficiency and inertness prior to, and then again after, extended use and heat treatment.

REFERENCES

1. M. J. E. Golay, in *Gas Chromatography 1958*, D. H. Desty, editor. Academic Press, New York, 1958, p. 36.

2. L. S. Ettre, *Open Tubular Columns in Gas Chromatography*. Plenum Press, New York, 1965.

3. D. H. Desty, J. N. Haresnape, and B. H. F. Whyman, *Anal. Chem.* **32,** 302 (1960).

4. M. Novotny and A. Zlatkis, *Chromatogr. Rev.* **14,** 1 (1971).

5. M. Verzele and P. Sandra, *J. High Resoln. Chromatogr./Chromatogr. Commun.* **2,** 303 (1979).

6. M. L. Lee and B. W. Wright, *J. Chromatogr.* **184,** 235 (1980).

7. W. Jennings, *Comparison of Glass and Fused Silica Columns for Gas Chromatography.* Huethig, Heidelberg, 1981.

8. B. W. Wright, B. E. Richter, and M. L. Lee, in *Recent Topics in Capillary Separation Techniques,* M. Novotny, editor, Wiley in press.

9. R. D. Dandeneau and E. H. Zerenner, *J. High Resoln. Chromatogr./Chromatogr. Commun.* **2,** 351 (1979).

10. R. Dandeneau, P. Bente, T. Rooney, and R. Hiskes, *Am. Lab.* **11**(9), 61 (1979).

11. R. H. Doremus, *Glass Science.* John Wiley and Sons, London, 1973, p. 28.

12. V. Pretorius and J. C. Davidtz, *J. High Resoln. Chromatogr./Chromatogr. Commun.* **2,** 703 (1979).

13. W. Eitel, *Silicate Science,* Vol. 11. Academic Press, New York, 1965, p. 349.

14. S. R. Lipsky, W. J. McMurray, M. Hernandez, J. E. Purcell, and K. A. Billeb, *J. Chromatogr. Sci.* **18,** 1 (1980).

15. C. Hishta and J. Bonstein, *Adv. Chromatogr.* **9,** 220 (1970).

16. M. L. Hair and A. M. Filbert, *Res. Dev.* **20**(11), 34 (1969).

17. A. V. Kiselev, *Russ. J. Phys. Chem.* **38,** 1501 (1964).

18. M. J. D. Low, M. Ramasubramanian, and V. V. Subba Rao, *J. Phys. Chem.* **71,** 1726 (1967).

19. P. C. Carman, *Trans. Faraday Soc.* **36,** 964 (1940).

20. L. T. Zhuravlev, A. V. Kiselev, V. P. Naidina, and A. L. Polyakov, *Russ. J. Phys. Chem.* **37,** 1216 (1963).

21. E. K. Lippincott and R. Schroeder, *J. Chem. Phys.* **23,** 1099 (1955).

22. V. Y. Davydov, L. T. Zhuravlev, and A. V. Kiselev, *Russ. J. Phys. Chem.* **38,** 1108 (1964).

23. M. L. Hair and W. Hertl, *J. Phys. Chem.* **73,** 4269 (1969).

24. M. L. Hair, *J. Non-Cryst. Solids* **19,** 299 (1975).

25. M. L. Hair, *Infrared Spectroscopy in Surface Chemistry.* Marcel Dekker, New York, 1967, p. 79.

26. R. K. Iler, *The Chemistry of Silica—Solubility, Polymerization, Colloid and Surface Properties, and Biochemistry.* John Wiley and Sons, New York, 1979.

27. B. W. Wright, M. L. Lee, and G. M. Booth, *Chromatographia* **15,** 584 (1982).

28. L. T. Zhuravlev and A. V. Kiselev, *Russ. J. Phys. Chem.* **39,** 236 (1965).

29. M. R. Basila, *J. Chem. Phys.* **35,** 1151 (1961).

30. W. Hertl and M. L. Hair, *J. Phys. Chem.* **72,** 4676 (1968).

31. O. M. Dzhigit, A. V. Kiselev, and G. G. Muttik, *Kolloid Zhur.* **23,** 504, 533 (1961); **24,** 241 (1962).

32. J. B. Peri and A. L. Hensley, Jr., *J. Phys. Chem.* **72,** 2926 (1968).

33. J. B. Peri, *J. Phys. Chem.* **70,** 2937 (1966).

34. J. Kunawicz, P. Jones, and J. A. Hockey, *Trans. Faraday Soc.* **67,** 848 (1971).

35. L. G. Ganichenko, V. F. Kiselev, K. G. Krasil'nikov, and V. V. Murina, *Russ. J. Phys. Chem.* **35,** 844 (1961).

36. J. M. Bather and R. A. C. Gray, *J. Chromatogr.* **122,** 159 (1976).

37. B. A. Morrow and I. A. Codey, *J. Phys. Chem.* **80,** 1995, 1998 (1976).

38. F. R. Aussenegg, *Appl. Spectrosc.* **32,** 587 (1978).

39. W. A. Zisman, *Ind. Eng. Chem.* **55,** 79 (1963); *Adv. Chem. Ser.* **43,** 1 (1964).

40. A. B. D. Cassie, *Disc. Faraday Soc.* **3,** 11 (1948).

41. K. D. Bartle, B. W. Wright, and M. L. Lee, *Chromatographia* **14,** 387 (1981).

42. R. Houwink, *Adhesion and Adhesives.* Elsevier, New York, 1965.

43. J. R. Dann, *J. Colloid Interface Sci.* **32,** 302, 321 (1970).

44. M. K. Bernett and W. A. Zisman, *J. Colloid Interface Sci.* **29,** 413 (1969).

45. D. A. Olsen and A. J. Osteraas, *J. Phys. Chem.* **68,** 2730 (1964).

46. F. I. Onuska, B. K. Afghan, and R. J. Wilkinson, *J. Chromatogr.* **158,** 83 (1978).

47. K. L. Ogan, C. Reese, and R. P. W. Scott, *J. Chromatogr. Sci.* **20,** 425 (1982).

48. D. H. Desty, *Chromatographia* **8,** 452 (1975).

49. V. Pretorius, J. C. Davidtz, and D. H. Desty, in *Fourth International Symposium on Capillary Chromatography,* R. E. Kaiser, editor. Huethig, Heidelberg, 1981, p. 201.

50. R. N. Wenzel, *Ind. Eng. Chem.* **28,** 988 (1936).

51. Y. Tamai and K. Aratani, *J. Phys. Chem.* **76,** 3267 (1972).

52. W. Jennings, *J. High Resoln. Chromatogr./Chromatogr. Commun.* **4,** 601 (1981).

53. W. Leibnitz and M. Mohnke, *Chem. Tech. (Berlin)* **14,** 753 (1962).

54. F. A. Bruner and G. P. Cartoni, *Anal. Chem.* **36,** 1522 (1964).

55. R. A. Hoolman, C. R. Green, and F. W. Best, *Anal. Chem.* **50,** 2157 (1978).

56. A. V. Kiselev, in *Gas Chromatography 1962,* M. van Swaay, editor. Butterworths, London, 1962, p. 34.

57. E. L. Ilkova and E. A. Mistryukov, *Chromatographia* **4,** 77 (1971).

58. G. Alexander and G. A. F. M. Rutten, *J. Chromatogr.* **99,** 81 (1974).

59. J. L. Marshall and D. A. Parker, *J. Chromatogr.* **122,** 425 (1976).

60. G. Alexander, G. Garzo, and G. Palyi, *J. Chromatogr.* **91,** 25 (1974).

61. J. J. Franken, G. A. F. M. Rutten, and J. A. Rijks, *J. Chromatogr.* **126,** 117 (1976).

62. H. T. Badings, J. J. G. van der Pol, and D. G. Schmidt, *Chromatographia* **10,** 404 (1977).

63. M. Novotny and K. Tesarik, *Chromatographia* **1,** 332 (1968).

64. J. D. Schieke, N. R. Comins, and V. Pretorius, *Chromatographia* **8,** 354 (1975).

65. J. D. Schieke, N. R. Comins, and V. Pretorius, *J. Chromatogr.* **112,** 97 (1975).

66. P. Sandra and M. Verzele, *Chromatographia* **10,** 419 (1977).

67. F. I. Onuska and M. E. Comba, *J. Chromatogr.* **126,** 133 (1976).

68. R. I. Onuska, M. E. Comba, T. Bistricki, and R. J. Wilkinson, *J. Chromatogr.* **142,** 117 (1977).

69. P. Sandra, M. Verstappe, and M. Verzele, *J. High Resoln. Chromatogr./Chromatogr. Commun.* **1,** 28 (1978).

70. M. Verzele, G. Redant, M. van Roelenbosch, M. Godefroot, M. Verstappe, and P. Sandra, in *Fourth International Symposium on Capillary Chromatography,* R. E. Kaiser, editor. Huethig, Heidelberg, 1981, p. 239.

71. F. I. Onuska and M. E. Comba, *Chromatographia* **10,** 498 (1977).

72. G. Schomburg, H. Husmann, and F. Weeke, *Chromatographia* **10,** 580 (1977).

73. K. Grob and G. Grob, *J. Chromatogr.* **125,** 471 (1976).

74. K. Grob, Jr., G. Grob, and K. Grob, *J. High Resoln. Chromatogr./Chromatogr. Commun.* **1,** 149 (1978).

75. K. Grob, J. R. Guenter, and A. Portmann, *J. Chromatogr.* **147,** 111 (1978).

76. G. Schomburg, H. Husmann, and H. Borwitzky, *Chromatographia* **12,** 651 (1979).

77. C. Watanabe and H. Tomita, *J. Chromatogr.* **121,** 1 (1976).

78. P. Sandra, M. Verstappe, and M. Verzele, *Chromatographia* **11,** 223 (1978).

79. R. C. M. deNijs, G. A. F. M. Rutten, J. J. Franken, R. P. M. Dooper, and J. A. Rijks, *J. High Resoln. Chromatogr./Chromatogr. Commun.* **2,** 447 (1979).

80. A. L. German and E. C. Horning, *J. Chromatogr. Sci.* **11,** 76 (1973).

81. R. S. Deelder, J. J. M. Ramaekers, J. H. M. van den Berg, and M. L. Wetzels, *J. Chromatogr.* **119,** 99 (1976).

82. C. N. Blakesley and P. A. Torline, *J. Chromatogr.* **105,** 385 (1975).

83. R. G. McKeag and F. W. Hougen, *J. Chromatogr.* **136,** 308 (1977).

84. W. Bertsch, F. Shunbo, R. C. Chang, and A. Zlatkis, *Chromatographia* **7,** 128 (1974).

85. E. D. Pellizzari, *J. Chromatogr.* **92,** 299 (1974).

86. C. A. Cramers, E. A. Vermeer, and J. J. Franken, *Chromatographia* **10,** 412 (1977).

87. E. Schulte, *Chromatographia* **9,** 315 (1976).

88.   G. A. F. M. Rutten and J. A. Luyten, *J. Chromatogr.* **74,** 177 (1972).

89.   C. A. Cramers, E. A. Vermeer, and J. J. Franken, *Chromatographia* **10,** 412 (1977).

90.   G. Marcelin, S. G. Traynor, W. Goins, Jr., and L. M. Hirschy, *J. Chromatogr.* **187,** 57 (1980).

91.   E. A. Mistryukov, A. L. Samusenko, and R. V. Golovnya, *J. Chromatogr.* **148,** 490 (1978).

92.   E. A. Mistryukov, A. L. Samusenko, and R. V. Golovnya, *J. Chromatogr.* **169,** 391 (1979).

93.   R. V. Golovnya, A. L. Samusenko, and E. A. Mistryukov, *J. High Resoln. Chromatogr./Chromatogr. Commun.* **2,** 609 (1979).

94.   D. A. Cronin, *J. Chromatogr.* **97,** 263 (1974).

95.   L. Blomberg and T. Wännman, *J. Chromatogr.* **148,** 379 (1978).

96.   M. Verzele, *J. High Resoln. Chromatogr./Chromatogr. Commun.* **2,** 647 (1979).

97.   M. A. Kaiser and D. B. Chase, *Anal. Chem.* **52,** 1849 (1980).

98.   R. F. Arrendale, R. F. Severson, and O. T. Chortyk, *J. Chromatogr.* **208,** 209 (1981).

99.   M. L. Lee, Ph.D. Dissertation, Indiana University, 1975.

100.   H. Borwitzky and G. Schomburg, *J. Chromatogr.* **170,** 99 (1979).

101.   K. Grob, G. Grob, and K. Grob, Jr., *Chromatographia* **10,** 181 (1977).

102.   K. Grob, G. Grob, and K. Grob, Jr., *J. High Resoln. Chromatogr./Chromatogr. Commun.* **2,** 31 (1979).

103.   K. Grob, G. Grob, and K. Grob, Jr., *J. High Resoln. Chromatogr./Chromatogr. Commun.* **2,** 527 (1979).

104.   M. L. Lee, D. L. Vassilaros, L. V. Phillips, D. M. Hercules, H. Azumaya, J. W. Jorgenson, M. P. Maskarinec, and M. Novotny, *Anal. Lett.* **12,** 191 (1979).

105.   B. W. Wright, M. L. Lee, S. W. Graham, L. V. Phillips, and D. M. Hercules, *J. Chromatogr.* **199,** 355 (1980).

106.   G. A. F. M. Rutten, C. C. E. van Tilburg, C. P. M. Schultjes, and J. A. Rijks, in *Proceedings of the Fourth International Symposium on Capillary Chromatography,* R. E. Kaiser, editor. Huethig, Heidelberg, 1981, p. 779.

107.   E. Dirkes, Jr., W. A. Rubey, and C. Pantano, *J. High Resoln. Chromatogr./Chromatogr. Commun.* **4,** 303 (1980).

108.   A. V. Kiselev, in *Gas Chromatography 1962,* M. van Swaay, editor. Butterworths, London, 1962, p. 3.

109.   M. Novotny, L. Blomberg, and K. D. Bartle, *J. Chromatogr. Sci.* **8,** 390 (1970).

110.   K. Grob, G. Grob, and K. Grob, Jr., *J. High Resoln. Chromatogr./Chromatogr. Commun.* **2,** 677 (1979).

111.   Th. Welsch, W. Engewald, and Ch. Klaucke, *Chromatographia* **10,** 22 (1977).

112.   K. Grob, G. Grob, and K. Grob, Jr., *J. High Resoln. Chromatogr./Chromatogr. Commun.* **2,** 31 (1979).

113.   M. Godefroot, M. Van Roelenbosch, P. Sandra, and M. Verzele, *J. High Resoln. Chromatogr./Chromatogr. Commun.* **3,** 337 (1980).

114.   K. D. Bartle and M. Novotny, *J. Chromatogr.* **94,** 35 (1974).

115.   K. Grob and G. Grob, *J. High Resoln. Chromatogr./Chromatogr. Commun.* **3,** 197 (1980).

116.   T. Welsch, R. Muller, W. Engewald, and G. Werner, *J. Chromatogr.* **241,** 41 (1982).

117.   F. O. Stark, O. K. Johannson, G. E. Vogel, R. G. Chaffee, and R. M. Lacefield, *J. Phys. Chem.* **72,** 2750 (1968).

118.   M. L. Hair and W. Hertl, *J. Phys. Chem.* **73,** 2372 (1969).

119. G. Schomburg, H. Husmann, and H. Borwitzky, *Chromatographia* **12**, 651 (1979).

120. T. J. Stark, R. D. Dandeneau, and L. Mering, 1980 Pittsburgh Conference, Abstract 002. Atlantic City, New Jersey, 1980.

121. B. W. Wright, P. A. Peaden, M. L. Lee, and T. Stark, *J. Chromatogr.* **248**, 17 (1982).

122. L. Blomberg, K. Markides, and T. Wännman, *J. High Resoln. Chromatogr./Chromatogr. Commun.* **4**, 527 (1980).

123. L. Blomberg, K. Markides, and T. Wännman, in *Proceedings of the Fourth International Symposium on Capillary Chromatography*, R. E. Kaiser, editor. Huethig, Heidelberg, 1981, p. 73.

124. I. Kari, A. Huhtikangas, J. Gynther, T. Vertiainen, and R. Hiltunen, *Chromatographia* **14**, 462 (1981).

125. V. Pretorius and D. H. Desty, *J. High Resoln. Chromatogr./Chromatogr. Commun.* **4**, 122 (1981).

126. V. Pretorius, J. W. du Toit, and J. H. Purnell, *J. High Resoln. Chromatogr./Chromatogr. Commun.* **4**, 344 (1981).

127. G. Dijkstra and J. de Goey, in *Gas Chromatography 1958*, D. H. Desty, editor. Academic Press, New York, 1958, p. 56.

128. T. Boogaerts, M. Verstappe, and M. Verzele, *J. Chromatogr. Sci.* **10**, 217 (1972).

129. K. D. Bartle, *Anal. Chem.* **45**, 1831 (1973).

130. L. Blomberg, *Chromatographia* **8**, 324 (1975).

131. J. Roeraade, *Chromatographia* **8**, 511 (1975).

132. D. A. Parker and J. L. Marshall, *Chromatographia* **11**, 533 (1978).

133. R. L. Levy, D. A. Murray, H. D. Gesser, and F. W. Hougen, *Anal. Chem.* **40**, 459 (1968).

134. J. P. J. van Dalen, *Chromatographia* **5**, 354 (1972).

135. M. L. McConnell and M. Novotny, *J. Chromatogr.* **112**, 559 (1975).

136. G. Schomburg, H. Husmann, and F. Weeke, *J. Chromatogr.* **99**, 63 (1974).

137. G. Schomburg and H. Husmann, *Chromatographia* **8**, 517 (1976).

138. F. Fairbrother and A. E. Stubbs, *J. Chem. Soc.*, 527 (1935).

139. M. Novotny, K. D. Bartle, and L. Blomberg, *J. Chromatogr.* **45**, 469 (1969).

140. L. Blomberg, *J. Chromatogr.* **138**, 7 (1977).

141. K. Grob and G. Grob, *Chromatographia* **4**, 422 (1971).

142. J. Bouche and M. Verzele, *J. Gas Chromatogr.* **6**, 501 (1968).

143. G. A. F. M. Rutten and J. A. Rijks, *J. High Resoln. Chromatogr./Chromatogr. Commun.* **1**, 279 (1978).

144. K. Grob, *J. High Resoln. Chromatogr./Chromatogr. Commun.* **1**, 93 (1978).

145. M. Giabbai, M. Shoults, and W. Bertsch, *J. High Resoln. Chromatogr./Chromatogr. Commun.* **1**, 277 (1978).

146. B. L. Goodwin, *J. Chromatogr.* **172**, 31 (1979).

147. K. Grob and G. Grob, *J. High Resoln. Chromatogr./Chromatogr. Commun.* **1**, 221 (1978).

148. A. Venema, L. G. J. v. d. Ven, and H. v. d. Steege, *J. High Resoln. Chromatogr./Chromatogr. Commun.* **2**, 69 (1979).

149. M. K. Cueman and R. B. Hurley, Jr., *J. High Resoln. Chromatogr./Chromatogr. Commun.* **1**, 92 (1978).

150. P. Sandra and M. Verzele, *Chromatographia* **11**, 102 (1978).

151. M. P. Maskarinec, personal communication.

152. C. H. Lochmuller and J. D. Fisk, *J. High Resoln. Chromatogr./Chromatogr. Commun.* **4**, 232 (1981).

153. K. Grob and G. Grob, *J. High Resoln. Chromatogr./Chromatogr. Commun.* **5**, 119 (1982).

154. E. L. Ilkova and E. A. Mistryukov, *J. Chromatogr. Sci.* **9**, 569 (1971).

155. E. L. Ilkova and E. A. Mistryukov, *J. Chromatogr.* **54**, 422 (1971).

156. W. G. Jennings, *Chromatographia* **8**, 690 (1975).

157. S. M. Volkov, V. M. Goryayev, V. I. Anikeyev, and V. A. Khripach, *J. Chromatogr.* **190**, 445 (1980).

158. B. W. Wright, P. A. Peaden, and M. L. Lee, *J. High Resoln. Chromatogr./Chromatogr. Commun.* **5**, 413 (1982).

159. S. L. Goren, *J. Fluid Mech.* **12**, 309 (1961).

160. R. K. Jain, I. B. Ivanov, C. Maldarelli, and E. Ruckenstein, in *Lecture Notes in Physics*, *105*, J. Ehlers, K. Hepp, R. Kippenhahn, H. A. Weidenmuller, J. Zittartz, W. Beiglbock, and T. S. Sorensen, editors. Springer-Verlag, New York, 1979, p. 140.

161. W. Noll, *Chemistry and Technology of Silicones*. Academic Press, New York, 1968, p. 464.

162. L. Blomberg, *J. High Resoln. Chromatogr./Chromatogr. Commun.* **5**, 520 (1981).

163. H. W. Fox, P. W. Taylor, and W. A. Zisman, *Ind. Eng. Chem.* **39**, 1401 (1947).

164. K. Grob and G. Grob, *J. Chromatogr.* **213**, 211 (1981).

165. R. G. Sinclair, E. R. Hinnenkamp, K. A. Boni, D. A. Berry, W. H. Schuller, and R. V. Lawrence, *J. Chromatogr. Sci.* **9**, 126 (1971).

166. K. Grob, *Helv. Chim. Acta* **51**, 729 (1968).

167. C. Madani, E. M. Chambaz, M. Rigaud, J. Durand, and P. Chebroux, *J. Chromatogr.* **126**, 161 (1976).

168. M. Rigaud, P. Chebroux, J. Durand, J. Maclouf, and C. Madani, *Tetrahedron Lett.* **44**, 3935 (1976).

169. C. Madani, E. M. Chambaz, M. Rigaud, P. Chebroux, and J. C. Breton, *Chromatographia* **10**, 466 (1977).

170. C. Madani and E. M. Chambaz, *Chromatographia* **11**, 725 (1978).

171. C. Madani and E. M. Chambaz, *J. Am. Oil. Chem. Soc.* **58**, 63 (1981).

172. L. Blomberg, J. Buijten, J. Gawdick, and T. Wännman, *Chromatographia* **11**, 521 (1978).

173. L. Blomberg and T. Wännman, *J. Chromatogr.* **168**, 81 (1979).

174. L. Blomberg and T. Wännman, *J. Chromatogr.* **186**, 159 (1979).

175. L. Blomberg, K. Markides, and T. Wännman, *J. Chromatogr.* **203**, 217 (1981).

176. L. Blomberg, J. Buijten, K. Markides, and T. Wännman, *J. Chromatogr.* **208**, 231 (1981).

177. M. Morton, editor, *Rubber Technology*. Van Nostrand-Reinhold, New York, 1973.

178. W. Noll, *Chemistry and Technology of Silicones*. Academic Press, New York, 1968.

179. K. Grob, G. Grob, and K. Grob, Jr., *J. Chromatogr.* **211**, 243 (1981).

180. K. Grob and G. Grob, *J. High Resoln. Chromatogr./Chromatogr. Commun.* **4**, 491 (1981).

181. K. Grob and G. Grob, *J. High Resoln. Chromatogr./Chromatogr. Commun.* **5**, 13 (1982).

182. P. Sandra, G. Redant, E. Schacht, and M. Verzele, *J. High Resoln. Chromatogr./Chromatogr. Commun.* **4**, 411 (1981).

183. L. Blomberg, J. Buijten, K. Markides, and T. Wännman, *J. High Resoln. Chromatogr./Chromatogr. Commun.* **4**, 578 (1981).

184. P. A. Peaden, B. W. Wright, and M. L. Lee, *Chromatographia* **15**, 335 (1982).

185. J. Buijten, L. Blomberg, K. Markides, and T. Wännman, *Chromatographia* **16**, 183 (1982).

186. K. Markides, L. Blomberg, J. Buijten, and T. Wännman, *J. Chromatogr.* **254**, 53 (1983).

187. B. E. Richter, J. C. Kuei, J. I. Shelton, L. W. Castle, J. S. Bradshaw, and M. L. Lee, *J. Chromatogr.* in press.

188. B. E. Richter, J. C. Kuei, N. J. Park, S. J. Crowley, J. S. Bradshaw, and M. L. Lee, *J. High Resoln. Chromatogr./Chromatogr. Commun.*, **6,** 371 (1983).

189. G. Schomburg, H. Husmann, S. Ruthe, and M. Herraiz, *Chromatographia* **15,** 599 (1982).

190. W. Bertsch, V. Pretorius, M. Pearce, J. C. Thompson, and N. G. Schnautz, *J. High Resoln. Chromatogr./Chromatogr. Commun.* **5,** 432 (1982).

191. W. J. Bobear, in *Rubber Technology*. Van Nostrand-Reinhold, New York, 1973, p. 371.

192. B. E. Richter, J. C. Kuei, J. I. Shelton, L. W. Castle, J. S. Bradshaw, and M. L. Lee, *Proceedings of the Fifth International Symposium on Capillary Chromatography*. Elsevier, Amsterdam, 1983, p. 108.

193. J. Buijten, L. Blomberg, K. Markides, and T. Wännman, *J. Chromatogr.* **237,** 465 (1982).

194. P. Sandra, M. Van Roelenbosch, I. Temmerman, and M. Verzele, *Chromatographia* **16,** 63 (1982).

195. K. Grob, Jr., G. Grob, and K. Grob, *J. Chromatogr.* **156,** 1 (1978).

196. M. J. Hartigan and L. S. Ettre, *J. Chromatogr.* **119,** 187 (1976).

197. M. Verzele and P. Sandra, *J. Chromatogr.* **158,** 111 (1978).

198. S. Cram, F. Yang, and A. Brown, *Chromatographia* **10,** 397 (1977).

199. B. W. Wright and M. L. Lee, unpublished results (1981).

200. L. S. Ettre and J. E. Purcell, in *Advances in Chromatography,* Vol. 10, J. C. Giddings and R. A. Keller, editors. Marcel Dekker, New York, 1974, p. 1.

201. D. H. Desty, A. Goldup, and W. T. Swanton, in *Gas Chromatography,* N. Brenner, J. E. Callen, and M. D. Weiss, editors. Academic Press, New York, 1962, p. 105.

202. J. H. Purnell, *J. Chem. Soc.,* 1268 (1960).

203. M. J. Hartigan, K. Billeb, and L. S. Ettre, *Chromatographia* **10,** 571 (1977).

204. R. E. Kaiser, *Optimierung in der HPLC*. Huethig, Heidelberg, 1979.

205. A. S. Said, *J. High Resoln. Chromatogr./Chromatogr. Commun.* **2,** 637 (1979).

206. M. J. E. Golay, *Anal. Chem.* **29,** 928 (1957).

207. J. C. Giddings, *J. Chromatogr.* **13,** 301 (1964).

208. D. C. Horne, J. H. Knox, and L. McLaren, *Separation Sci.* **1,** 531 (1966).

209. I. Brown, *Chromatographia* **12,** 265 (1979).

210. T. W. Smuts, T. S. Buys, T. G. du Toit, J. W. du Toit, and P. A. Torline, *J. High Resoln. Chromatogr./Chromatogr. Commun.* **4,** 385 (1981).

211. R. E. Kaiser and R. Rieder, *Chromatographia* **10,** 455 (1977).

212. J. C. Giddings, *Anal. Chem.* **36,** 741 (1964).

213. C. A. Cramers, F. A. Wijnheijmer, and J. A. Rijks, *Chromatographia* **12,** 643 (1979).

214. L. S. Ettre, *Chromatographia* **8,** 291, 355 (1975).

215. R. E. Kaiser and R. Rieder, *Chromatographia* **8,** 491 (1975).

216. R. E. Kaiser, *Chromatographia* **9,** 463 (1978).

217. W. Jennings and K. Yabumoto, *J. High Resoln. Chromatogr./Chromatogr. Commun.* **3,** 177 (1980).

218. T. A. Rooney and M. J. Hartigan, *J. High Resoln. Chromatogr./Chromatogr. Commun.* **3,** 416 (1980).

219. J. Krupcik, J. Garaj, G. Guiochon, and J. M. Schmitter, *Chromatographia* **14,** 501 (1981).

220. K. Grob, Jr. and K. Grob, *J. Chromatogr.* **207,** 291 (1981).

221. A. I. M. Keulemans, *Gas Chromatography,* 2nd ed. Reinhold, New York, 1959, p. 124.

222. A. Klinkenberg, in *Gas Chromatography 1960,* R. P. W. Scott, editor. Butterworths, London, 1960, p. 182.

223. K. Grob, Jr. and K. Grob, *Chromatographia* **10,** 250 (1977).

224. G. Goretti and A. Liberti, *J. Chromatogr.* **161,** 89 (1978).

225. J. D. Schieke and V. Pretorius, *J. Chromatogr.* **132,** 217 (1977).

226. L. S. Ettre, *J. Chromatogr. Sci.* **13,** 354 (1975).

227. K. Grob, G. Grob, and K. Grob, Jr., *J. Chromatogr.* **219,** 13 (1981).

228. M. Verzele and P. Sandra, *J. High Resoln. Chromatogr./Chromatogr. Commun.* **2,** 303 (1979).

229. G. Schomburg, *J. High Resoln. Chromatogr./Chromatogr. Commun.* **2,** 461 (1979).

230. K. Grob and G. Grob, *J. High Resoln. Chromatogr./Chromatogr. Commun.* **1,** 302 (1978).

231. R. C. M. de Nijs and R. P. M. Dooper, *J. High Resoln. Chromatogr./Chromatogr. Commun.* **3,** 583 (1980).

232. H. Saito, *J. Chromatogr.* **243,** 189 (1982).

233. K. Grob, *J. High Resoln. Chromatogr./Chromatogr. Commun.* **3,** 505 (1980).

234. G. Schomburg, R. Dielmann, H. Borwitzky, and H. Husmann, *J. Chromatogr.* **167,** 337 (1978).

235. J. Edge and F. F. Oldfield, *J. Soc. Glass Tech.* **42,** 227T (1958).

236. A. Venema, L. G. J. v. d. Ven, and H. v. d. Steege, *J. High Resoln. Chromatogr./Chromatogr. Commun.* **2,** 405 (1979).

237. A. Venema and J. B. Beltman, in *Proceedings of the Fourth International Symposium on Capillary Chromatography*, R. E. Kaiser, editor. Huethig, Heidelberg, 1981, p. 91.

FOUR

INSTRUMENTATION

4.1 BASIC COMPONENTS

A schematic flow diagram of a basic gas chromatograph for use in open tubular column gas chromatography is shown in Figure 4.1. The five basic components are: (a) a carrier gas flow control system that regulates carrier gas flow rates; (b) a low dead volume sample injection system which allows injection of reproducible sample size and representative sample composition; (c) a column oven that can be used for isothermal and temperature-programmed analysis; (d) a low dead volume and high-sensitivity detector for high-speed GC and trace analysis; and (e) a short time constant recorder or a chromatographic data system with high analog-to-digital clock rates for accurate recording of the peaks eluted from open tubular columns.

High-purity carrier gas from the outlet of a two-stage pressure regulator at a high-pressure gas supply cylinder is fed through a combination of carrier gas purification traps including molecular sieve 5A or 13X, activated charcoal, and an oxygen scrubber (for ECD operation) to a high-quality flow controller or a pressure regulator for precise flow-rate control. The carrier gas from the flow controller or the pressure regulator is then fed into the sample injector. The sample is introduced into the injector (or column inlet) via a syringe or a sampling valve. The sample is then vaporized in the injector vaporization chamber or at the inlet of the column (on-column injection) and is carried downstream to the outlet of the column by the carrier gas.

The chromatographic separation process for the sample components occurs during their migration through the column. The sample components eluted from the column are then detected by chromatographic detector(s). The output signal of the detector is either displayed on a recorder or input

100

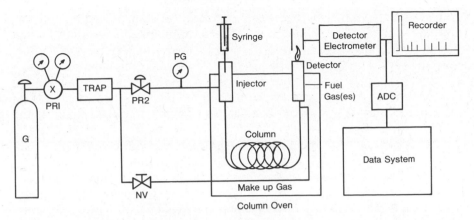

FIGURE 4.1. Schematic diagram of a basic gas chromatograph for open tubular column GC. Key: G, high-pressure gas supply cylinder; PR1, two-stage pressure regulator; PR2, pressure regulator or flow controller; PG, pressure gauge for the column inlet pressure; NV, needle valve for adjusting the make-up gas flow rate. For operation in which the make-up gas is different from the carrier gas, a separate flow regulation system is required for the makeup gas.

into a chromatographic data system through an analog-to-digital converter (ADC) for further data reduction.

4.2 CARRIER GAS FLOW CONTROL AND MEASUREMENT

Carrier gas flow rates in open tubular column GC are substantially lower than those used in packed-column GC. A flow control system that maintains column flow rates below a few cubic centimeters per minute is essential for operating open tubular columns at their full potential.

Four commonly used pneumatic flow control systems are: (a) pressure regulator; (b) flow controller; (c) needle valve or flow restrictor; and (d) combination of flow controller and back-pressure regulator. The performance characteristics and requirements of each system in conjunction with open tubular column GC are discussed below.

Pressure Regulator

A pressure regulator is the most widely utilized system for carrier gas flow regulation in open tubular column GC. This is because of the need in split sampling for easy flow-rate regulation for a wide split ratio. The pressure regulator maintains a constant downstream pressure by the adjustment of the range spring and controller diaphragm. By connecting the column inlet to the outlet of the pressure regulator, a constant column inlet pressure is obtained. Carrier gas flow rates are then dependent on the pressure gradient across the column and the viscosity of the carrier gas.

The advantages of using a pressure regulator for open tubular column GC are the following: (a) It allows a wide range of operating flow rates and an instantaneous change from low to high (e.g., from 1 to 500 cm³ min⁻¹) flow rate and vice versa. (b) The carrier gas flow rate in the column remains constant at a given temperature if the pressure gradient across the column is constant. (c) A leak upstream of the column inlet will not affect column carrier gas flow rate and therefore a built-in septum purge and split sampling are permitted. (d) It allows the use of wide-bore columns in which low column inlet pressures are employed as well as microbore columns in which several hundred bar column inlet pressures are required.

The drawbacks of using a pressure regulator for open tubular column GC are the following: (a) Pressure regulators are very sensitive to upstream pressure changes. Constant upstream pressure regulated by a two-stage pressure regulator at the carrier gas supply cylinder is required. Operation at high detector sensitivities and chromatographic retention time reproducibilities requires both pressure regulators PR1 and PR2 (as shown in Figure 4.1) to be thermostated because pressure regulators are known to vary markedly with ambient temperature fluctuations. (b) Poor retention time reproducibility in isothermal analysis may result from the variation in the column outlet pressure (ambient pressure). Since the stability of the column carrier gas flow rate depends on the maintenance of the pressure gradient across the column, both column inlet and outlet pressures need to be maintained at the set pressures. The pressure regulator, if thermostated, can accurately control the column inlet pressure in the normal GC operating pressure range to 0.02–1%. However, the column outlet pressure, if equal to the ambient pressure, can vary as much as 50 mm Hg during a day. A 50 mm Hg pressure variation represents a 3% change in the pressure required for operating a 25 m × 0.2 mm i.d. column at 2-bar column inlet pressure. In consequence, poor retention time reproducibility can be expected. (c) In column temperature-programming operation, the carrier gas flow rate in the column is greatly reduced since the carrier gas viscosity increases with the rise in column temperature. The change in flow rate ($F_{Z,T}$) at any point Z along the column with respect to column temperature (T) may be approximated by[1]

$$F_{Z,T} = \frac{F_{o,T_0}}{[1 + (P^2 - 1)(1 - Z/L)]^{1/2}} \left(\frac{T_0}{T}\right)^{1.7} \tag{4.1}$$

Here F_{o,T_0} is the carrier gas flow rate at the column outlet and at the initial column temperature (T_0). P is the ratio of column inlet to outlet pressures: $P = P_i/P_o$. The carrier gas flow-rate variation with respect to column temperature change can be written as

$$F_{o,T} = F_{o,T_0} \left(\frac{T_0}{T}\right)^{1.7} \tag{4.2}$$

For a normal temperature-programmed analysis from 40 to 300°C, the column outlet carrier gas flow rate decreases to 36% of the initial flow rate. This decrease results in long analysis times and poor detection of well-retained sample components. In addition, if a concentration-dependent detector is used for solute detection, a very significant change in response, linearity, and baseline stability due to the carrier gas flow-rate change may occur. Quantitative analysis, under such circumstances, requires extreme caution. (d) A gas leak or a broken column in the oven gives no immediate and obvious indication or warning to the operator. As a result, poor baseline stability, retention time reproducibility, and quantitative reproducibility and accuracy are often obtained. Furthermore, when hydrogen is used as the carrier gas, a severe leak allows hydrogen to escape into the column oven or the laboratory at a high rate with potential explosion hazard. (e) The inherent problems of the pressure regulator such as hysteresis, permanent deformation, and outgassing of the diaphragm are more critical in open tubular column GC due to the requirements of low flow rates and high detector sensitivities. In terms of outgassing, stainless steel and Viton are preferred over Fairprene and Buna-N for diaphragm materials. However, stainless steel diaphragms may have greater hysteresis and, thus, a rapid flow-rate change from low to high (e.g., from 1 to 500 cm^3 min^{-1}) or vice versa may result in an offset in the pressure setting and baseline shift at the detector. The diaphragms made of organic polymers such as Viton and Fairprene have less hysteresis and better pressure regulation characteristics. However, because of the high outgassing rates of these types of controller diaphragms, it is necessary to bake-out the regulator at a temperature (e.g., 50–60°C with continuous He gas purge) higher than the working temperature (e.g., 40°C) before use. In addition, the working temperature should also be maintained constant in a pneumatic oven such that constant flow rate and outgassing rate are maintained: the baseline stability at high detector sensitivity is then not affected.

Flow Controller

Two types of flow controllers are conventionally used for carrier gas flow regulation. The first type regulates the flow rate by maintaining a fixed pressure drop across a needle valve by adjusting the needle-valve orifice size. The second type has a fixed needle-valve orifice size and regulates the flow by adjusting the spring force exerted on a controller diaphragm to change the pressure drop across the needle valve. For both types of flow controller, once the differential pressure drop across the needle valve and the orifice size are set, a constant carrier gas flow rate is maintained regardless of changes in the downstream pressure and flow restriction.

The advantages of using a flow controller in open tubular column GC are the following: (a) It maintains constant mass flow rate at the column outlet regardless of the column flow resistance change due to ambient pressure, carrier gas viscosity, and column changes. In column temperature-programming operations, flow control allows a constant column outlet flow rate that is especially advantageous when a concentration-sensitive detector is used. (b) It maintains the preset carrier gas flow rate regardless of the column used. (c) With a pressure gauge at the column inlet, a restriction to flow or a leak downstream of the flow controller, developed in routine operation, can be easily noticed from a change in the pressure gauge reading. (d) A severe gas leak downstream of the flow controller such as a broken column only allows carrier gas to escape at the preset rate (normally ≤ 10 cm^3 min^{-1}). This is particularly important when hydrogen is the carrier gas. (e) In temperature-programming operation, flow control normally gives better chromatographic performance in terms of speed of analysis, peak resolution, peak shape, and peak detection than that obtained with a pressure regulator system under the same initial chromatographic conditions.

Flow control, in principle, is preferred over pressure regulation for the control of carrier gas flow rates in open tubular column GC. However, due to the limitation of present flow controller technology and the inherent properties of the flow controller, a number of drawbacks prevent their widespread use. The difficulties include sensitivity to temperature and upstream pressure regulation, long stabilization periods, large minimum pressure drop, and hysteresis.

Needle Valve and Flow Restrictor

The use of a needle valve for carrier gas flow control is satisfactory if the upstream pressure of the needle valve is high and constant.[2] The required upstream pressure (P_N) of the needle valve for a fractional change (θ) of column carrier gas flow rate during a temperature-programming operation can be calculated from Equation (4.3):

$$P_N = \left(\frac{P_{i,T}^2 - (1 - \theta)P_{i,T_0}^2}{\theta} \right)^{1/2} \tag{4.3}$$

Here P_{i,T_0} and $P_{i,T}$ are the column inlet pressures at the column initial and final program temperatures, T_0 and T, respectively.

Equation (4.3) indicates that to maintain a constant carrier gas flow rate during a temperature-programmed analysis, a high pressure upstream of the needle valve is essential. For example, if the column is temperature programmed from 40 to 300°C with H$_2$ carrier gas, the required upstream pressure required to maintain better than 1% column flow stability is 18.2 bar. Here, the column inlet pressures at 40 and 300°C are assumed to be 2.0 and 2.7 bar, respectively.

The advantages of using a needle valve for flow control are the relative simplicity, low cost, and absence of polymeric elastomer exposed to the carrier gas flow stream which precludes outgassing or contamination. The disadvantages are the requirement of high upstream pressure and the susceptibility to plugging by particulate contaminants. In addition, the stability of the carrier gas flow rates depends on the stability of the upstream pressure and the valve temperature. Needle valves are not normally used to control carrier gas flow rate but are often used for make-up gas and detector fuel gas control.

Combined Flow Controller and Back-Pressure Regulator

A-combined flow controller and back-pressure regulator system is shown in Figure 4.2. Here, the flow controller is used to maintain a constant carrier gas flow rate to the injector regardless of column inlet pressure. Column carrier gas flow rate is controlled by the back-pressure regulator which maintains constant column inlet pressure whatever the downstream pressure of the regulator. This system allows both split and splitless sampling modes. In the split sampling mode, the total volumetric carrier gas flow from the flow controller is split at the inlet of the column. A small fraction of the carrier gas (or sample stream) is directed into the column, while the major fraction is vented through the back-pressure regulator.

In the splitless sampling mode, the solenoid valve upstream of the back-pressure regulator is switched to allow a major portion of carrier gas to flow through the "splitless sampling bypass" to the splitter vent. As a result, only the column flow is allowed to pass into the injector vaporization chamber and the column.

Because the column carrier gas flow rate is controlled by the pressure gradient across the column as a pressure-regulated system, the column flow rate is affected by the column temperature change in temperature-programmed operations. It has the same characteristics and drawbacks as the pressure-regulated system previously discussed. However, in the case of severe gas leakage, the leak rate is limited to the controlled flow rate of the flow controller. In addition, severe gas leakage can be immediately noticed

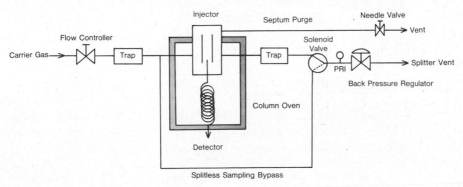

FIGURE 4.2.   Basic flow diagram of a flow controller and back-pressure regulator combined flow regulation system.

from the drop in pressure reading at the gauge, PR1, upstream of the back-pressure regulator.

In the design of the system, a flow controller with a wide range flow span $(0-600 \text{ cm}^3 \text{ min}^{-1})$, accurate flow regulation, and good flow reproducibility should be used to allow a wide split ratio operation range. A contaminant trap downstream of the flow controller and upstream of the injector should be used to prevent contamination by diaphragm outgassing. An additional trap should be used between the injector splitter vent and upstream of the back-pressure regulator so that the vented sample can be trapped before reaching the regulator. Additional design considerations similar to those presented in the previous sections of pressure regulators and flow controllers should also be taken into account.

4.3   COLUMN TEMPERATURE CONTROL

Open tubular column GC requires critical consideration of the effects of column temperature stability and temperature program linearity on the chromatographic performance of the low thermal mass glass or fused-silica open tubular columns. The requirements of column temperature stability for open tubular column GC can be considered from: (a) the dependence of the carrier gas flow rate on column temperature; and (b) the effect of column temperature variation on the retention time reproducibility. The former was discussed in Section 4.2. The latter effect can be considered from the dependence of the solute–stationary liquid phase partition coefficient and, therefore, the net retention volume (or time) on column temperature. The solute net retention volume (V'_g) can be related to the molar enthalpy of solution in the stationary phase (ΔH_n) and the entropy of solution $(\mathscr{R} \ln A)$.

$$\ln V'_g = \frac{\Delta H_n}{\mathscr{R}T} + \ln A \qquad (4.4)$$

By using Equation (4.4) and the data adapted from Baumann et al.,[3] the percent change in net retention volume $(\% \Delta V'_g/V'_g)$ or net retention time $(\% \Delta t'_R/t'_R)$ with respect to the change in column temperature (ΔT) is plotted in Figure 4.3 for solutes with ΔH_n of 6 and 12 kcal mol^{-1} and $\ln A$ of -5.37 and -9.97, respectively.

As shown by Figure 4.3, a 1°C column temperature change may alter the net retention volume (or time) of sample compounds by 5%. The change in retention volume may be as high as 20% if the column temperature strays 5°C from 100°C. To obtain better than 0.1% net retention volume (or time) reproducibility, the column temperature should be maintained to within $\pm$ 0.02°C in this example.

The apparatus conventionally used for high-precision column temperature control consists of a temperature-controlled liquid bath or a heated air convection oven. The forced-air convection oven is the most convenient and

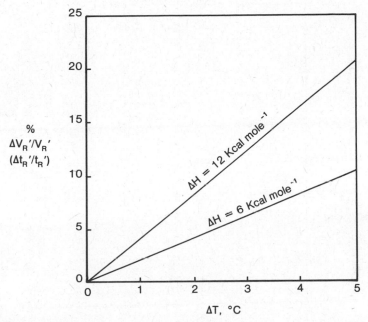

FIGURE 4.3. Relationship between column temperature stability and percent net retention volume (time) variation. A column temperature of 100°C was assumed.

the most popular column temperature control system available. It is used in all commercial gas chromatographs because it allows wide operating temperatures, fast heat-up and cool-down, and good temperature stability (± 0.1°C).

In the design of a column oven, it is important to ensure a uniform temperature throughout the column coil region. Temperature uniformity depends on the geometry of the oven, the location of the heater and sensor, and the pattern of mixing and circulation of air. Direct radiative heating of the column by the heating element should also be minimized when using low thermal mass glass or fused-silica columns. The combination of direct radiative heating and heater element temperature fluctuation could cause flow instability and detector baseline noise.

Temperature Programming

Commercial gas chromatographs equipped with air oven convective heating systems in general allow both ballistic and linear temperature programming. Ballistic programming results from full power heating to raise the column temperature nonlinearly. For linear temperature programming, program rates in 0.1 and 1°C min^{-1} increments are normally provided. Multiple-step linear temperature programming is also common in modern GC.

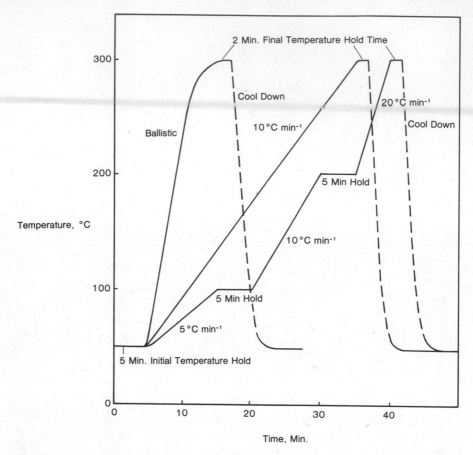

FIGURE 4.4. Examples of ballistic, single-step, and multiple-step linear temperature programs.

Figure 4.4 shows examples of typical ballistic, single-step, and multiple-step linear temperature program profiles from 50 to 300°C. The initial column temperature, initial temperature hold time, program rate, final temperature, and final temperature hold time are all adjustable and are selected according to the needs of each individual analytical problem. The optimization of column temperature-programming parameters has been treated by Harris and Habgood.[1]

Rapid heat-up and cool-down of the column oven temperature are essential for fast analysis. Although the thermal mass of glass or fused-silica columns is low, the actual heat-up and cool-down rates are limited by the power supply and the thermal mass of the oven.

At high heat-up rates, a substantially long period of initial temperature lag and final temperature overshoot are normally measured, as shown in Figure 4.5. As a result, the speed of the analysis is limited. A modern microprocessor-controlled instrument utilizes a proportional-integral-differential control algorithm[4] that has power feed-forward capability to minimize the initial temperature lag and final temperature overshoot. A temperature program profile obtained from this instrument is given in Figure 4.5.

Rapid cool-down of the column oven is achieved by opening the oven door to allow heat dissipation either manually or with an automated system.[5] The latter approach is incorporated into most of the modern gas chromatographs and allows automatic GC analysis in conjunction with an automatic sample injection system.

The cool-down rate (dT/dt) of an air oven may be approximated by Equation (4.5) using a single lumped model:[6]

$$\frac{dT}{dt} \simeq - \frac{\mathcal{J}A(T_i - T_f)}{C_p m} \tag{4.5}$$

where $\mathcal{J}$ is the average heat transfer coefficient, A is the surface area of the oven, and T_i and T_f are the column initial and final temperatures in the cool-down cycle, respectively. C_p and m are the specific heat and the mass of the oven, respectively. $C_p m$ is then the heat capacity of the oven.

Equation (4.5) indicates that a large oven surface area and heat transfer coefficient will allow rapid temperature cool-down. A lower thermal mass results in a faster cool-down rate. The optimization of oven thermal mass to allow rapid cool-down as well as high temperature stability is important in the design of the column oven for temperature-programming operation.

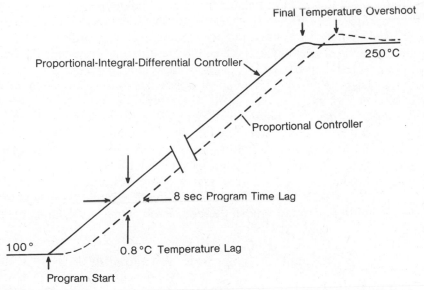

FIGURE 4.5. Comparison of the programmed temperature profiles using proportional controller and proportional-integral-differential controller. Temperature-program rate of 5°C min^{-1}.

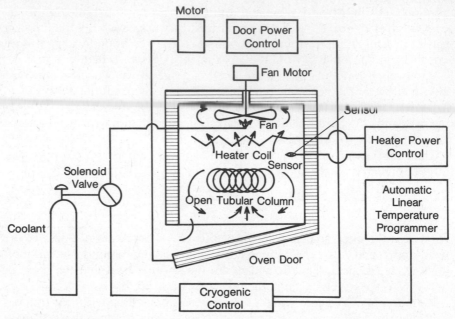

FIGURE 4.6. Schematic diagram of an automatic cryogenic temperature-programmable gas chromatograph.

Resistance Heating

The low thermal mass of the flexible fused-silica columns[7] may allow in the future the utilization of direct electric resistance heating or thermoelectric heating and cooling for column temperature control. The direct electric resistance heating of a metallic coated thin-wall flexible fused-silica column could offer significant advantages such as ultrarapid heating of the column, no thermal lag problem, and minimum power consumption. Rapid cool-down could also be expected because of its low thermal mass. It could also facilitate thermal focusing, multidimensional GC, and the miniaturization of open tubular column gas chromatographs.

In addition to producing a uniform temperature along the length of the column as found in normal temperature-controlled GC, it would also be feasible to produce a temperature gradient along the column by a gradient coating of the metallic layer. This approach could provide the advantages of the column temperature gradient effects as obtained in the technique of "chromathermography."[8–14]

Subambient Temperature Programming

An open tubular column gas chromatograph equipped with subambient temperature equipment allows column oven temperature cool-down to as low as $-99°C$ with liquid nitrogen and $-50°C$ with solid CO_2. A typical subambient system[5] is shown in Figure 4.6. It consists of a cryogenic control and coolant source in addition to a heater and a forced-air fan as found in normal temperature-programmed GC. When the column temperature is below a preset coolant turn-on temperature (normally near 60°C) and is above the set column temperature, the solenoid valve is switched on. This allows coolant to spread from a supply nozzle to the fan blades and to mix and cool a large quantity of surrounding air. The cool air is then circulated at a high velocity along the walls of the oven enclosure and allows a rapid column oven cooldown.

In open tubular column subambient temperature operation, it is particularly important to arrange the fan and the air circulation patterns to prevent direct spread of the coolant droplets onto the column.

4.4 INJECTION SYSTEMS

The design and performance of the injection systems in open tubular column GC are vital to the chromatographic performance in terms of peak shape, resolution, and quantitation. In recent years, significant progress in the development of injection systems has been reported. Some commonly used injection systems are the splitter injector, splitless injector, microflash vaporizer direct injector, cold on-column injector, purge-and-trap (precolumn) injection system, and pyrolyzer sampling system. The utility and, hence, the selection of an injector depend on the sample concentration, volatility, boiling point, and thermal and chemical stabilities. No universal injection system can be constructed to handle every complex sample encountered. It is important to recognize the application ranges and the limitations of the different sampling systems. The selection of the injector and the optimization of its performance are essential to success in open tubular column GC analysis.

In this section, state-of-the-art open tubular column GC sampling systems are described. Special emphasis will be placed on the design requirements, and the utility and performance in terms of peak broadening effects and quantitation.

Splitter Injector

The splitter injector is a flash vaporization injector that allows a small fraction of the volatilized sample to enter the open tubular column, while the major portion of the sample is vented to waste. The basic splitter injector was introduced in 1959 by Desty et al.[15] Descriptions of many other splitter injectors and their performances have been reported.[16-22] Excellent reviews are available on the splitter injectors as well as other sampling processes.[23]

Figure 4.7 shows a cross-sectional view of a conventional injector of this

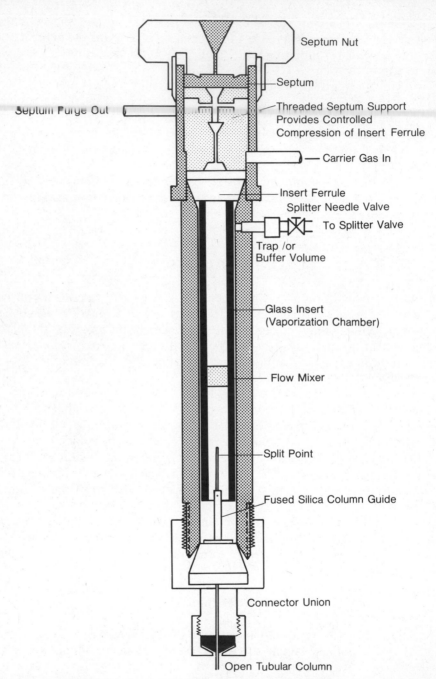

Septum Nut

Septum

Septum Purge Out

Threaded Septum Support
Provides Controlled
Compression of Insert Ferrule

Carrier Gas In

Insert Ferrule
Splitter Needle Valve

To Splitter Valve

Trap /or
Buffer Volume

Glass Insert
(Vaporization Chamber)

Flow Mixer

Split Point

Fused Silica Column Guide

Connector Union

Open Tubular Column

FIGURE 4.7. Cross-sectional view of a typical injector splitter.

type. Carrier gas, controlled by a pressure regulator or a combination of a flow controller and a back-pressure regulator, enters the injector near the septum. The carrier gas flow is first divided into two streams at the inlet of the vaporization chamber. The low flow-rate stream flows upward to purge the inlet and the septum of the injector. The high flow-rate carrier gas stream enters the vaporization chamber where the sample is vaporized and mixed with carrier gas. The sample and carrier gas mixed stream is then split at the column inlet, and a small fraction of the sample is allowed to enter the column. A needle valve in the high-temperature injector block is used to adjust the split ratio. A buffer volume or filter trap downstream from the split point is employed to contain the sample during the sampling process so that no sample stream reaches the needle valve to change the split ratio due to viscosity change of the flow stream. The column inlet is used as the split point and is located in the heated zone of the injector to prevent sample recondensation before splitting. The column-to-injector connection should have low thermal mass so that the column inlet temperature can rapidly track the oven temperature to avoid uneven temperature zones at the column inlet.

Design requirements for a splitter injector are: (a) minimum extra column band spreading due to the injector; (b) maximum chemical inertness; (c) wide operation range of split ratios; (d) minimum contamination; (e) linear and reproducible sample splitting; and (f) high-temperature operating range.

Minimum extra-column band spreading from the splitter injector may be achieved by minimizing the sample residence time in the injector. The two easiest approaches are (a) to increase the total carrier gas flow rate in the injector, and (b) to decrease the vaporization time of the sample. A long sample residence time in the injector could result in severe loss of column efficiency. For example, for an assumed effective volume of a vaporization chamber of 1 cm^3, a 1 cm^3 sample vapor volume (equal to about 2 μL of liquid sample injected) transferred at a carrier gas flow rate of 100 cm^3 min^{-1} could contribute 0.36 s^2 to the column peak variant of 0.018 s^2 (assumed W_h = 1 s), or a 21-fold increase in the column plate height. Low flow-rate split sampling may be used in the case where a cold trap or solvent effect[17,24] is applied to the inlet of the column for minimizing the initial sample band width.

The vaporization time, in general, contributes no significant band spreading in split sampling because of the small sample size injected, and a sufficient amount of heat has already been transferred by the preheated carrier gas for rapid vaporization. However, in cases where less-volatile samples are analyzed, high injector temperatures are desirable for preventing slow vaporization of the sample.

The sample residence time in the injector may also depend on the syringe needle residence time in the heated injector. During the time between insertion and withdrawal of the syringe needle, the sample can be continuously vaporized from the needle and transferred downstream to the column as a broad initial sample band. Rapid injection, as demonstrated by Gaspar et al.,[25] is important for achieving maximum column efficiency.

The absence of irreversible adsorption, and catalytic and thermal degradation of the sample compounds in the heated vaporization chamber are essential for trace sample quantitation. An all-glass or fused-silica sample vaporization chamber and sample transfer line should be used for open tubular column GC. Deactivation of the glass or fused-silica vaporization chamber may be required for extremely difficult samples.

A wide split-ratio operation range allows flexible utilization of the splitter injector with a wide variety of column dimensions and sample concentrations and volumes.

Contamination effects such as ghost peaks from the injector are often noticed, especially when high sensitivity detectors are used. It is important to prevent organic residuals, such as cutting oil, and organic-based materials, such as rubber O-ring material, from entering the heated zone in the injector. The septum may require a continuous positive purge to eliminate contamination. Linear and reproducible sample splitting is the most important and the most difficult performance requirement of a splitter injector. Linear splitting implies nondiscriminative splitting against varying sample component volatilities, viscosities, and polarities. Linearity demands a vaporization chamber with appropriate geometry, proper vaporization, uniform mixing of sample with carrier gas before splitting, and proper selection of injector temperature, split ratio, and injection speed. Vaporization chambers made of 2–4 mm i.d. glass tubes of 5–10 cm length with baffle, glassbead frit, mixing cup, or glasswool packing for enhancing vaporization and effective mixing are generally used. Glasswool packing inside the vaporization chamber gives excellent mixing and linear splitting.[16] However, there may be difficulties resulting from bleeding, washing by solvent of deactivated glasswool, and ghost peak effects due to trapping of nonvolatile compounds and septum fragments. Frequent replacement of the vaporization chamber (glass insert) is important to prevent ghost peaks and loss in separation efficiency, and is particularly desirable when dirty samples are analyzed.

The selection of injector temperature, injection speed, and split ratio depend on the sample. Linearity of split sampling of a particular mixture may be obtained by proper selection of these parameters. A detailed discussion of this is given in Chapter 5.

Linear and reproducible splitting of samples containing components with wide ranges of volatilities, polarities, viscosities, and concentrations is, in practice, difficult to achieve with a splitter injector, but nonetheless important. To achieve reproducible splitting, sampling parameters such as injector and column temperatures, split ratio, injection speed, sample size, flow rate, and so on, all have to be maintained constant. Autosampler injection is highly desirable for achieving reproducible splitting of wide volatility range samples.

Splitless Injector

A splitless injector (Figure 4.8) differs from a splitter injector only in the geometry of the sample vaporization chamber and the provision of a solenoid valve downstream of the splitter vent. The splitter can be shut off for a period of time (normally 30–90 s, depending on the sample size) while the sample is injected, vaporized, and transferred to the column.

A small-bore open vaporization chamber (normally a 2 mm i.d. × 10 cm

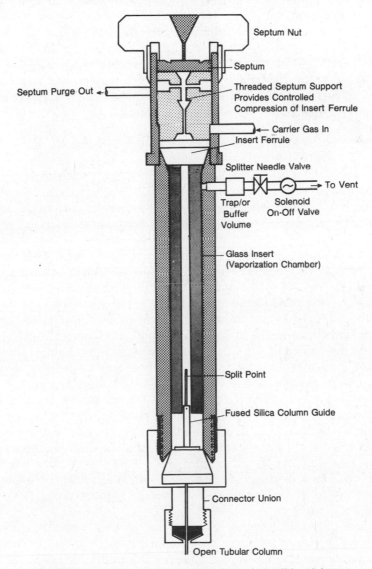

FIGURE 4.8. Cross-sectional view of a typical splitless injector.

glass or quartz tube) is required to allow minimum mixing so that the sample vapor can be transferred as a plug with minimum dilution into the column. Because of the large vaporization chamber volume used in the splitless injector and the exponential dilution that occurs in the splitless sampling process, this sampling method requires a rapid purge of the injector by switching on the splitter carrier gas flow via the solenoid valve after the splitless sampling time.

The splitless sampling time is normally longer than 30 s, which is unacceptable for conventional high-resolution open tubular column GC. The initial sample band has to be refocused at the inlet of the column by using either cold trapping or the solvent effect technique.[17,24] Splitless sampling demands both reproducible sample size and sampling time.[26] Automation of splitless sampling improves both retention times and quantitative reproducibility for the analysis of wide volatility range samples.[26] A detailed discussion of splitless sampling parameters and techniques is given in Chapter 5.

Microflash Vaporizer Direct Injector

The flash vaporizer direct injector is routinely used in packed-column GC for quantitative analysis. The use of the microvolume flash vaporizer for direct sample injection without splitting in open tubular column GC has been reported.[27-33] A conventional microflash vaporizer direct injector[28] is shown in Figure 4.9.

A microflash vaporizer direct injector designed for high-resolution open tubular column GC requires minimum sample backflush to ensure quantitative sample transfer to the column, and minimum injector contribution to the initial sample bandwidth to preserve column efficiency. The first goal cannot be achieved by increasing the volume of the vaporization chamber as is prescribed, in general, in the literature. The sample capacity of an injector is not limited by the volume of the vaporization chamber, but rather by its ability to transfer dynamically the entire sample downstream into the column and to prevent sample backflush. One effective approach is to use a narrow-bore vaporization chamber with a flow controller to allow the establishment of a positive pressure gradient along the annular flow path between the tip of the syringe needle and the top of the chamber during sample injection. Figure 4.9 shows the integration of a standard plunger-in-needle 5 μL syringe with an o.d. of 0.66 mm and a glass vaporization chamber of 0.7 mm i.d. The pressure gradient across a 4 cm annular path allows large sample sizes up to 5 μL to be slowly injected without sample loss due to backflushing.

The goal of having minimum injector contribution to band spreading can be achieved by using a narrow-bore vaporization chamber. For example, the calculated contribution to peak variance from a 0.7 mm i.d. × 10 cm long vaporizer is 0.6 s^2 at 1 cm^3 min^{-1} carrier gas flow rate. This corresponds

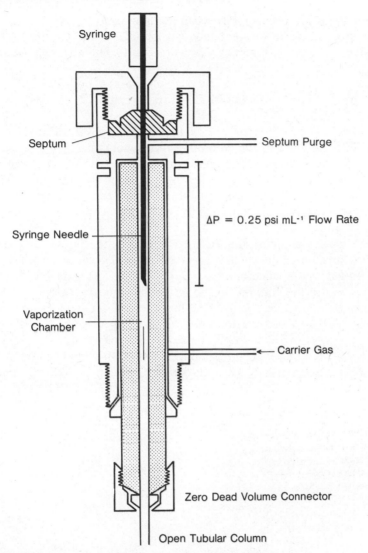

Syringe

Septum

Septum Purge

$\Delta P = 0.25$ psi mL^{-1} Flow Rate

Syringe Needle

Vaporization
Chamber

Carrier Gas

Zero Dead Volume Connector

Open Tubular Column

FIGURE 4.9. Cross-sectional view of a microflash vaporizer direct injector.

to a conventional 4 mm i.d. × 5 cm long vaporizer splitter at a split ratio of 80:1 (81 cm^3 min^{-1} injector carrier gas flow rate). The contribution to initial sample band spreading can be reduced by a factor of 4 if the flow rate in the injector is doubled. For the use of wide-bore columns where high carrier gas flow rates (2–5 cm^3 min^{-1}) are normally used, the direct injector contribution to the initial sample band spreading is negligible.

The microflash vaporizer direct injector allows small sample size (<1 μL) injections for isothermal or temperature-programmed operation. For trace analysis where sample volumes larger than 1 μL are required, the effect of sample volume on band spreading can be minimized by using a cold trap or solvent effect to focus the solute zone at the inlet of the column. A detailed discussion of the direct sampling parameters and techniques is given in Chapter 5.

Cold On-Column Injector

The cold on-column injector[16,34–36] allows the injection of a liquid sample directly into the inlet of the column, and thus eliminates the possibility of sample loss during vaporization and transfer from the vaporizer to the column. Excellent quantitative precision and accuracy for thermally labile compounds and wide volatility range samples have been reported.[16,36–38]

Figure 4.10 shows a syringeless on-column injector[16] for open tubular column GC. Schomburg et al.[16] have discussed both the macro and micro version of direct sampling. In the macro version, the liquid sample is contained in a small crucible which is inverted over the column inlet. The sample is then forced by the injector carrier gas pressure in combination with capillary action into the column. In the micro version, the sample is placed in a micropipette which can be inserted into the column. This system requires no syringe and is a true carrier gas pressure-lock system. Excellent quantitative precision and accuracy have been reported.[16,38]

The use of a syringe for on-column injection on wide-bore open tubular columns was first described by Zlatkis and Walker.[39] The technique was

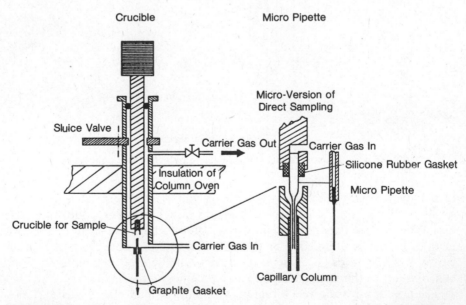

FIGURE 4.10. Direct sampling of liquids into capillary columns, macro and micro versions. (Reproduced with permission from ref. 16. Copyright Elsevier Scientific Publishing Company.)

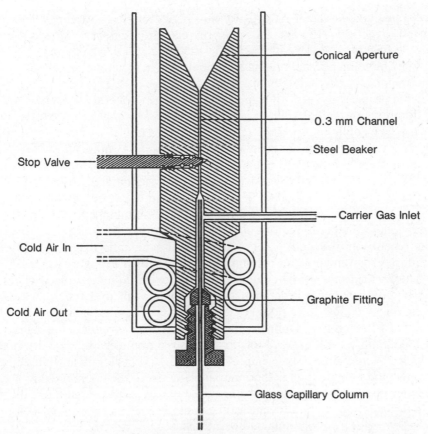

Conical Aperture

0.3 mm Channel

Steel Beaker

Stop Valve

Carrier Gas Inlet

Cold Air In

Graphite Fitting

Cold Air Out

Glass Capillary Column

FIGURE 4.11. Cross-sectional view of an on-column injector. (Reproduced with permission from ref. 35. Copyright Elsevier Scientific Publishing Company.)

later adapted by Grob and Grob[35] for smaller diameter open tubular columns (≤0.3 mm i.d.) using a syringe needle of 0.23 mm o.d., 0.1 mm i.d., and 85 mm length. A cross-sectional view of the Grob on-column injector is shown in Figure 4.11. With the use of fused-silica tubing for the syringe needle, a column i.d. as small as 0.2 mm may be used. The minimum sample volume that can be quantitatively injected is about 0.3 μL with a stainless steel needle, and about 0.2 μL with a narrow-bore fused-silica needle.

This on-column injector has a conical syringe needle guide and a syringe needle passage channel on top. The diameter of the passage channel must be narrow in order to match the outside diameter of the syringe needle such that no measurable pressure drop or carrier gas leakage occurs when the stop valve is open and the syringe needle is inserted into the injector. Gas leakage could result in the loss of high-volatility solutes. Cool air is forced through the bottom of the injector body to cool the column inlet, thus allowing on-column injection of the sample in the liquid state. The carrier gas flow rate can be controlled by a pressure regulator or a flow controller after sample injection. However, a pressure regulator is normally used during sample injection because of leakage in the needle passage channel.

The above mentioned on-column injectors use column oven temperature programming to initiate the chromatographic process. It is important to select the proper solvent with respect to the initial temperature of the oven.[36] In addition, the initial column inlet zone has to be properly thermostated (secondary cooling) during the sampling process in order to prevent sample discrimination.[36] With the injection of large sample volumes (e.g., ≥ 1 μL), peak splitting[40] of the high boiling point components due to droplet formation and the nonuniform and excessive spreading of the initial distribution profile of the injected sample at the inlet of the column may often be observed.

To improve the initial sample profile, an on-column injector which allows independent temperature control of the column inlet from the column oven was recently introduced.[41] A cross-sectional schematic diagram of a temperature-programmable on-column injector is shown in Figure 4.12.

The temperature of the column inlet where the liquid sample is injected can be controlled and programmed from subambient to 350°C by an automatic linear temperature program (ALTP). Low programming rates can be selected for thermally labile compounds, and high programming rates (up to 180°C min^{-1}) can be used for rapid vaporization of wide volatility range samples to minimize initial sample bandwidth and peak tailing. In addition, because the column inlet is thermally isolated from the column oven, it can be controlled at temperatures below the column oven temperature during sample injection, thus allowing isothermal as well as high initial column-oven temperature-programming operations.

The on-column injector shown in Figure 4.12 uses a flow controller instead of the pressure regulator used in other on-column injectors for carrier gas flow regulation, and thus allows a constant carrier gas flow rate throughout sample injection and temperature-programmed operation. The use of hydrogen carrier gas is also much safer because of the controlled low flow rate (0–10 cm^3 min^{-1}).

The use of a flow controller for the carrier gas requires the system to be totally leak free during the process of valve opening when the syringe needle is inserted through the valve passage to the column inlet for on-column injection. One elegant approach is to use a narrow-bore fused-silica or stainless steel needle sheathed with a 26 gauge stainless steel tubing for protection and to achieve a pressure-lock system with a spring loaded O-ring seal at the top of the injector body. A pressure-lock carrier gas leak-tight system is important for the quantitation of volatiles that elute near the solvent peak. The backflush of solvent containing these volatiles during the insertion of the syring needle into the column inlet resulting from a surge of carrier gas pressure or a leak through the valve opening could result in discrimination and poor quantitation of these sample components.

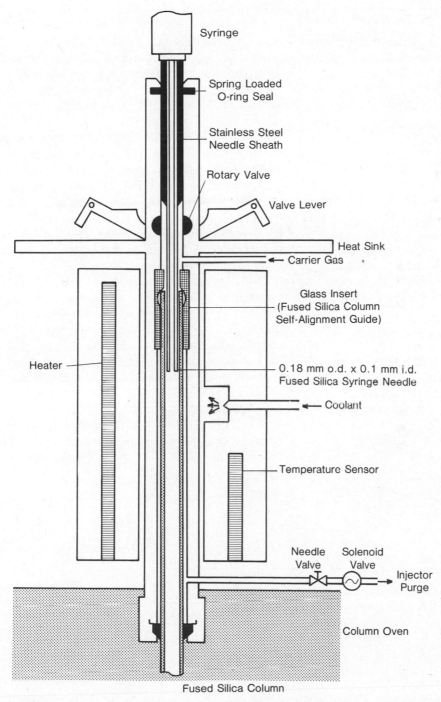

Syringe

Spring Loaded
O-ring Seal

Stainless Steel
Needle Sheath

Rotary Valve

Valve Lever

Heat Sink

← Carrier Gas

Glass Insert
(Fused Silica Column
Self-Alignment Guide)

Heater

0.18 mm o.d. x 0.1 mm i.d.
Fused Silica Syringe Needle

← Coolant

Temperature Sensor

Needle   Solenoid
Valve   Valve

Injector
Purge

Column Oven

Fused Silica Column

FIGURE 4.12.   Cross-sectional view of a temperature-programmable on-column injector.

Purge-and-Trap Sampling System

Gas phase organic volatiles associated with a wide range of samples including food and flavors, biological fluids, air samples, water pollutants, and polymers are, in general, complex mixtures with a wide range of concentrations. The use of purge-and-trap sampling techniques for sample concentration prior to open tubular column GC is important in the analysis of these sample types. Gas phase analytical techniques with open tubular column GC for qualitative and quantitative analysis of trace organics present in air and other gaseous media have been reported.[42-63] A simple purge-and-trap sampling system which includes a headspace sample chamber (or a packed-bed precolumn), a cold trap, and an open tubular column is shown in Figure 4.13. The operation of the system for the purge-and-trap sampling mode, normal analysis mode, and backflushing precolumn mode are discussed below.

In the purge-and-trap sampling mode, carrier gas, which is controlled by a pressure regulator or a flow controller, is directed by a three-way solenoid valve at position 1 to the sample chamber (headspace or packed precolumn sampling accessory) where the volatile sample is purged downstream to the cold trap. The cold trap is maintained at a suitable cold temperature using liquid N_2 or dry ice. After the sample concentration step, the flow is shut off at V_3 and the cold trap heater is turned on. Once thermal equilibrium has been reached in the cold trap, the flow is switched on at V_3 and the three-way solenoid valve, V_2, is switched from valve position 3 to 4. The desorbed sample is then split at the inlet of the open tubular column in the splitter injector, and normal analysis proceeds.

If cold trap backflush is desired, the carrier gas flow can be diverted by switching the solenoid valve, V_1, from position 1 to 2. The needle valve, NV_2, is used to adjust the pressure drop between position 2 of V_1 and the injector to be the same as that between position 1 of V_1 and the injector.

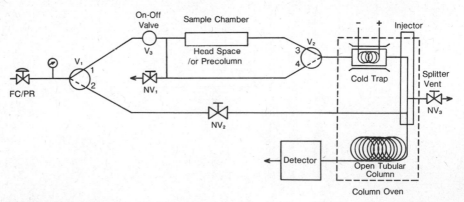

FIGURE 4.13. Schematic flow diagram of a purge-and-trap sampling system in conjunction with an open tubular column.

The other needle valve, NV_1, allows the adjustment of the backflush flow rate of the cold trap.

The precolumn used in the purge-and-trap sampling mode can be an open tubular column or a short packed column (5 cm $\times$ 2 mm i.d.). Selection of the adsorbent depends on the nature of the compounds to be analyzed. The use of adsorbents such as silica gel, alumina, and activated charcoal is normally limited because of irreversible adsorption and decomposition. The temperature limitations and bleeding levels should also be considered in the selection of the adsorbent. Porous polymers such as Tenax GC and particles coated with nonpolar liquid stationary phases such as OV-1, OV-101, SE-30, and Dexsil 300 are normally used in precolumns.

Pyrolyzer Sampling System

Pyrolysis sampling in conjunction with open tubular column GC for the analysis of polymers, paints, drugs, textiles, biological fluids, microorganisms, coal and related materials, and so on, has been discussed in a number of publications.[64-74] The nature and quantity of pyrolysis fragments reflect the elemental and structural characteristics of the parent molecules. The earliest applications of pyrolysis gas chromatography were in the establishment of fingerprint pyrograms of nonvolatile organics. However, because of the complexity of thermal degradation products, pyrolysis GC usually employs open tubular columns because of their tremendous separation power and sensitivity. In addition, small sample sizes are required to ensure rapid heating to well-defined temperatures and to prevent the occurrence of secondary reactions in pyrolysis.

Since the pyrolyzer functions as an open tubular column GC inlet system, the design of the pyrolyzer must exclude any dead volume and ensure a sharp plug sample injection profile. Two major design types are continuous-mode and pulse-mode pyrolyzers.

The continuous-mode pyrolyzer is also called the reactor chamber pyrolyzer. As the name implies, the system includes a low dead-volume reactor chamber that can be heated externally. The sample can be placed in a moveable boat or cup that can be brought into the uniformly heated reaction zone. In the case of a liquid sample, direct sample injection can also be used. The volatile pyrolysis products are then split or carried by the carrier gas directly into the column. Figure 4.14 is a schematic diagram of a continuous-mode pyrolyzer–GC system.[70] The splitter injector is heated by a separate heater and controller to prevent the formation of cold spots in the interface and to prevent condensation of volatile pyrolysis fragments.

Pulse-mode pyrolyzers may also be called flash pyrolyzers. The two most commonly employed are resistively heated ribbon or filament pyrolyzers and Curie-point pyrolyzers using radio-frequency induction heating.

Resistively heated filament or ribbon type pyrolyzers are commonly used and commercially available for a wide variety of applications. They typically

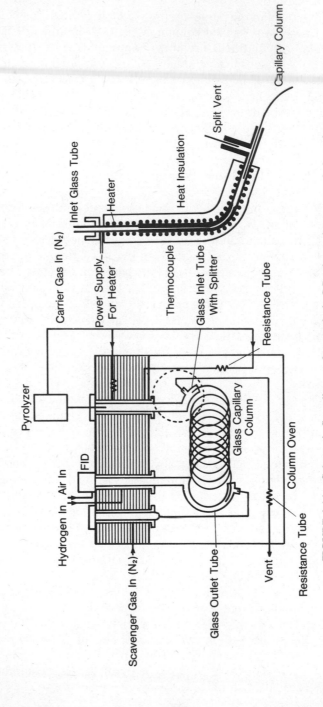

FIGURE 4.14. Schematic flow diagram for pyrolysis GC using a glass open tubular column. (Reproduced with permission from ref. 70. Copyright American Chemical Society.)

consist of an inert filament or ribbon (made of platinum or platinum-rhodium) heated by a high current power supply. The heating rate and temperature are reproducibly controlled by a power and temperature proportional controller. Filament temperatures as high as 1500°C and a linear heating rate as fast as 20,000°C s^{-1} can be obtained. A schematic diagram of a typical filament pyrolyzer is shown in Figure 4.15.

The Curie-point pyrolyzer differs from the resistive heating type pyrolyzer in that the ferromagnetic heater element is heated inductively to its Curie point in a radio-frequency field. The Curie-point temperature of a ferromagnetic material (the temperature at which the alloy becomes paramagnetic and ceases to adsorb energy) depends on the composition of the alloy (Table 4.1) and is highly reproducible. A typical Curie-point pyrolyzer is also shown in Figure 4.15.

In pulse-mode pyrolysis, the sample is applied to the filament as a thin and uniform coating by wetting with a solution in a volatile and nonde-

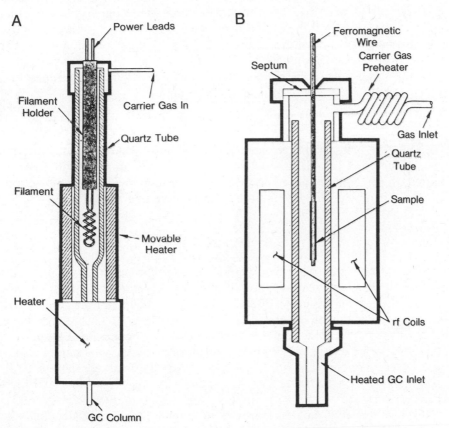

FIGURE 4.15. Schematic diagrams of (A) filament and (B) Curie-point pyrolyzers. (Reproduced with permission from ref. 75. Copyright American Chemical Society.)

TABLE 4.1. Curie-point temperatures for
different ferromagnetic alloys[a]

Alloy composition	Curie-point temperature (°C)
100% Ni	358
18:51:1% Fe:Ni:Cr	440
49:51% Fe:Ni	510
40:60% Fe:Ni	590
30:70% Fe:Ni	610
67:33% Fe:Co	660
100% Fe	770
40:60% Ni:Co	900
50:50% Fe:Co	980

[a] Taken from ref. 75.

structive solvent such as CS_2. In the case of insoluble samples, the filament or ribbon can be bent or shaped to accommodate the sample. For example, hollow ferromagnetic tubes have been used with powder samples, and a helically shaped wire to hold a quartz sample capillary tube containing the sample is also often employed.

Practical considerations of the selection of pyrolyzer type, temperature control, and sample size are discussed in Chapter 5.

Autosampling

The sampling systems discussed above can be improved in their quantitative performance and sample throughput if the systems are automated. Automation, in principle, requires precisely time-programmed syringe injections or valve opening cycles for unattended repetitive operations.

4.5   DETECTORS

Because of the high separation efficiency of open tubular columns, selection and use of detectors require consideration of detector response times and extra-column band spreading effects. The requirements in terms of detector time constants and cell volumes were discussed in Chapter 2 and by many other authors.[76-84] Modern GC detectors having response time constants of 50 ms or less are compatible with state-of-the-art open tubular columns. The detector cell volume band spreading effect can be effectively minimized by using make-up gas at the end of the column for high-resolution and high-speed analysis.

The usefulness of the detector in open tubular column GC depends also on its sensitivity, stability, detection limit, linearity, and reactivity. The definitions and implications of these characteristics are discussed below.

Sensitivity

Detector sensitivity (S) is defined as the change in the response with the change in detected quantity. For a mass flow-rate-dependent detector, the quantity detected is the sample mass $(\dot{M})$ in the carrier gas. The sensitivity for such a detector is therefore expressed as

$$S = \frac{A}{M} \qquad (4.6)$$

where A is the integrated peak area. The sensitivity for a mass flow-rate-dependent detector remains unchanged regardless of packed or open tubular column applications, because the same peak areas are measured for the same mass flow rates.

For a concentration-dependent detector where the response is proportional to the concentration of the sample in the carrier gas, the sensitivity is expressed as

$$S = \frac{h_t}{C} \qquad (4.7)$$

where h_t is the peak height and C is the concentration.

Baseline Stability

Short-term and long-term detector baseline stabilities are measured by noise and drift, respectively. Low noise is an important detector performance parameter and allows the detector to be operated at high sensitivity. Small baseline drift allows analyses that require extended time periods. For a given detector, baseline stability may be obtained by using high-purity gases and by maintaining constant gas flow rates and detector temperatures.

Minimum Detectable Quantity (Q_0)

Q_0 is defined as the mass flow rate (for mass flow-rate-dependent detectors) or the concentration (for concentration-dependent detectors) of the detected sample that yields a response equal to twice the noise level.

Detectors with low Q_0 are essential for use with thin-film open tubular columns and for trace analysis. High Q_0 implies the necessity of large sample size which in turn requires high column sample capacity. As a consequence, either a long column and thus long analysis time, or a wide-bore column with poor resulting separation efficiency must be used.

With the use of high-resolution open tubular columns, the detector Q_0 can be reduced due to the narrow and sharp elution peak shape. It is obvious that the higher the column efficiency, the narrower the peak width and thus the lower the Q_0.

Linearity and Linear Dynamic Range

The linearity is normally defined as the range of mass flow rates, or concentration of the detected sample in the carrier gas, over which the response of the detector is constant. The linear dynamic range is defined as the incremental change in detected sample which produces an incremental change in detector response to within $\pm 5\%$ of linearity.

Open tubular columns, in general, operate in the low mass flow rate or concentration range of the detector, which is a compromise between detector noise and column sample capacity. In many cases, this operation range is narrower than the available detector linear dynamic range. For those detectors in which the column operation range is greater than the detector linear dynamic range, the nonlinear range of the detector response should be avoided in order to prevent peak distortion and error in quantitation.

Reactivity

Reactivity in terms of adsorption and catalytic decomposition of labile compounds in the detector is detrimental to quantitation. An all-glass or quartz system may be required for the analysis of chemically unstable and highly polar samples, especially when only a trace amount of the sample is analyzed.

A comparison of the properties of some commonly used detectors is given in Table 4.2. The apparatus, requirements, advantages, and limitations of these detectors are discussed below.

Flame Ionization Detector (FID)

The flame ionization detector is the most commonly used detector in open tubular column GC because: (a) it is highly sensitive to organic carbon containing compounds; (b) it has a 10^7 linear range and excellent baseline stability; (c) it is highly reliable and easy to use; (d) it is relatively insensitive to small column flow-rate changes in temperature-programmed operation; and (e) it has low detector dead volume effects and a fast response time constant.

The limitations of the FID are: (a) it has little or no response to compounds[85] such as N_2, O_2, CO, CO_2, H_2O, H_2S, CS_2, COS, HCN, NH_3, NO, N_2O, NO_2, N_2O_3, CCl_4, $SiCl_4$, CH_3SiCl_3, SiF_4, and all noble gases; (b) it is a destructive detector (chemical and physical properties of the sample are irreversibly changed); (c) its response is strongly dependent on the structure of the sample and on the presence of heteroatoms (e.g., the presence of O, S, and halogens decreases the response of the FID); (d) a large injection of solvents such as CS_2 may produce a negative peak as a result of their combustion properties which cause a momentary separation of the flame from the polarized flame tip (the detector background current is thus low

TABLE 4.2. Properties of some commonly used detectors in open tubular column GC.

Parameters	FID	FPD	TSD	PID	ECD
Operation mode	Mass flow rate	Mass flow rate	Mass flow rate	Concentration	Concentration
Selectivity	Organic carbon	Sulfur/phosphorus	Nitrogen/phosphorus	Organic hydrocarbon	Electron affinity
Sensitivity	0.02 C g^{-1}	32 C g^{-1} (P) 0.32 C g^{-1} (S)	0.8 C g^{-1} (P) 0.4 C g^{-1} (N)	0.3 C g^{-1}	—
Noise $Q_0{}^a$	0.01 pA 0.5 pg s^{-1}	8 pA 0.5 pg s^{-1} (P) 50 pg s^{-1} (S)	0.01 pA 25 fg s^{-1} (P) 50 fg s^{-1} (N)	0.01 pA 0.5 pg μL^{-1}	5 Hz (constant-current) 1 pg μL^{-1}
Linearity	10^7	10^5 (P) 10^3 (S)	10^4 (P) 10^5 (N)	10^7	10^4
Function	Destructive	Destructive	Destructive	Nondestructive	Nondestructive

^{a} Q_0 values are measured for packed-column GC. Q_0 for open tubular column operation can be calculated by dividing the packed column Q_0 by the enhancement factor, E, due to the column according to Equations (2.76) and (2.77).

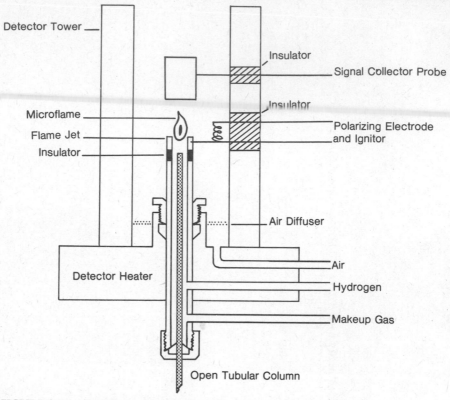

Detector Tower

Insulator

Signal Collector Probe

Microflame

Insulator

Flame Jet

Polarizing Electrode and Ignitor

Insulator

Air Diffuser

Detector Heater

Air

Hydrogen

Makeup Gas

Open Tubular Column

FIGURE 4.16. Simplified schematic diagram of a flame ionization detector.

during CS_2 elution); and (e) chlorinated solvents such as CH_2Cl_2 and $CHCl_3$ produce soot and black smoke in the hydrogen-rich flame which causes spikes and detector instablity. A hotter flame with lower hydrogen content can prevent this incomplete combustion of chlorinated solvents.

The design requirements and performance characteristics of the FID have been thoroughly investigated.[86-92] A schematic diagram of the FID is shown in Figure 4.16.

As is shown, the end of the column is directly interfaced to the detector housing via a connector union and a Vespel or graphite ferrule. The thermal mass of the interface connector should be small so that the column end can rapidly track the column oven temperature. Make-up gas at the end of the column is used for both minimizing band broadening from the connector volume and the detector cell volume and preventing back diffusion of sample into the interface. Make-up gas may also be required to optimize the detector response and stability. The ratio of the total carrier gas flow rate to that of the hydrogen fuel gas affects the FID flame temperature, flame geometry, and ionization and collection efficiencies.

Figure 4.17 shows the effect of the total carrier gas flow rate (He carrier gas flow rate fixed at 0.5 cm^3 min^{-1}, plus variable make-up gas flow rate) on the sensitivity of an FID measured as the peak height for *n*-alkanes eluted from an OV-101, 25 m × 0.25 mm i.d., glass open tubular column.[93] As the total carrier gas flow rate approaches that of the hydrogen fuel gas at 30 cm^3 min^{-1}, the best FID sensitivity is obtained. Figure 4.17 also indicates that, in the optimum detector sensitivity range, a change in the carrier gas flow rate of as much as 10 cm^3 min^{-1} does not significantly affect the response of the FID. For column temperature-programmed operation, using a pres-

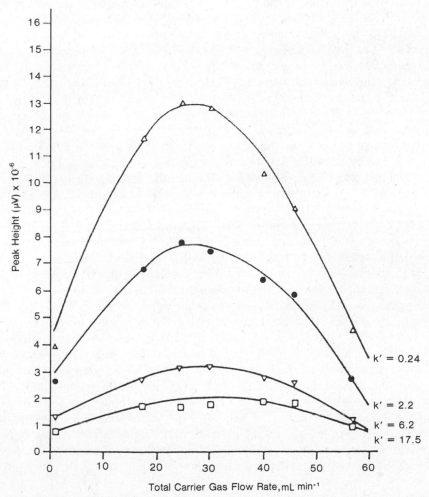

FIGURE 4.17. Effect of the total carrier gas flow rate on FID sensitivity. Column was 25 m × 0.25 mm i.d. coated with OV-101; isothermal at 130°C. (Reproduced with permission from ref. 93. Copyright Dr. Alfred Huethig Publishers.)

sure regulator for carrier gas flow control, it is important to adjust the total carrier and make-up flow rate to that of the hydrogen fuel gas flow rate so that the sensitivity of the FID is not significantly affected by variations in the column carrier gas flow rate. If hydrogen is used for both carrier and detector fuel gases, the FID response may change in temperature-programmed operation due to changes in the column carrier gas flow rate. Frequent calibration or the use of a flow controller is required for the main tenance of a constant hydrogen carrier gas flow rate for quantitative measurements.

The use of nitrogen as a make-up gas can double the detector sensitivity over that obtained with helium. All gases introduced to the FID should be preheated to the detector temperature to maintain a steady flame temperature and ionization efficiency.

An air-to-hydrogen fuel gas ratio of approximately 10 was demonstrated[85,89] to give good flame stability and ionization efficiency. Air and hydrogen flow rates on the order of 300 and 30 $cm^3 min^{-1}$, respectively, are normally employed. The flame temperature and background current of the FID are on the order of 2000–2200°C and 10^{-14} A, respectively. The use of a gas filter and electrolytic grade hydrogen are required for operating the detector at high sensitivity and with low noise.

The end of the column should be positioned just below the flame tip to prevent pyrolysis of the liquid stationary phase or the external coating (in the case of fused silica). This also minimizes the detector dead volume and the possibility of adsorption or catalytic decomposition due to contact with the high-temperature flame tip inner surface.

The flame tip inner diameter can affect the sensitivity of the FID. Experimental results[93] indicate that a 0.025 cm i.d. flame tip produces almost double the sensitivity of that obtained with a 0.05 cm i.d. flame tip. Flame tip materials commonly chosen are stainless steel and quartz. A quartz flame tip is advantageous for open tubular column GC because of its greater chemical inertness and, therefore, low resultant detector noise. However, the sensitivity of the FID with a quartz flame tip is usually lower than that measured with a stainless steel flame tip[94] due to the poor heat conduction and low heat capacity of the former, and, thus, low ionization efficiency.

The geometry of the collector electrode has been extensively investigated.[95–99] In general, the ion collection efficiency is affected by the density and the distribution of the electric field between the collector electrode and the flame burner electrode. The geometry of the collector electrode defines the distribution of the electrical field and the flow pattern of the gas stream between two electrodes. A symmetrical and cylindrical collector electrode has shown better ion collection efficiency and linearity than a ring electrode.[93]

FID polarization potentials between 100 and 300 V are normally used, depending on the saturation current of the detector.[99] The collector electrode is generally selected as the anode and the flame tip as the cathode, due to the more efficient collection of the low mobility cations generated near the flame tip as a result of their short migration distance.[89]

Although thermostating the FID is not required, the detector tower should be maintained at above 100°C to prevent water vapor condensation and, thus, noise and extinguishing of the flame.

Electron Capture Detector (ECD)

The electron capture detector is generally used for the analysis of compounds that have high electron affinities such as pesticides, insecticides, drugs and their metabolites, and so on. The detection mechanism and response characteristics of the ECD in coulometric, constant current, and constant frequency modes have been reviewed;[100-108] Table 4.3 lists typical properties. Coulometric ECD offers the best detection limits but limited linearity. The response in the coulometric mode is proportional to the number of solute molecules that react with electrons. It is independent of the carrier gas flow rate and detector temperature[103] if an extremely clean carrier gas and low column bleed system is used. The constant-frequency mode offers the maximum linear dynamic range, but the worst detection limits among the three compared. In modern ECD, the constant-current mode is often used because of its superior linearity (10^4) and subpicogram detection limits. It should be noted that in both constant-current and constant-frequency modes, the ECD responses are dependent on sample concentrations and their electron affinities. Accurate calibration using standard samples is necessary for proper quantitative analysis.

Since the ECD is a concentration-dependent detector, a small detector cell volume is desirable for optimizing the detector sensitivity and minimizing the dead volume. However, because of the large differences in the spe-

TABLE 4.3. The ECD properties in coulometric, constant-frequency (with long pulse period), and constant-current (with pulse frequency between 10^3 to 10^6 Hz) modes. Assumed peak width at half height is 5 s and sample molecular mass is 150 daltons.

Operation modes	Ionization source	Noise (pA)	Linearity	$Q_0{}^a$ (pg)	References
Coulometric	^{63}Ni (15 mCi)	1.5	5×10^2	0.02	101
	^{3}H (500 mCi)	3	5×10^2	0.05	
Constant-frequency	^{63}Ni (15 mCi)	1.5	10^7	2	101
	^{3}H (500 mCi)	3	2×10^7	4	
Constant-current	^{63}Ni (7.5 mCi)	5 Hz	10^5	0.1	105
	^{3}H (250 mCi)	—	10^4	0.05^b	109

^{a} Measured using packed column.

^{b} Measured using open tubular column.

cific activities of the radioactive emission sources used for the ECD, a small detector cell volume can only be obtained with a high specific activity source. The two most commonly used radioactive sources, ^{63}Ni and ^{3}H (tritium), have specific activities of 5 and 9800 Ci g^{-1}, respectively. Using Ti^{3}H and Sc^{3}H, detector cell volumes as small as 140 μL were reported[109] for open tubular column GC applications. The state-of-the-art ECD using a ^{63}Ni source has a detector cell volume of 300 μL.[103] A small flow cell ECD using ^{63}Ni in a unique displaced-coaxial-cylindrical design is shown in Figure 4.18.

The temperature limits for use of Ti^{3}H and Sc^{3}H, as set by the U.S. Atomic Energy Commission, are 220°C and 325°C, respectively. Above these maximum operating temperatures, tritium leakage and contamination can occur. The ^{63}Ni source, however, allows analysis up to 350°C and heating up to 420°C for short periods. ^{55}Fe was also suggested as a high-temperature source,[110] but no further development has been reported.

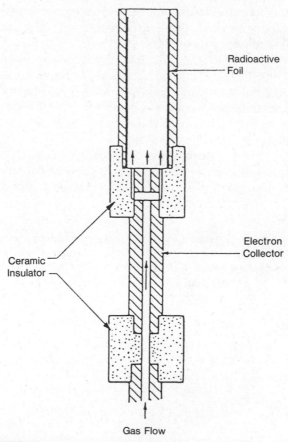

FIGURE 4.18. Schematic diagram of a displaced-coaxial-cylinder electron capture cell. (Reproduced with permission from ref. 105. Copyright Elsevier Scientific Publishing Company.)

The ECD stability in terms of noise and drift depends on the stability of the carrier gas flow rate, impurities in the carrier gas, oxygen leakage, temperature fluctuation, detector pressure stability, and the radiation source. Carrier gas flow fluctuations can significantly affect the standing current and thus the stability of the baseline and the response of the detector. In open tubular column ECD operation, constant carrier gas flow control is preferred. If a pressure regulator is used, addition of a make-up gas at the end of the column is desirable for minimizing the effects of column carrier gas flow-rate reduction due to the increase in column temperature during temperature programming.

Impurities in the carrier gas, such as water and oxygen, can decrease the concentration of free electrons. Competition for electrons between the eluted solute molecules and impurities may occur.[111] In the constant-current mode, impurities in the carrier gas increase the base frequency and the noise level of the detector. As a result, the detection limits and linearity range are affected. It is necessary to use a high-purity carrier gas and to filter the gas carefully using an oxygen scrubber or adsorbents such as molecular sieves and activated charcoal before it enters the column and the ECD. Stationary liquid phases of high electron affinity such as trifluoropropylmethylsilicone (OV-202, OV-210, OV-215) should not be used with the ECD.

It is also important to operate the ECD above the column oven temperature to minimize condensation of column effluents. In addition, a thermostated temperature control is required because the detector temperature affects the number of electrons emitted from the radioactive source, their energies, and the electron capture mechanism. Experimental data indicate that a marked increase in the ECD noise level occurs unless the detector temperatures are controlled to better than $\pm 0.1°C$.[112-114] To obtain better than 1% accuracy, control of the ECD temperature to better than $\pm 0.3°C$ was also suggested.[114,115]

The effect of pressure[116,117] on the ECD response is shown in Figure 4.19. The minimum detectable quantity is improved by increasing the pressure in the detector cell. A linear relationship between response and cell pressure was measured for cell pressures higher than 2 bar. The pressure control in that pressure range is important for quantitative analysis. However, as is also shown in Figure 4.19, the detector response at 1 bar is less dependent on cell pressure, and the cell pressure control may not be critical for normal ECD operation.

The effect of the carrier gas on the performance of the ECD requires careful consideration in selecting the carrier gas for open tubular column GC. Theoretically, the standing current increases with an increasing ionization cross-section of the carrier gas. The high-energy β-ray ionization cross-sections for nitrogen, hydrogen, and helium are 7.04, 1.23, and 1.64, respectively. Nitrogen is apparently the best choice among the three gases commonly used in open tubular column GC for the optimum performance of the ECD. In practice, the ECD sensitivity is not greatly affected by using

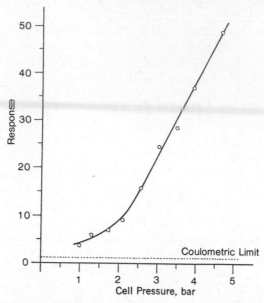

FIGURE 4.19. Electron capture detector response at different pressures (bar) to 1 pg of tec-nazene, measured in faradays peak area per mole of analyte, at optimum voltage conditions. (Reproduced with permission from ref. 117. Copyright Elsevier Scientific Publishing Company.)

helium or hydrogen at a low flow rate, if nitrogen or a mixture of methane–argon (10% v/v) at a flow rate ≥ 20 cm^3 min^{-1} is used as the make-up gas. A high volumetric flow rate of hydrogen (>5 cm^3 min^{-1}) carrier gas could significantly affect the sensitivity and linearity of the ECD due to the low ionization cross-section of hydrogen. Hydrogen carrier gas should not be used if a tritium source is used. It was found[118] that hydrogen molecules are exchanged for tritium molecules, with resulting decrease in the source activity and detector lifetime.

Flame Photometric Detector (FPD)

The flame photometric detector has had widespread application[93,119–128] for the analysis of pollutants in air and water, pesticides, and coal hydrogenation products. A number of reviews[102,129–132] have appeared. The FPD is mainly used for selective detection of sulfur, phosphorus, halogens, and/or nitrogen-containing compounds. Some applications for the detection of metals such as tin,[133–135] chromium,[136,137] selenium,[137] and germanium[138] have also been reported.

The advantages of using the FPD in high-resolution open tubular column GC are that it is highly sensitive and selective to sulfur- and phosphorus-containing compounds. Detection limits as low as 0.5 pg s^{-1} phosphorus and 50 pg s^{-1} sulfur were reported,[122] and one phosphorous atom produces the same response as 10^6 carbon atoms, while one sulfur atom produces the response equivalent to 10^3 carbon atoms at low sulfur content and 10^6 carbon atoms at high sulfur content. The latter variation arises because the carbon interference depends linearly on the number of carbon atoms while the sulfur response depends on the second power of the number of sulfur atoms.

The FPD is a destructive detector and has a 2–3 decade linear dynamic range for sulfur detection, and a 4–5 decade linear dynamic range for phosphorus detection.[122] The design considerations and the operation of the FPD are given below.

The two most widely used FPD designs are the single-burner FPD as first described by Brody and Chaney,[120] and the dual-flame FPD reported by Patterson and co-workers.[122,123] In the case of the single-burner FPD, a diffusion flame containing an excess of H_2 to obtain a cool flame[139,140] is used for reducing detector noise and enhancing phosphorus and sulfur responses. Because the cool single flame is used for both sample decomposition and phosphorus or sulfur excitation, a large sample size can change the flame temperature and geometry, and adversely affect light emission. In more extreme cases, solvent flameout[122,141] can occur due to momentary starving the flame of oxygen as the solvent peak elutes from the column, causing it to be extinguished. In addition, the sulfur and phosphorus light emission can be severely quenched if water[142] or hydrocarbons[121,143–146] are coeluted with the sample. Furthermore, the sulfur response of the single-flame FPD varies from a first-order to a second-order response[132,147–152] depending on the heteroatom environment.

The dual-flame FPD employs two longitudinally separated air–H_2 flames. The lower flame, shown in Figure 4.20, converts the sample molecules into combustion products containing relatively simple molecules, such as S_2 and HPO. The upper flame generates light-emitting excited species such as S_2^* and HPO* which are viewed photometrically through the viewing window adjacent to the upper flame. The dual-flame FPD was demonstrated to give no solvent flameout and no hydrocarbon quenching effect. A true second-order sulfur response was reported for a dual-flame FPD.[122,123]

The FPD sensitivity depends on the emitted light intensity on the basis of chemiluminescence. The intensity of the light emission increases with decreasing flame temperature and with excess hydrogen in the diffusion flame. The flame temperature decreases with a decreasing amount of fuel gases and with the use of a high thermal conductivity carrier gas such as He or H_2. If the fuel gas flow rates are decreased, the FPD background current and noise are also reduced and a better signal-to-noise ratio is obtained. It was suggested that the FPD response was proportional to the cube of the hydrogen concentration. For this reason, the FPD is normally operated at high H_2 flow rates under accurate flow control. The maximum hydrogen flow rate is limited by the fact that a flame with excess hydrogen is unstable and is easily extinguished during the elution of the solvent or major sample components.

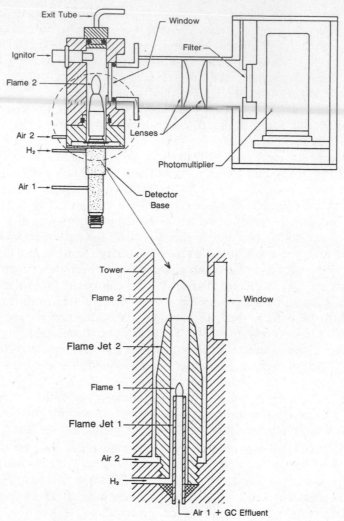

FIGURE 4.20. Schematic diagram of the FPD dual-flame burner. (Reproduced with permission from ref. 123. Copyright American Chemical Society.)

When hydrogen is used as the carrier gas, it is important to maintain the flow rate constant. A high-quality flow controller is important for quantitative measurements. Helium is preferred over nitrogen as carrier gas because the FPD response increases with a reduction in the flame temperature and flame background noise.

The FPD temperature can affect the background current and noise level of the detector. With an increasing detector temperature, the FPD response is reduced.[153] In quantitative work, it is therefore important to control the detector temperature.

Thermionic Specific Detector (TSD)

The thermionic nitrogen/phosphorus specific detector has also been called the alkali flame ionization detector (AFID), nitrogen–phosphorus detector (NPD), or thermionic ionization detector (TID). The AFID employs an alkali metal salt reservoir or a salt pellet that is placed near the burner of a hydrogen–air flame. The NPD, TID, or TSD is the second generation of the N/P detector which employs a flameless design. The alkali metal salt is either coated onto, or uniformly mixed into, a glass or ceramic bead which is placed above the detector burner tip. The flameless TSD is generally easier to operate than the flame-based AFID because better baseline stability, response, and detector operating lifetime can be achieved.

The TSD allows the most sensitive detection for compounds containing nitrogen (50 fg s^{-1} nitrogen in azobenzene) and phosphorus (25 fg s^{-1} phosphorus in malathion).[154] The widespread use of the TSD in gas chromatography is evident from the numerous applications[155–166] in the areas of pesticides, petroleum, fossil fuels, shale oil, food and flavors, clinical samples, drug and drug metabolites, and so on. The TSD has 10^4–10^5 times the nitrogen or phosphorus selectivity relative to the response of organic carbon.

The TSD is a destructive and mass flow-rate-dependent detector. A typical TSD for open tubular column GC is shown in Figure 4.21. The basic design of the TSD is identical to the FID except that an alkali metal salt source is placed between the burner tip and the collector. An alkali metal such as Na, Rb, or Cs is normally contained in a matrix of silica or ceramic beads.[167–171] A uniform distribution of the alkali metal salt in a high-density ceramic matrix was reported to give relatively long operation lifetime and baseline stability.[154,169] The alkali source is electrically heated, and the bead temperature is controlled by the input current. The temperature of the alkali source determines the vapor pressure and the thermal energy of the alkali metal and thus affects the detector sensitivity, background current, and lifetime. The alkali source temperature should be kept as low as possible for stable response and long lifetime. It should also be optimized for the best response for the sample analyzed.

Two TSD operation modes, namely flame ionization (AFID) and flameless thermionic ionization (TID), are normally used. In the flameless thermionic ionization mode, the sample decomposition and ionization processes occur on the boundary layer of the hot surface of the alkali ceramic/glass bead.[169] A small hydrogen flow rate (≤ 6 cm^3 min^{-1}) is normally required. The bead temperature is controlled by electrical heating and is little affected by a small variation in the hydrogen flow rate. However, accurate control of the hydrogen flow rate is important due to the fact that the TSD response depends on the concentration of the H atoms in the gaseous boundary layer of the bead. An increase in the hydrogen flow rate from 3.1 to 6.3 cm^3 min^{-1} increases the phosphorus response and decreases the specificity for nitrogen with respect to hydrocarbons and phosphorus.[169]

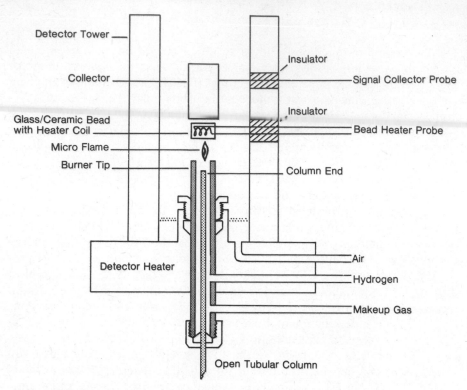

FIGURE 4.21.   Schematic diagram of a typical thermionic specific detector.

If a flame is employed for sample decomposition and ionization, the background current is strongly dependent on the flow rate of the detector hydrogen. A decrease in hydrogen flow rate may decrease the response. It has been reported[172] that a change in the hydrogen flow rate of 0.05% leads to a change in the ionization current of an AFID of 1%. In consequence, a highly stable hydrogen flow rate is required for the AFID.

The response and background current of the TSD depend also on air and carrier gas flow rates; the response decreases, in general, with increasing air and carrier gas flow rate.[154,157,168]

The selection of the carrier gas is also important. When helium is used, the TSD response may decrease to only 10% of that measured using nitrogen[154,168,173] because of increased cooling of the alkali source and incomplete decomposition of the sample.

Photoionization Detector (PID)

Significant improvements[174-178] in the design of the photoionization detector in the past few years have eliminated the deficiencies of the windowless PID.[179-190] Recent developments allow the PID ionization chamber to be operated at atmospheric pressure with high stability and sensitivity. A PID designed for open tubular column GC was recently reported.[175,178,191]

The application of the PID for open tubular column effluent detection has the following advantages: (a) Since it is a concentration-dependent detector, the enhancement in peak detection due to peak sharpening can further extend the detection limits for trace analysis. (b) The PID has high sensitivity (0.3 C g^{-1} carbon), low noise ($\sim 10^{-14}$ A), and excellent linearity (10^7),[178] and thus allows trace analysis. (c) The PID is a nondestructive detector and can be used in series with a second detector for more selective detection. (d) The PID can be operated in a universal or a selective detection mode by simply changing the photon energy of the ionization source.

The development of multiphoton detection methods such as two-photon photoionization[175] for open tubular column GC could further extend the potential of the PID for: (a) improving ionization efficiency to about three orders of magnitude higher than for one-photon ionization; (b) improving the detection limits to subfemtogram levels using advanced laser sources; (c) decreasing detector cell volumes to the order of a nanoliter utilizing a tightly collimated laser beam; and (d) increasing spectral selectivity to allow identification of solutes based on their spectral specificity.

A typical PID for open tubular column operation is shown in Figure 4.22. The basic design of the PID is the same as that of the FID, except that a photon source is located between the end of the column and the ion collector electrode. An optically transparent window (made of LiF, MgF_2, NaF, or sapphire) separates the detection compartment from the discharge com-

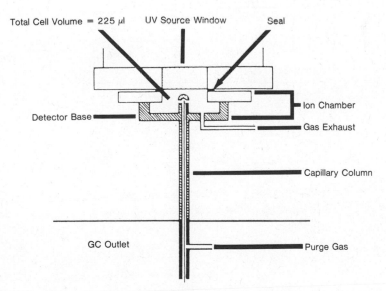

FIGURE 4.22.   Schematic diagram of a PID flow cell for open tubular column GC. (Reproduced with permission from ref. 178. Copyright Dr. Alfred Huethig Publishers.)

partment which is filled with a suitable discharge gas such as argon, hydrogen, helium, neon, krypton, or xenon at a pressure of 0.1–1 mm Hg. The photon flux and energy depend on the shape of the discharge tube, the gas in the discharge tube, and the window material. Table 4.4 lists the optical characteristics of some crystals and discharge gases used in the PID.[108] By proper selection of the discharge gas (such as He) and the optical window, a high photoionization energy source can be obtained for universal detection of organic hydrocarbons which have ionization energies lower than the source of photon energy. Selective detection can also be obtained by lowering the photon energy of the PID, for example, by using a xenon or krypton discharge gas or using a low photon energy transparent crystal such as BaF_2 or sapphire. A major design difficulty of the PID for high-resolution open tubular column GC is the minimization of the detector cell volume due to the requirements in photon flux concentration and the physical volume of the collector and the polarization electrodes. A large detector cell volume affects both detection limits and peak broadening. As discussed earlier (Chapter 2), A submicroliter detector cell volume is required for no more than a 1% loss in efficiency of the narrow-bore open tubular column with an i.d. <0.25 mm. The use of a pulsed laser with nanosecond to microsecond pulse duration and a 100 mm focal length fused-silica lens could result in an effective multiphoton focal volume of a few nanoliters. This extremely small

TABLE 4.4. Photon energy limits for some optically transparent crystals and discharge gases.[a]

Substances	Photon energy or energy transparence limits (eV)
Crystals	
LiF	11.9
MgF_2	11.1
CaF_2	10.3
BaF_2	9.2
Sapphire	8.7
Discharge gases	
He	11.3–20.3
H_2	6.9
Ar	7.3–11.6
Xe	5.6–8.5
Kr	6.9–10.0
Ne	12.4–16.7

[a] Taken from ref. 108.

laser focal volume, when used with on-column detection,[192] could result in an effective detector volume on the order of a nanoliter.

A carrier gas with an ionization potential higher than the photon energy of the PID must be used. Hydrogen, helium, and nitrogen with ionization potentials of 15.43, 24.48, and 15.58 eV, respectively, are suitable for use as carrier gases with a conventional PID which consists of a sealed UV lamp that emits the Lyman α line of hydrogen at 121.6 nm (10.2 eV) through a MgF_2 window into the ionization chamber. Carrier gas impurities such as water and oxygen have no effect on the PID response if their ionization potentials are higher than the PID photon energy. Furthermore, the PID is very stable to variations in the carrier gas flow rate and therefore is suitable for open tubular column GC operation with a pressure regulation system.

Mass Spectrometer (MS)

Open tubular column gas chromatography–mass spectrometry (GC-MS) is the most powerful tool available for the chemical analysis of volatile samples because of its inherent high selectivity and sensitivity. It combines a high-performance separation method with a high-performance measuring technique. The mass spectrometer can be used as a selective detector in the selected-ion monitoring mode for quantitative analysis with a detection limit on the order of picograms, or in the scanning mode for qualitative analysis where it has the highest information content of all identification and structure elucidation methods for organic compounds.[193] In the second case, the mass range must be scanned significantly rapidly (over several seconds only) in order to minimize mass spectral discrimination due to scanning on the upward or downward edge of the peak where concentrations change rapidly, especially with open tubular columns.

Open tubular columns have often been criticized, in connection with mass spectrometry, for low sample capacity and dynamic range. The dynamic range of sample quantities in GC-MS is limited by the signal-to-noise (S/N) ratio of the MS detector at the lower end, and by column nonideality, or detector limitations (nonlinear response, memory effects, etc.) at the higher end. Eyem[194] compared the S/N ratios and dynamic ranges of MS combinations with three different column types: open tubular columns, SCOT columns, and packed columns. This information is tabulated in Table 4.5. As can be seen, the minimum sample quantities that can be analyzed with full sensitivity of the MS are at least 100 times and 10 times higher on packed and SCOT columns, respectively, than on wall-coated open tubular columns. The difference is even more pronounced when considering the effects of adsorption on porous supports used in packed and SCOT columns. Since the detection limit of the MS in the selected-ion mode is (in a favorable case) approximately 10^{-12} g, the dynamic range for an open tubular column with maximum sample capacity of 1 μg is, therefore, 10^6.

The major problem encountered in interfacing a mass spectrometer to a

TABLE 4.5. Characteristic S/N data for different column types.[a]

Characteristics	WCOT		SCOT	Packed
Length (m)	25		25	2.5
Efficiency (N)	70,000		25,000	3500
Maximum sample quantity (μg)	1[b]		10[c]	≥100
Peak area	669		669	669
Capacity factor (k)	5		12.4[b]	60
Retention time (s)	750	750[c]	1675[b]	750[d]
He velocity (cm s^{-1})	20	45[b]	20[c]	20.3
He flow (cm^3 min^{-1})	0.9	4.7	2.1	14
Peak height	100	56[b]	25[b]	21
S/N × 10^{-3}	500	41	41	4.9
Relative S/N	102	8.4	8.4	1

[a] Taken from ref. 194.
[b] Calculated.
[c] Defined.
[d] Adjusted.

gas chromatograph is providing for the reduction of pressure from 1 bar at the end of the chromatographic column to approximately 10^{-5} torr or lower in the ion source of the mass spectrometer. This must be accomplished by some means that minimizes the simultaneous reduction of sample available for identification. The criteria of a perfect coupling can be summarized as follows: (a) The GC resolution obtained with the GC-MS system should be the same as is obtained with conventional GC detection. (b) No compound should have its chemical nature altered after leaving the column, until ionization within the ion source of the MS. (c) By introducing as much of the column effluent as possible into the ion source for highest sensitivity, the operation of the MS should in no way deteriorate. (d) The chromatographic profile obtained using total-ion-current detection should resemble as closely as possible the profile obtained with conventional GC detection for easy comparison. (e) The coupling of the chromatographic column to the MS should be easily and quickly accomplished in order to facilitate rapid change of columns. (f) The performance and sensitivity of the mass spectrometer should remain constant with changes such as column diameter, column length, carrier gas flow rate, and GC temperature programming.

The various ways of connecting gas chromatographic columns to mass spectrometers which satisfy the previous requirements to some degree can be classified into three types: direct connection, molecular separator, and open-split connection. Although most couplings of packed columns to mass spectrometers involve molecule separators, they have found little use in open tubular column GC-MS.

The best results using a molecular separator have been obtained with a single-stage jet separator,[195,196] although compound adsorption and thermal degradation have been observed with this separator, especially when constructed of metal. Other separators are generally not suitable for capillary columns: they often have large dead volumes, low efficiencies, and/or memory effects. Furthermore, they are somewhat inflexible in the choice of different types of columns or optimal separation parameters.

In recent years, most interest has centered on direct and open-split interfacing. In a direct connection, the end of the chromatographic column is is joined directly to the MS ion source by a vacuum tight seal. Such a connection was first described in 1964.[197] The major advantages of this type of connection are the quantitative transfer of the sample into the MS and the lack of dead volume which reduced the separating power of the open tubular column.

The problem encountered initially in direct connection was the limitation on carrier gas flow rates (~ 1 cm^3 min^{-1}) which could be handled by ion source pumps. However, as early as 1971,[198] the 150 L s^{-1} oil diffusion pump on the ion source of a Varian MAT CH·5 MS was replaced with a 600 L s^{-1} pump which gave an ion source pressure of 7×10^{-5} torr for a helium flow rate of 8 cm^3 min^{-1}. Soon after, Henderson and Steel[199] described a system that could handle up to 20 cm^3 min^{-1} of carrier gas. With modern pumping technology, there is no reason to consider open tubular column flow rates as a limiting factor in direct GC-MS coupling.

A more immediate problem associated with direct coupling is that the end of the column is at an undefined pressure, always lower than atmospheric pressure. Consequently, the direct comparison of chromatograms obtained from a gas chromatograph alone with those from the GC-MS combination can often be difficult. Furthermore, it is impossible to adopt parameters of the separation which had been optimized independently of the GC-MS system.

There has been some debate in the past as to the influence of vacuum at the column outlet on separation efficiency. GC-MS operation with reduced column outlet pressure has been claimed to cause no loss of column efficiency for long (~ 100 m) open tubular columns.[200,201] However, careful measurements have shown that vacuum at the outlet causes a decrease in resolution which increases with lower outlet pressure (up to 25% loss).[202] This decrease in separation efficiency is accounted for by the pressure gradient correction terms originally derived by Giddings et al.[203] In most cases, the loss in column efficiency is not large enough to affect significantly the separation desired.

An advantage of reduced column outlet pressures is a shift of the optimum gas velocity to a higher value.[202] This results in increased analysis speeds, narrower chromatographic peaks, and increased sensitivity.[204]

Similar results (i.e., a 30% reduction in column efficiency and a shift of the optimum velocity to higher values) were obtained using short (12 and 23 m) open tubular columns connected directly to the MS ion source. Under these conditions, the entire column was operated under subatmospheric pressure conditions.

The ideal direct interface, in terms of performance, from the open tubular column to the MS ion source is the column itself. Unfortunately under these conditions column change becomes a laborious and time-consuming task. For this reason, an intermediary capillary between the GC oven and the MS ion source is normally preferred. The materials initially used (stainless steel and glass-lined tubing) for interface lines, although mechanically strong, were impossible to deactivate completely, and other materials were sought. The proper treatment of glass provided excellent solute transfer[205-207] but the mechanical stability was a problem. More recently, platinum–iridium[208-210] and fused-silica capillaries have been used. The flexibility and mechanical stability of these materials simplify the interfacing problems, but platinum–iridium may under certain conditions lead to the catalytic destruction or transformation of certain compounds.[205,210]

The open-split coupling[211-215] to the MS is represented in Figure 4.23. The outlet tubing from the GC and the inlet tubing of the MS are separated by a narrow slit, which is scavenged by helium gas. In such an arrangement, the inlet flow rate of the mass spectrometer is fixed for a given inlet capillary dimension, temperature, and carrier gas. The total sample is fed to the MS only if the inlet flow rate of the MS is equal to or below the column outlet flow rate. For larger column outlet flow rates, a part of the sample is split off to waste, and only a fraction, controlled by the split ratio, enters the ion source. If the column outlet flow rate is smaller than the fixed MS inlet flow

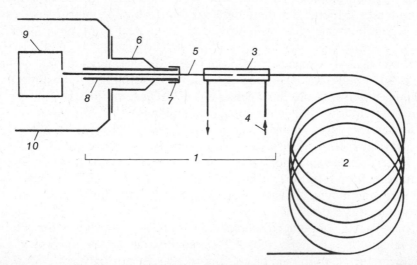

FIGURE 4.23. Schematic diagram of the open-split GC-MS interface. Key: 1, interface; 2, column; 3, splitting device; 4, scavenger gas flow; 5, capillary restriction; 6, connection flange; 7, graphite seal; 8, heated guide tube; 9, ion source; 10, source housing. (Reproduced with permission from ref. 214. Copyright Elsevier Scientific Publishing Company.)

rate, the sample is diluted by helium and the sample concentration at the MS inlet decreases.

The advantages of the open-split connection are as follows:[211] (a) Even in separations with about 100,000 effective plates, no peak broadening caused by the splitting device is observed. (b) The end of the GC column is at atmospheric pressure so that the chromatograms obtained from the GC-MS combination are directly comparable with separate GC runs. (c) Time-consuming GC optimizations can be made remote from the GC-MS combination. (d) Change of columns is rapid and without difficulty because no vacuum sealing is involved. (e) The whole connection device is simple and reliable. (f) The splitting device allows the use of any type of column with no or only minor adaptation. In order to reduce the possibility of catalytic degradation and adsorption, all-glass open-split interfaces have been constructed.[216,217]

The optimization of the open-split GC-MS coupling with respect to GC resolution and MS sensitivity has been discussed by Huber et al.[215] MS sensitivity depends on the ion production rate in the ion source which is determined for a given sample concentration by the inlet flow rate. The latter influences the sensitivity because it determines the mass flow of the sample to the ion source as well as the pressure in the ion source. For each ion source, an optimum inlet flow rate exists for which the sensitivity is a maximum. This optimum inlet flow rate can be fixed by means of an inlet tube having the proper flow resistance. In order to maximize the transfer to, and concentration of the sample in, the ion source, the outlet flow rate of the column must match the inlet flow rate of the ion source.

There are two ways of adjusting the column outlet flow rate: by adjusting the carrier gas flow velocity or the cross-section area of the column. Since the column efficiency depends on the flow velocity, the maximum sensitivity and minimum detection limit can be obtained by operating at the flow velocity which gives the minimum theoretical plate height, and adjusting the column cross-sectional area such that the outlet flow rate matches the fixed inlet flow rate of the MS. For a fixed MS inlet flow rate of 2.20 cm^3 min^{-1}, a 0.3 mm i.d. open tubular column was found to be optimum.[215]

The majority of GC-MS systems are equipped with a low-resolution MS. In many instances, high-resolution MS would be useful for assigning elemental compositions to each mass fragment. Meili et al.[218] have described the successful coupling of a high-resolution GC and a high-resolution MS.

Monitoring of the gas chromatographic effluent with a universal detector such as the FID in parallel with the MS is often useful for interrelating mass spectral data with chromatographic profiles, especially in the case of chemical ionization MS.[219,220] By splitting the column effluent between the FID and MS, and calibrating both the FID and total-ion-current (TIC) records with an event marker, accurate peak assignments can be made. Figure 4.24 shows simultaneous FID and TIC chromatograms of a synthetic mixture of approximately equal amounts of aliphatic and aromatic compounds with

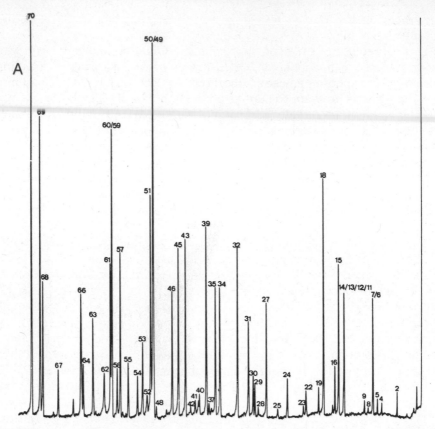

FIGURE 4.24. Chromatograms showing comparison of (A) total ion current and (B) flame ionization detection for a synthetic mixture. Note: 50 m Emulphor ON 870 column, temperature program from 50 to 170°C at 2°C min^{-1}, m/e 60–350 mass range scan, isobutane reagent gas. Selected peak assignments: (18) cyclohexyl methyl ketone; (26) n-tetradecane; (37) phenylcyclohexane; (50) 2-bromobenzaldehyde; (63) 2-bromophenol; (70) 2-aminoacetophenone. (Reproduced with permission from ref. 220. Copyright Elsevier Scientific Publishing Company.)

isobutane as the chemical ionization reagent. The chromatographic profiles are significantly different.

Infrared (IR) Detector

The indirect recording of IR spectra of compounds eluted from SCOT columns has been reported:[221] the vapor was trapped in a special cell, but microgram quantities were still required—quantities inconsistent with efficient columns. Early methods for the direct (i.e., on-line) recording of IR spectra of GC effluents depended on rapid dispersion scanning,[222] but periods of ~30 s were involved for a complete scan, so that the method was scarcely applicable to open tubular column work. However, by the use of

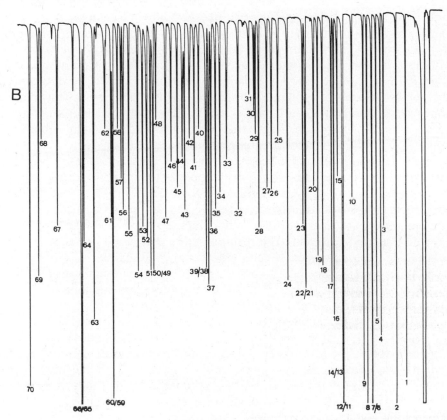

FIGURE 4.24.   (*Continued*)

multiplex IR with Fourier-transform (FT) decoding of the signals, all resolution elements of the spectrum are viewed concurrently and spectrum acquisition can be very rapid.[223] There is also an increase in the signal-to-noise ratio from the higher optical throughput of the FT-IR spectrometer.[223] Scan repetition can allow averaging of accumulated signals to improve sensitivity further.

FT-IR spectra of GC effluents are recorded by transferring the gas stream to a reflecting light pipe (typically 1–2 mm in diameter and 20–80 cm long) which is the sample cell. This large volume inevitably broadens peaks, and make-up gas may be passed[224] into the light pipe to offset the effect. Interferograms are accumulated at a rate of approximately one per second, so that, in principle, IR spectra can be fairly simply obtained on-the-fly over the elution of an open tubular column chromatographic peak. The peaks may be located by the use of a separate detector with effluent splitting, or a (nondestructive) katharometer.

Alternatively, the chromatogram may be reconstructed from the FT-IR measurements alone. At least three approaches have been made to this end.[225,226] Firstly, the interferograms may be Fourier transformed individually, and a search made for nonbackground spectra. In a second method, small (512 point) interferograms are transformed to give low (32 points cm^{-1}) resolution spectra.[227] The total absorbance is calculated for various spectral windows (e.g., of the carbonyl absorption region)[228] and graphed against time to produce a "chemigram" or chemical-functional group chromatogram.[228,229] A third procedure has been recommended,[225] both as more economical in computer time than transformation of all individual interferograms and as affording much greater resolution than the small interferogram method. The Gram–Schmidt process is used as a basis for describing the instrumental characteristics and to orthogonalize the interferograms: the orthogonal component is a function of the total IR absorbance, and the reconstructed chromatogram is plotted as the variation of this with time. A complete software package for a minicomputer has been described[230] to execute the data treatment. The resulting chromatograms are analogous to those from total-ion-current detection in coupled GC-MS.

In spite of sophisticated computer methods, the inherent sensitivity of FT-IR detection in GC is low. The possibilities for summing many interferograms recorded for the same peak[229] are small for narrow peaks from open tubular columns. Nonetheless, IR spectra may be recorded on-the-fly for as little as 10 μg of single compounds.[231,232] While this is feasible for relatively simple mixtures, or for previously separated mixtures, on efficient wall-coated open tubular columns, more complex mixtures may require the increased capacity of thick-film, wide-bore open tubular columns or even SCOT columns to avoid the problems of overloading. Because of the high information content of IR spectra, there is every reason to expect the FT-IR detection will become much more important in the future as sensitivities are improved.

Other Detectors

The preceding discussions have been focused on the most commonly used detectors in open tubular column GC. A brief discussion of other detectors including the thermal conductivity detector (TCD), the electrolytic conductivity detector (ELCD), the optical absorbance or fluorescence detector, and the microwave inductive plasma detector for open tubular column effluent detection is given below.

The thermal conductivity detector (TCD) is one of the most popular packed-column GC detectors today. However, its application in high-resolution open tubular column GC is quite limited. The main reason is that most of the column effluents from open tubular columns can be easily detected by ionization detectors now available. The advantage of the TCD lies in detection of gases such as CS_2, COS, H_2S, SO_2, CO, NO, NO_2, CO_2, and so on, which are usually not analyzed by open tubular column GC (low capacity ratios at temperatures above ambient). In addition, conventional

TCD detectors have large detector cell volumes and require the use of make-up gas to minimize the detector dead volume. As a result, a large sample concentration is required to ensure an adequate detector response. The minimum concentration detectable by a conventional TCD is about 0.5 ng cm^{-3}. For a column with a 50 ng cm^{-3} maximum sample capacity, the useful sample concentration range is between 0.5 and 50 ng cm^{-3} or a linear operation range of 10^2. This relatively narrow range limits the use of the TCD for samples with wide concentration ranges, particularly in trace analysis. The use of a microflow cell in the TCD cannot be easily realized in practice, due to the detector's poor sensitivity and relatively unstable baseline. New technology may be needed for the TCD design in order to allow this universal detector to be useful in high-resolution open tubular column GC.

The electrolytic conductivity detector (ELCD) was first reported by Piringer and Pascalau[233] in 1962, for detection of organic column effluents by measuring the conductivity of the combustion product, CO_2, in aqueous solution. Today, the ELCD has been developed into a highly sensitive, selective, and linear detector[234-239] for packed-column effluent detection. Several reviews have been reported in the literature.[108,240,241]

The ELCD offers picogram per second detection of nitrogen and sulfur, and subpicogram per second chlorine detection limits. It also has greater than 10^4 linearity range and excellent selectivity for hydrocarbons. The ELCD is an important detection tool in the monitoring of environmental pollutants, and its application for open tubular column effluent detection has been demonstrated.[242] However, due to the difficulties in its routine operation, column to detector interfacing, and the loss of sensitivity due to the use of a solvent purge for minimizing connector, reactor, and scrubber dead volume, the ELCD is not routinely used with high-resolution open tubular columns.

Gas phase optical detectors such as UV absorbance and fluorescence detectors have also been investigated for GC column effluent detection. Novotny et al.[243] constructed a 50 μL flow cell and demonstrated UV detection limits of the order of nanograms per cubic centimeter for naphthalene eluted from an open tubular column. This poor detector sensitivity limits the useful sample concentration range to about 10^2, comparable to that obtainable using a TCD. The advantages of the UV detector over the TCD, however, are its selectivity and specificity. By using multiple wavelength detection in combination with peak area (or height) ratio calculations, chromatographically unresolved sample components may be identified and quantified.

Gas phase fluorescence detectors achieve subnanogram per cubic centimeter detectivity[244-248] and may reach picogram per cubic centimeter detectivity using laser-induced fluorescence techniques[249,250] such as pulsed nitrogen lasers and flash-lamp pumped dye lasers. The use of on-column detection[192] with high-intensity stable laser sources and a suitable optical arrangement (e.g., using fiber optics) could provide selective, sensitive, and low dead volume optical detectors for high-resolution open tubular column GC.

In addition to the detectors discussed above, the microwave inductive plasma detector (MIPD) has also been demonstrated to have great potential in specific element detection and qualitative determination of the empirical formulas of unknown compounds. The MIPD utilizes microwave plasma excitation and a grating or multiple wavelength diode array spectrometer for selective element detection of sample components. Column effluents are fed continuously to a high-temperature helium (or argon) plasma where they are completely atomized, and the excited atoms emit their characteristic line spectra.

The use of the MIPD for open tubular column effluent detection has been demonstrated.[251-255] The combination of open tubular column GC with MIPD is particularly attractive because of the small sample volume required. This could prevent disruption of the helium plasma discharge due to an excessive solvent plug or hydrogen from dissociated hydrocarbons. For large sample volumes, an interface between the column and the MIPD which provides for solvent venting is required to prevent disruption of the discharge by the solvent.[252] It also requires special attention to avoid sample decomposition and to minimize dead volume.[251] An electrically heated transfer line using fused-silica or glass capillary tubing is normally employed. Carrier gas (helium) and scavenger gas of high purity should be used in order to operate at high sensitivity. In addition, all gases to the detector should be precisely controlled to achieve low background noise and reproducible peak measurements.

The MIPD is a very valuable tool for selective and quantitative analysis of components containing metal atoms as well as oxygen-, halogen-, nitrogen-, or phosphorus-containing volatile compounds. However, its application in open tubular column GC is currently limited by its linearity (2–4 decades), and the possibility of mutual element influences which require careful calibration.[251] In addition, the total quenching of the helium plasma discharge due to large sample volume injection and the buildup of carbonaceous deposits on the wall of the discharge tube is a problem when operated at reduced pressure. The future development of the direct current atmospheric pressure argon plasma,[256,257] the inductive coupled plasma,[258,259] and the atmospheric pressure active nitrogen afterglow plasma detectors[260-262] may be important for element specific detection of open tubular column effluents.

4.6   DATA SYSTEMS

The advent of microelectronic technology in recent years has allowed low-cost microcomputers to be integrated into chromatographs and data systems.[263-265] The use of microprocessors in gas chromatography allows data acquisition and reduction as well as real-time control of the entire instrument including sample injector, column temperature control, temperature programming, valve switching, detector and attenuation selections and programming, and so on. The advantages of microprocessor integrated GC are as follows: (a) It can be fully automated and can thus eliminate random analytical errors and improve chromatographic performance in terms of retention time and sample size reproducibility. (b) It increases system utility, capability, and sample throughput. (c) It allows highly precise and intelligent data handling, thus improving quantitative precision and accuracy.

Instrument Control

A general bus structure diagram of a microprocessor-based gas chromatograph is shown in Figure 4.25. The keyboard allows user interfacing with the microprocessor for instrument control and methods development. The instrument conditions are displayed on the cathode ray tube (CRT). Detector control involves detector operation parameters including cell current and power supply optimization and time-programmed control of the detector amplifier range and attenuation. The modern GC also has the detector auto-zero control and allows corrections for background change due to column conditions, detector leakage, and amplifier drift. Valve control involves relay switch open/closure, and allows time-programmed gas sampling or multidimensional switching valve operations. The microprocessor control valve switching allows a fast valve open/closure time constant of a few milliseconds. In general, the valve switching time constant is limited by the mechanical open/closure time constant of the valve.

A high-resolution printer–plotter can be used as a recorder and display for the information such as chromatographic conditions, methods, error messages, peak areas, retention times, peak widths, and peak resolution. A state-of-the-art printer–plotter with a fixed thermal printer head is superior to the conventional strip chart recorder for accurate recording of chromatographic peaks. The printer–plotter, in general, has fast full-scale response time and has no mechanical moving parts. It does not have the overshoot, dead-band,

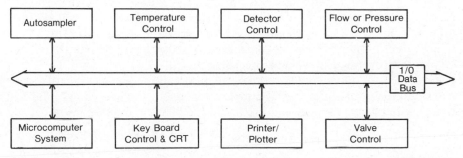

FIGURE 4.25.   General bus structure diagram of a microprocessor-based gas chromatograph.

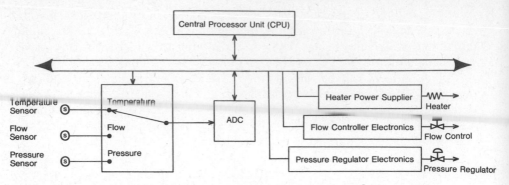

FIGURE 4.26. Schematic diagram of a closed-loop real-time feedback control system.

and damping errors associated with a strip chart recorder and moving head devices.

Microprocessor-controlled temperature zones and flow and pressure regulation are generally achieved by a closed-loop real-time feedback control scheme as shown in Figure 4.26.

As an example, in the case of temperature control (such as those of injector, detector, and column oven), the central processing unit (CPU) requests the temperature sensor to take temperature measurements at a given time interval defined by the software control algorithm. The temperatures registered in analog values are converted to the corresponding digital numbers via an analog-to-digital converter (ADC) and analyzed by the CPU according to the control algorithm. By the use of a control algorithm such as the proportional-integral-differential,[265] the power supply to the heater is controlled and regulated to allow precise temperature programming to obtain minimum initial temperature lag and final temperature overshoot. In Figure 4.5, programmed column temperature profiles from 100 to 250°C at 5°C min^{-1} are compared for a microprocessor controller and a conventional analog proportional controller. The disadvantages of the classical proportional controller are evident in that a 0.8°C temperature lag behind the set-point temperature is measured. In addition, there is a nonlinear programmed temperature profile due to an 8 s temperature lag, and a temperature overshoot of almost 1 min measured for the proportional controller. The microprocessor controller also minimizes the effect of powerline voltage fluctuation to less than an isothermal oven temperature variation of ±0.1°C.

The microprocessor can be designed to allow high-precision (0.1% or better) temperature, flow, and pressure control. The microprocessor can also be used for autosampler control and automation. The autosampler control might include the number of sample injections, sample volume, sampling time, sample tray labeling, and method sequencing and automation.

Data Acquisition and Reduction

In addition to instrument control, the other major function of the central processor unit (CPU) is the processing of the chromatogram and the reduction of the chromatographic data. Accurate processing of chromatographic data is essential for quantitative and qualitative measurements. It is important to have a sufficiently fast ADC clock rate for fast open tubular column peak acquisition; a clock rate higher than 20 Hz (20 data points per second) is required for measuring a peak with plate height, H, of 0.3 mm (Figure 4.27).

The algorithm for peak detection, in general, utilizes the first derivative of the signal with respect to elution time. The start of a peak is registered when the first derivative exceeds a given threshold value (e.g., twice signal-to-noise) (Figure 4.28).

With a conventional first derivative algorithm, the higher the threshold range and the greater the peak tailing, the larger the deviation in the measured peak start and peak end from the true peak values. A better peak detection algorithm should allow time forward and backward from the detected peak start and peak end to give more accurate determinations.

Figure 4.28 also shows that the smaller the threshold range the more accurate will be the peak detection using the first derivative algorithm. The threshold range is limited by the noise in the detector signal; too low a threshold value, however, could result in the detection of irrelevant noise and drift as an apparent chromatographic peak. It is therefore important to

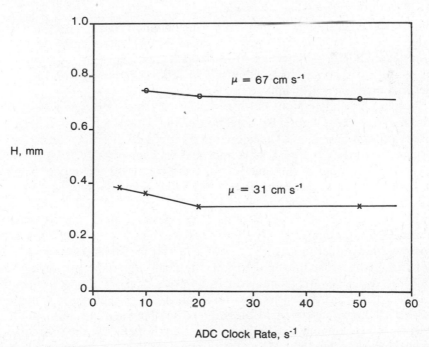

FIGURE 4.27. ADC clock rate required for high-resolution open tubular column data acquisition. Column is 10 m × 0.2 mm i.d. coated with SE-30. The peak measured has a partition ratio of 0.35.

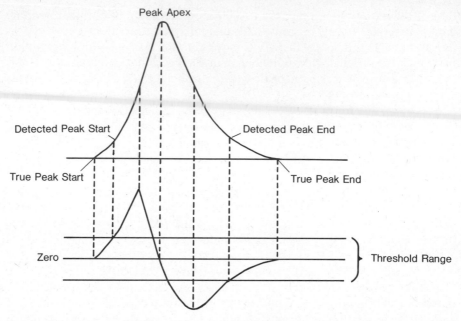

FIGURE 4.28.   Practical basis for peak detection from the first derivative of the signal.

reduce the detector noise in the signal. One common signal processing technique involves digitally filtering the signal by the "bunching" technique which integrates n numbers of digital points from the ADC and improves the signal-to-noise ratio by a factor $n^{1/2}$. n can be optimized with respect to the peak width at half height.

Accurate peak detection also depends on the baseline stability. A drifting baseline may cause imprecise detection of the peak start, peak maximum, peak end, and baseline. As is shown in Figure 4.29, it is impossible to determine accurately the peak start and peak end for peak B using the first-derivative algorithm or the time forward and backward technique. The use of a nonlinear curve fit, or a second-derivative algorithm may improve the accuracy of the peak detection.

The peak detection error in terms of the detected peak start, peak end, and peak baseline can seriously affect peak area measurements (Figure 4.29). The area integrated for peak B, as indicated by the shaded area, is certain to be in gross error compared to the true peak area. The use of the time forward and backward technique can improve peak detection by locating the true peak start and peak end for peak A. In this case, only the area between the true baseline and the straight line between the true peak start and peak end is not included in the peak A area. For peak B, the time forward and backward technique can locate the true peak end but not the peak start. The error is the area added between the true baseline (curved line) and the

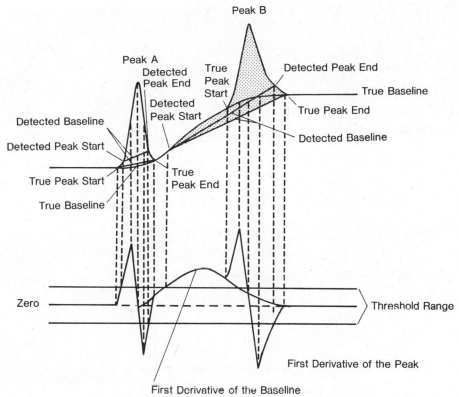

FIGURE 4.29. Peak detection errors due to drifted baseline.

straight line between the detected peak start and the true peak end for peak B.

The precision of peak quantitation also depends on peak resolution: complete resolution of adjacent peaks is essential. A small shift in retention time can result in a variation in peak detection and thus peak area measurement for fused peaks. Location of the true baseline is difficult in the case of drift or of fused peaks. It is generally best to subtract the baseline signal from the chromatogram in order to obtain a straight and simple baseline.[264]

Figure 4.30 compares the uncorrected chromatogram of Figure 4.29 with the baseline corrected chromatogram. It is apparent that peak detection and quantitation is relatively easy after baseline subtraction.

The use of the baseline subtraction technique is important in open tubular column GC where complex mixtures and trace components are to be analyzed using high column temperatures (high column bleed), high detector sensitivity, and low detection threshold ranges.

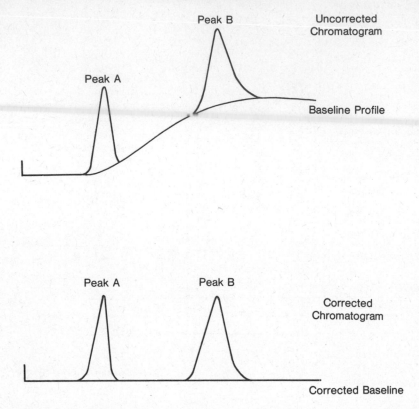

FIGURE 4.30.   Comparison between uncorrected and baseline corrected chromatograms.

4.7   MULTIDIMENSIONAL SYSTEMS

Advances in column technology in recent years (see Chapter 3) have led to the production of columns of extremely high resolving power for the separation of complex and difficult samples. However, in many instances, where highly complex sample matrices are involved such as petrochemicals, environmental pollutants, biological fluids, cigarette smoke, foods and flavors, and so on, the selectivity and resolving power of a single column may not be adequate, and the use of one or more additional columns either in series or in parallel may be necessary. A gas chromatographic system designed for the use of a combination of columns (packed–packed, packed–open tubular, and open tubular–open tubular) to achieve improved separation efficiency and selectivity is named multidimensional column GC. This is a powerful technique for the separation of complex samples, trace enrichment, and qualitative and quantitative analysis. It is also useful for sample cleanup with a precolumn, or with an on-line or off-line high-performance liquid chromatographic system prior to open tubular column separation. In this section, the practical aspects of multidimensional column GC and techniques involving GC-GC and on-line LC-GC systems are discussed.

Multidimensional GC-GC systems

Multidimensional GC-GC systems involving packed–packed, packed–open tubular, and open tubular–open tubular columns have been reviewed and described in detail in the literature.[266–280] The modes of operation, applications, and hardware aspects are briefly discussed below.

A multidimensional system, depending on its complexity, may be used in a combination of operational modes such as the following:

1. *Backflushing.*   Backflushing reverses the flow in the precolumn to remove the retained fraction of sample and prevents these retained components and contaminants from entering the analytical column. As a result, the analysis times of samples with wide polarity or volatility range can be reduced. In addition, backflushing minimizes contamination of the analytical column and the detector. It also minimizes sample preparation time and the net migration of the stationary phase from the precolumn to the analytical column. Because of the elimination of the highly retained components, the analytical column can be operated at lower column temperatures so that column lifetime is extended, and column bleed is minimized.

2. *Solvent flush.*   Solvent eluted from the precolumn is vented to waste so that dilute samples containing water, chlorinated solvents, polar solvents, and derivatization agents that are not compatible with columns or detectors can be analyzed. Solvent flushing is very important when using open tubular columns downstream of the precolumn. Water and chlorinated solvents which are not tolerated either by the thin film phase or the ECD have to be eliminated before entering the analytical column. The solvent flush can also be used to prevent the main components from entering the analytical column, so that overloading is prevented.

3. *Heart cutting.*   Heart cutting is a switching technique that allows quantitative transfer of fractions of the column effluent from a precolumn into the analytical column. The principle applications of heart cutting are: (a) Optimization of chromatographic resolution: here, nonresolved sample components eluted from the first column are selectively diverted to a second column of higher efficiency or better selectivity. (b) Separation of trace components coeluted with major components: the technique allows the transfer of a narrow cut containing the trace component peak into the second column, thus reducing the interference of the major peak during the subsequent separation in the second column. (c) Trace component enrichment: multiple runs of the precolumn with heart cutting and cold trapping allow concentration of the trace component for later analysis.

4. *Column selection.*   The sample stream from the precolumn or injector can be selectively diverted to either one or both columns downstream

from the precolumn. The column selection technique allows the determination of retention time data from two selective columns (normally one polar and one nonpolar) and thus provides additional information for peak identification.

5. *Sample injection.* A sample stored in a sample loop or eluted from a precolumn can be selectively injected into an analytical column. This mode allows highly precise and reproducible sample injection, as well as automation.

Multidimensional Column GC Instrumentation

The operational modes discussed above can either be implemented by using an in-line low dead volume and high-temperature rotary switching valve, or by using off-line solenoid valves for alternating the line pressure at various times to direct or change the carrier gas flow direction. The advantages of using an in-line rotary valve are that it is simple to set-up and operate, allows temperature programming for the second column, and is easy to automate. The disadvantages are potential dead volume and adsorption effects, limited maximum operating temperature, potential gas leakage or flow path plugging due to repeated temperature cycling and valve switching, and limited flexibility.

The advantages of valve switching (Dean switching) are firstly that no valve and moving parts are in either the sample flow path or the high-temperature zone, so that none of the disadvantages of the rotary valve system apply to valve switching, and secondly its versatility and flexibility. Multioperation modes can be utilized in the valve switching system, but the second column oven can only be operated isothermally, and extremely precise and high-quality valves and pressure regulators must be used. Microprocessor-controlled multidimensional column systems have greatly reduced the difficulties, however.

Design requirements for off-line solenoid valve switching in open tubular column multidimensional GC are: carrier gas velocities in both columns (precolumn and analytical column) must be independent of the valves being opened or closed; there must be no back diffusion of sample into the transfer line and no memory effects; sample transfer should be quantitatively accurate and reproducible; there should be no dead volume; and the valve switching should be rapid even for a low pressure difference. The basic components of a simple multidimensional column system based on the principle of Dean switching are shown in Figure 4.31.

The system consists of two low-temperature-coefficient, low-pressure-coefficient, and fast-response pressure regulators A and B. A controls the inlet pressure of column A, which is indicated on the pressure gauge, P_A. Three (normally open) rapid (time constant <20 ms) on–off solenoid valves, C, D, and E, are employed for redirecting the carrier gas flow. A flow restrictor, R_1, bypasses the solenoid valve C for purging upstream of the in-

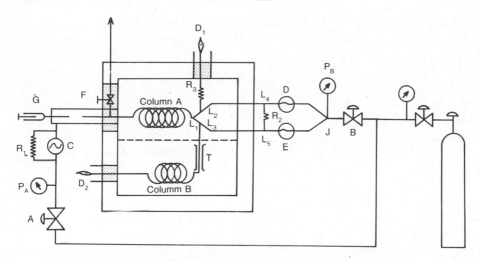

FIGURE 4.31. Schematic diagram of a multidimensional GC system.

jector and for preventing backflushing of the sample into the carrier gas line. A high-temperature needle valve, F, is used for split and backflush operations, and the injector, G, for split injection. Both valve and injector are heated to eliminate solute condensation in the flow path and the valve. The end of column A is connected to the center leg, L_1, of the microvolume interface union. The two legs, L_2 and L_3, closest to L_1 are connected in one of the combination options (detector–column B, detector–trap–column B, column B and column C, and detector B and detector C). If a detector is mounted on L_2, restrictive interface tubing is required to maintain the required pressure at the union junction. The additional two legs, L_4 and L_5, are connected to two rapid solenoid on–off valves D and E (normally open). A low flow resistance bypass, G ($\Delta P = 0.1$ psi), is employed to purge L_3 and L_4 and to prevent sample diffusion into the flow paths during the switching operation. The inlet end of the three-way solenoid valve, J, is connected to the carrier gas, which is regulated by a fast-response pressure regulator, B, and the pressure setting is indicated at gauge P_B. The inlet ends of both regulators A and B are connected to the outlet of a two-stage pressure regulator at the carrier gas supply tank.

This design allows L_1, L_2, L_3, and L_4 to be maintained at the same pressure even at different column-oven temperatures. Temperature programming of the second oven containing column B does not affect the pressure gradient across column A. Columns A and B can be placed in the same oven or in two separate ovens. A cold trap, K, can also be employed for minimizing initial sample bandwidth at the inlet of the second column.

The design requirements for in-line rotary valve (or sliding valve) switching are as follows: (a) Carrier gas flow velocities in both columns (in series) must be independent of the valve being open or closed, although this re-

quirement poses some difficulties in open tubular column operations. (b)
There must be only a small interface volume and no unswept volume in the
valve. (c) A high-temperature valve is required for high-temperature operation, because the valve needs to be placed inside the column oven. (d) The
valve material should be chemically inert and should have no adsorption
properties (e) No flow path plugging or gas leakage must occur at the operating pressure. The basis flow diagram using an eight port rotary valve for
some basic modes of multidimensional operation are given in Figure 4.32.
The carrier gas can be controlled either by a pressure regulator or a flow
controller, PC, at an inlet pressure, P_1. A cold trap, T, may be installed at
the inlet of the column B for sample enrichment or for minimizing initial
sample zone spreading. A column effluent splitter may also be installed at
the end of column A such that the effluent from column A can be monitored
by a second detector to allow more intelligent valve switching operation. In
the mode of operation shown in Figure 4.32, a sample injected into the
injector is analyzed by column A (precolumn for sample cleanup) and column
B. By switching the valve to the other position, the flow direction is altered
to allow backflushing of column A while analysis continues in column B.
To utilize the same eight-port valve for solvent removal, the auxiliary carrier
gas should be connected to port number 3 while a detector, D_2, with a
restrictor, R_2, is connected to port number 4. R_2 should have the same
restriction as column B.

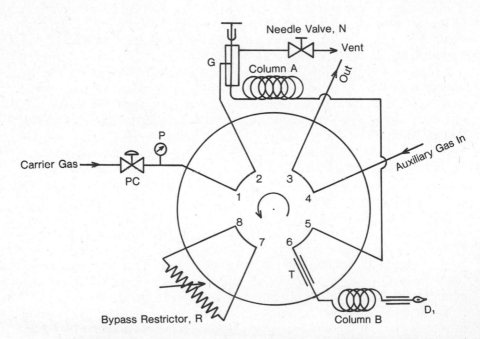

FIGURE 4.32.   In-line rotary valve switching for two columns in series.

Multidimensional LC-GC System

The combination of liquid chromatography (LC) and open tubular column GC in multidimensional operation offers a powerful approach to the solution of very complex separation problems such as the analysis of oils, natural products, and environmental pollutants. The system allows the use of the full capability of both LC and GC columns and detectors. The optimum combination of LC and GC columns and detectors can be selected for the required separation. For example, normal phase, reversed phase, and size exclusion LC columns can be selected for separation, and in GC, both packed and high-resolution open tubular columns can be used. The use of GC allows more sensitive and selective detection than that normally achievable using LC.

The multidimensional LC-GC technique is carried out using either an off-line or an on-line system. The off-line LC-GC technique,[281-288] in general, uses LC as a cleanup or preliminary separation technique for gas chromatography. The sample eluted from the LC column is then injected into the GC column for further separation and analysis.

The off-line LC-GC technique has the advantage of concentrating the solute if the concentration is too low for direct GC analysis. In addition, solvent incompatibility problems can be easily overcome, that is, a salt buffer solvent in LC can be exchanged for a GC column compatible solvent. However, the off-line LC-GC technique suffers difficulties in automation and quantitation. It also results in a greater chance of sample loss due to adsorption, oxidation, and sample handling.

The on-line LC-GC method has recently become possible due to the advances in microprocessor-controlled instrumentation. A fully automated multidimensional LC-GC system was first reported by Cram and co-workers.[289-292] They used an LC-GC interface that is automated by a time-programmed autosampler module for LC column effluent switching and injection. The LC column effluent is normally vented through an autosampler

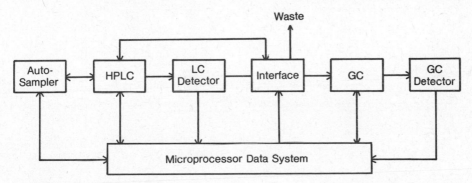

FIGURE 4.33.   Schematic diagram of the control and communication of an automated on-line LC-GC multidimensional system.

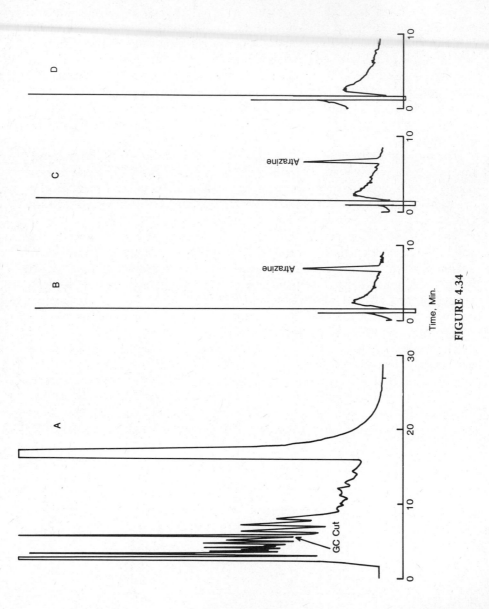

FIGURE 4.34

syringe to the waste container. When the sample of interest has filled the autosample syringe, the syringe is disengaged from the waste container and injects the preset volume into the GC column. A schematic flow diagram of an on-line LC-GC system[289] is shown in Figure 4.33. An example of an on-line LC-GC application in the trace quantitation of atrazine[289,292] is shown in Figure 4.34.

A comparison of chromatograms A and B indicates the quantitative accuracy of the automated LC-GC multidimensional technique for trace analysis. In chromatogram D, the absence of atrazine in a control sample is clear. The on-line LC-GC technique has greater chemical selectivity and broader application than GC-GC multidimensional systems or high-resolution open tubular GC alone. Through automation, less sample handling is required and high sample throughput is obtained. However, the technique does also have same obvious limitations. Samples that can be analyzed by the LC-GC system are limited to volatile samples which can be eluted from GC columns. Solvents are limited to those compatible with GC columns and detectors. Buffer solutions may not be tolerated by open tubular columns and GC detectors. Chemically bonded phases are best suited for LC-GC operation in which large amounts of LC solvents are directed into the open tubular column in order to achieve adequate detection.

REFERENCES

1. W. E. Harris and H. W. Habgood, *Programmed Temperature Gas Chromatography*. John Wiley & Sons, New York, 1966.

2. D. R. Rushneck, W. D. Dencker, E. T. Parker, and S. Rich, *J. Chromatogr. Sci.* **10,** 123 (1972).

3. F. Baumann, R. F. Klaver, and J. F. Johnson, *Gas Chromatography*, M. van Swaay, editor. Butterworth, London, 1962, p. 152.

4. W. Nickel, D. C. Guidinger, A. C. Brown III, K. R. Iwao, and G. E. Marshall, "A High-Performance Microprocessor-Controlled Gas Chromatograph," Varian Instrument Division Technical Bulletin, No. 81-101, 1981.

5. K. R. Iwao, *Control System For A Chromatography Apparatus Oven Door*, U.S. Patent No. 4,186,295, January 1980.

6. F. Kreith, *Principles of Heat Transfer*, 3rd ed. IEP, Dun-Donnelley, New York, 1973.

FIGURE 4.34. Chromatograms of atrazine in sorghum: (A) HPLC chromatogram of sorghum extract containing atrazine (15 cm × 4 mm column containing cyano bonded phase, 2 cm³ min⁻¹ flow rate, 2% isopropanol in hexane mobile phase, column was flushed for 1 min with 100% isopropanol at 18 min mark to remove strongly retained components); (B) GC chromatogram of HPLC/GC cut from chromatogram A (25 m OV-101 glass capillary column, 8 μL splitless injection, 200°C isothermal column temperature, nitrogen-selective thermionic detector); (C) GC chromatogram of atrazine standard carried through entire HPLC/GC procedure (0.2 ppm atrazine, conditions same as for chromatogram B); and (D) GC chromatogram of sorghum control carried through entire HPLC/GC procedure (conditions same as for chromatogram B). (Reproduced with permission from ref. 292. Copyright Preston Publications, Inc.)

7. R. Dandeneau and E. H. Zerenner, *J. High Resoln. Chromatogr./Chromatogr. Commun.* **2**, 351 (1979).

8. N. M. Turkel'taub, V. P. Shvartsman, T. V. Georgievskaya, O. V. Zolotareva, and A. I. Karymova, *Zhur. Fiz. Khim.* **27**, 1827 (1953).

9. N. M. Turkel'taub, O. V. Zolotareva, A. G. Latukhova, A. I. Karymova, and E. Kalnima, *Zhur. Anal. Khim.* **11**, 159 (1956).

10. A. A. Zhukhovitskii, B. A. Kazanskii, O. D. Sterligov, and N. M. Turkel'taub, *Dokl. Akad. Nauk. SSSR* **123**, 1037 (1958).

11. A. A. Zhukhovitskii, N. M. Turkel'taub, and V. A. Sokolov, *Dokl. Akad. Nauk. SSSR* **88**, 859 (1953).

12. A. A. Zhukhovitskii, O. V. Zolotareva, V. A. Sokolov, and N. M. Turkel'taub, *Dokl. Akad. Nauk. SSSR* **77**, 435 (1951).

13. A. P. Tudge, *Can. J. Phys.* **40**, 557 (1962).

14. R. W. Ohline and D. D. Deford, *Anal. Chem.* **35**, 227 (1963).

15. D. H. Desty, A. Goldup, and B. A. F. Whymann, *J. Inst. Petroleum* **45**, 287 (1959).

16. G. Schomburg, H. Behlau, R. Dielmann, F. Weeke, and H. Husmann, *J. Chromatogr.* **142**, 87 (1977).

17. K. Grob and K. Grob, Jr., *J. Chromatogr.* **94**, 53 (1974).

18. W. G. Jennings, *J. Chromatogr. Sci.* **13**, 185 (1975).

19. S. F. Spencer, *Am. Lab.,* Oct., 69 (1977).

20. M. J. Hartigan and L. S. Ettre, *J. Chromatogr.* **199**, 187 (1976).

21. J. D. Green, *J. Chromatogr.* **211**, 25 (1981).

22. S. P. Cram, R. N. McCoy, R. L. Howe, and K. R. Iwao, Pittsburgh Conference on Analytical Chemistry and Applied Spectroscopy, paper No. 223. Cleveland, Ohio, 1977.

23. G. Schomburg, in *Proceedings of the 4th International Symposium on Capillary Chromatogr,* R. E. Kaiser, editor. Huethig, Heidelberg, 1981, p. 371.

24. K. Grob and G. Grob, *J. Chromatogr. Sci.* **7**, 584 (1969).

25. G. Gaspar, P. Arpino, and G. Guiochon, *J. Chromatogr. Sci.* **15**, 256 (1977).

26. F. J. Yang, A. C. Brown, III, and S. P. Cram, *J. Chromatogr.* **158**, 91 (1978).

27. H. Kern and B. Brander, *J. High Resoln. Chromatogr./Chromatogr. Commun.* **2**, 313 (1979).

28. F. J. Yang, R. L. Howe, E. Freitas, and S. P. Cram, Pittsburgh Conference on Analytical Chemistry and Applied Spectroscopy, paper No. 495. Cleveland, Ohio, 1979.

29. P. M. J. VandenBerg and Th. P. H. Cox, *Chromatographia* **5**, 301 (1972).

30. C. A. Cramers and E. A. Vermeer, *Chromatographia* **8**, 479 (1975).

31. L. Ettre, E. Cieplinski, and W. Averill, *J. Gas Chromatogr.* **1**, 7 (1963).

32. M. J. Hartigan and J. E. Purcell, *Chromatogr. Newslett.* **3**, 1 (1974).

33. S. M. Sonchik and J. Q. Walker, *J. Chromatogr. Sci.* **17**, 227 (1979).

34. K. Grob, Jr. and H. P. Neukom, *J. Chromatogr.* **189**, 109 (1980).

35. K. Grob and K. Grob, Jr., *J. Chromatogr.* **151**, 311 (1978).

36. M. Galli, S. Trestianu, and K. Grob, Jr., *J. High Resoln. Chromatogr./Chromatogr. Commun.* **2**, 366 (1979).

37. K. Grob, Jr., *J. Chromatogr.* **178**, 387 (1979).

38. G. Schomburg, H. Husmann, and R. Rittmann, *J. Chromatogr.* **204**, 85 (1981).

39. A. Zlatkis and J. Q. Walker, *J. Gas Chromatogr.* **1**, 9 (1963).

40. K. Grob, Jr., *J. Chromatogr.* **213**, 4 (1981).

41. J. V. Hinshaw, Jr. and F. J. Yang, *J. High Resoln. Chromatogr./Chromatogr. Commun.* **6** (1983).

42. A. Zlatkis, H. A. Lichtenstein, and A. Tishbee, *Chromatographia* **6**, 67 (1973).
43. K. Grob, *J. Chromatogr.* **84**, 255 (1973).
44. R. Teranishi and T. R. Mon, *Anal. Chem.* **44**, 18 (1972).
45. M. Novotny, M. L. Lee, and K. D. Bartle, *Chromatographia.* **7**, 333 (1974).
46. J. D. Green, *J. Chromatogr.* **210**, 25 (1981).
47. M. E. Parrish, C. T. Higgins, D. R. Douglas, and D. C. Watson, *J. High Resoln. Chromatogr./Chromatogr. Commun.* **2**, 551 (1979).
48. K. Grob and G. Grob, *J. Chromatogr.* **90**, 303 (1974).
49. K. Grob, K. Grob, Jr., and G. Grob, *J. Chromatogr.* **106**, 299 (1975).
50. R. E. Sievers, R. M. Barkley, G. A. Eiceman, R. H. Shapiro, H. F. Walton, K. J. Kolonko, and L. R. Field, *J. Chromatogr.* **142**, 745 (1977).
51. J. Roeraade and C. R. Engell, *J. Agr. Food Chem.* **20**, 1039 (1972).
52. A. Zlatkis, C. F. Poole, R. Brazell, K. Y. Lee, and S. Singhawangcha, *J. High Resoln. Chromatogr./Chromatogr. Commun.* **2**, 428 (1979).
53. W. J. Kirsten, P. E. Mattosons, and H. Alfons, *Anal. Chem.* **47**, 1974 (1975).
54. P. Sandra, T. Saeed, G. Redant, M. Grodefroot, M. Verstappe, and M. Verzele, *J. High Resoln. Chromatogr./Chromatogr. Commun.* **3**, 107 (1980).
55. D. Nurok, J. W. Anderson, and A. Zlatkis, *Chromatographia* **11**, 188 (1978).
56. R. S. Brazell and M. P. Maskarinec, *J. High Resoln. Chromatogr./Chromatogr. Commun.* **4**, 1044 (1981).
57. A. Rapp and W. Knipser, *Chromatographia* **13**, 698 (1980).
58. W. Bertsch, R. C. Chang, and A. Zlatkis, *J. Chromatogr. Sci.* **12**, 175 (1974).
59. W. E. May, S. N. Chesler, S. P. Cram, B. H. Gump, H. S. Hertz, D. P. Enagonio, and S. M. Dyszel, *J. Chromatogr. Sci.* **12**, 535 (1975).
60. T. Ramstad and T. J. Nestrick, *Anal. Chim. Acta* **121**, 345 (1980)
61. D. Kalman, R. Dills, C. Perera, and F. Dewalle, *Anal. Chem.* **52**, 1993 (1980).
62. R. Otson, *Anal. Chem.* **53**, 929 (1981).
63. M. Novotny and R. Farlow, *J. Chromatogr.* **103**, 1 (1975).
64. K. Grob, *Chromatographia* **8**, 423 (1975).
65. K. Grob and G. Grob, *J. Chromatogr.* **125**, 471 (1976).
66. G. Schomburg and H. Husmann, *Chromatographia* **8**, 517 (1975).
67. G. Alexander and G. A. F. M. Rutten, *Chromatographia* **6**, 213 (1973).
68. J. P. Schmid, P. P. Schmid, and W. Simon, *Chromatographia* **9**, 597 (1976).
69. K. Wasserfallen and F. Rinderknecht, *Chromatographia* **11**, 128 (1978).
70. Y. Sugimura and S. Tsuge, *Anal. Chem.* **50**, 1968 (1978).
71. H. L. C. Meuzelaar, H. G. Ficke, and H. C. den Harink, *J. Chromatogr. Sci.* **13**, 12 (1975).
72. J. W. deLeeuw, W. L. Maters, D. VanderMeent, and J. J. Boon, *Anal. Chem.* **49**, 1881 (1977).
73. O. Mlcjnek, *J. Chromatogr.* **191**, 181 (1980).
74. J. D. Twibell, J. M. Home, and K. W. Smalldon, *Chromatographia* **14**, 366 (1981).
75. C. J. Wolf, M. A. Grayson, and D. L. Fanter, *Anal. Chem.* **52**, 349A (1980).
76. S. P. Cram and T. H. Glenn, Jr., *J. Chromatogr.* **112**, 329 (1975).
77. S. P. Cram and T. H. Glenn, Jr., *J. Chromatogr.* **119**, 55 (1976).
78. G. Gaspar, R. Arpino, C. Vidal-Madjar, and G. Guiochon, *Anal. Chem.* **50**, 1512 (1978).
79. J. C. Sternberg, in *Advances in Chromatography,* Vol. 2, J. C. Giddings and R. A. Keller, editors, Marcel Dekker, New York, 1966.

80. J. C. Giddings, *J. Gas Chromatogr.* **1,** 12 (1963).

81. I. G. McWilliam and H. C. Bolton, *Anal. Chem.* **41,** 1755 (1969).

82. I. G. McWilliam and H. C. Bolton, *Anal. Chem.* **41,** 1762 (1969).

83. V. Maynard and E. Gruska, *Anal. Chem.* **44,** 1427 (1972).

84. F. J. Yang and S. P. Cram, Pittsburgh Conference on Analytical Chemistry and Applied Spectroscopy, paper No. 115, Atlantic City, New Jersey, March 1981.

85. R. D. Condon, R. P. Scholly, and W. Averill, in *Gas Chromatography 1960,* R. P. W. Scott, editor. Butterworths, Washington, D. C., 1960, p. 30.

86. J. Harley, W. Nel, and V. Pretorius, *Nature* **181,** 177 (1958).

87. I. G. McWilliam and R. A. Dewar, *Nature* **181,** 760 (1958).

88. D. H. Desty, C. J. Geach and A. Goldup, in *Gas Chromatography 1960,* R. P. W. Scott, editor. Butterworths, London, 1960, p. 46.

89. L. Ongkiehong, in *Gas Chromatography 1960,* R. P. W. Scott, editor. Butterworths, London, 1960, p. 7.

90. A. Fowlis, R. J. Maggs, and R. P. W. Scott, *J. Chromatogr.* **15,** 471 (1964).

91. R. A. Dewar, *J. Chromatogr.* **6,** 312 (1961).

92. I. G. McWilliam, *J. Chromatogr.* **6,** 110 (1961).

93. F. J. Yang and S. P. Cram, *J. High Resoln. Chromatogr./Chromatogr. Commun.* **2,** 487 (1979).

94. S. F. Micheletti and G. T. Bryan, *Anal. Chem.* **48,** 51 (1969).

95. J. Novak, P. Bocek, L. Reprt, and J. Janak, *J. Chromatogr.* **51,** 385 (1970).

96. R. Nannikhoven, *Fresenius Z. Anal. Chem.* **236,** 79 (1968).

97. R. W. McCoy and S. P. Cram, *J. Chromatogr. Sci.* **7,** 17 (1969).

98. B. Q. Prescott, H. L. Wise, and D. A. Chestnut, U. S. Patent No. 3,451,780 (1969).

99. H. Bruderrek, W. Schneider, and I. Halasz, *Anal. Chem.* **36,** 461 (1964).

100. J. Connor, *J. Chromatogr.* **200,** 15 (1980).

101. J. Connor, *J. Chromatogr.* **210,** 193 (1981).

102. S. O. Farwell, D. R. Gage, and R. A. Kagel, *J. Chromatogr. Sci.* **19,** 358 (1981).

103. W. A. Aue and S. Kapila, *J. Chromatogr. Sci.* **11,** 255 (1973).

104. A. Zlatkis and D. C. Fenimore, *Rev. Anal. Chem.* **2,** 317 (1975).

105. P. L. Patterson, *J. Chromatogr.* **134,** 25 (1977).

106. E. P. Grimsrud and S. W. Warden, *Anal. Chem.* **52,** 1842 (1980).

107. W. A. Aue and K. M. W. Siu, *Anal. Chem.* **52,** 1544 (1980).

108. J. Sevcik, *Detectors in Gas Chromatography,* Elsevier, Amsterdam, 1976.

109. B. Brechbühler, L. Gay, and H. Jaeger, *Chromatographia* **10,** 478 (1977).

110. D. J. Dwight, E. A. Lorch, and J. E. Lovelock, *J. Chromatogr.* **116,** 257 (1976).

111. F. W. Karasek and D. M. Kane, *Anal. Chem.* **45,** 576 (1973).

112. P. Devaux and G. Guiochon, *J. Gas. Chromatogr.* **5,** 341 (1967).

113. H. J. VandeWiel and P. Tommassen, *J. Chromatogr.* **71,** 1 (1972).

114. E. D. Pellizzari, *J. Chromatogr.* **98,** 323 (1974).

115. R. J. Maggs, P. L. Joynes, A. J. Davies, and J. E. Lovelock, *Anal. Chem.* **43,** 1966 (1971).

116. M. Scolnick, *J. Chromatogr. Sci.* **7,** 300 (1969).

117. S. Kapila and W. A. Aue, *J. Chromatogr.* **118,** 233 (1976).

118. H. C. Hartmann, *Anal. Chem.* **45,** 733 (1973).

119. F. J. Yang, *J. Chromatogr. Sci.* **19,** 523 (1981).

120. S. S. Brody and J. E. Chaney, *J. Gas Chromatogr.* **4,** 42 (1966).

121. W. E. Rupprecht and T. R. Phillips, *Anal. Chim. Acta* **47**, 439 (1969).

122. P. L. Patterson, *Anal. Chem.* **50**, 345 (1978).

123. P. L. Patterson, R. L. Howe, and A. Abu-Shumays, *Anal. Chem.* **50**, 339 (1978).

124. E. R. Adlard, L. F. Creaser, and P. H. D. Matthews, *Anal. Chem.* **44**, 64 (1972).

125. P. J. Groenen and L. J. Van Gemert, *J. Chromatogr.* **57**, 239 (1971).

126. A. D. Horton and M. R. Guerin, *J. Chromatogr.* **90**, 63 (1974).

127. W. P. Cochrane and R. Greenhalgh, *Chromatographia* **9**, 255 (1976).

128. G. Goretti and M. Possanzini, *J. Chromatogr.* **77**, 317 (1973).

129. S. O. Farwell and R. A. Rasmussen, *J. Chromatogr. Sci.* **14**, 224 (1976).

130. H. M. McNair and C. D. Chandler, *J. Chromatogr. Sci.* **11**, 454 (1973).

131. D. F. S. Natusch and T. M. Thorpe, *Anal. Chem.* **45**, 1185A (1973).

132. C. H. Hartmann, *Anal. Chem.* **43**, 113A (1971).

133. W. A. Aue and G. C. Flinn, *J. Chromatogr.* **142**, 145 (1977).

134. S. Kapila and C. R. Vogt, *J. Chromatogr. Sci.* **18**, 144 (1980).

135. W. A. Aue and G. C. Flinn, *Anal. Chem.* **52**, 1537 (1980).

136. R. Ross and T. Shafik, *J. Chromatogr. Sci.* **11**, 46 (1973).

137. G. C. Flinn and W. A. Aue, *J. Chromatogr.* **153**, 49 (1978).

138. G. C. Flinn and W. A. Aue, *J. Chromatogr.* **186**, 229 (1979).

139. R. M. Dagnall, K. C. Thompson, and T. S. West, *Analyst* **92**, 506 (1967).

140. R. M. Dagnall, K. C. Thompson, and T. S. West, *Analyst* **93**, 72 (1968).

141. C. A. Burgett and L. E. Green, *J. Chromatogr. Sci.* **12**, 356 (1974).

142. H. Okali, P. L. Splitstone, and J. J. Ball, *J. Air Pollut. Contr. Assoc.* **23**, 514 (1973).

143. W. L. Crider and R. W. Slater, Jr., *Anal. Chem.* **41**, 531 (1969).

144. T. Sugiyama, Y. Suzuki, and T. Takeuchi, *J. Chromatogr.* **80**, 61 (1973).

145. C. D. Pearson and W. J. Hines, *Anal. Chem.* **49**, 123 (1977).

146. D. A. Clay, C. H. Rogers, and R. H. Jungers, *Anal. Chem.* **49**, 126 (1977).

147. A. I. Mizany, *J. Chromatogr. Sci.* **8**, 151 (1970).

148. R. K. Stevens, J. D. Mulik, A. E. O'Keefe and K. S. Krost, *Anal. Chem.* **43**, 827 (1971).

149. J. G. Eckhardt, M. B. Denton, and J. L. Moyers, *J. Chromatogr. Sci.* **13**, 133 (1975).

150. C. H. Burnett, D. F. Adams, and S. O. Farwell, *J. Chromatogr. Sci.* **15**, 230 (1977).

151. J. F. McGaughey and S. K. Gangwal, *Anal. Chem.* **52**, 2079 (1980).

152. D. G. Greer and T. J. Bydalek, *Environ. Sci. Technol.* **7**, 153 (1973).

153. R. E. Pescar and C. H. Hartmann, *J. Chromatogr. Sci.* **11**, 492 (1973).

154. P. L. Patterson, *J. Chromatogr.* **167**, 381 (1978).

155. D. R. Schulz, *Bull. Environ. Contam. Toxicol.* **5**, 6 (1970).

156. D. K. Albert, *Anal. Chem.* **50**, 1822 (1978).

157. M. J. Hartigan, J. E. Purcell, M. Novotny, M. L. McConnell, and M. L. Lee, *J. Chromatogr.* **99**, 339 (1974).

158. G. Norheim and J. Rygge, *J. Chromatogr.* **154**, 291 (1978).

159. W. Vogt, K. Jacob, and K. Knedel, *J. Chromatogr. Sci.* **12**, 658 (1974).

160. Y. Hoshika, *J. Chromatogr. Sci.* **19**, 444 (1981).

161. H. Dekirmenjian, J. I. Javaid, B. Duslak, and J. M. Davis, *J. Chromatogr.* **160**, 291 (1978).

162. P. I. Jotlow and D. N. Bailey, *Clin. Chem.* **21**, 1918 (1975).

163. C. Bianchetti and P. L. Morselli, *J. Chromatogr.* **153**, 203 (1978).

164. J. I. Javaid, H. Dekirmenjian, U. Liskevyck, and J. M. Davis, *J. Chromatogr. Sci.* **17**, 666 (1979).

165. J. I. Javaid, H. Dekirmenjian, U. Liskevyck, R. L. Lin, and J. M. Davis, *J. Chromatogr. Sci.* **19**, 439 (1981).

166. H. A. McLead, A. G. Butterfield, D. Lewis, W. E. J. Philips, and D. E. Coffin, *Anal. Chem.* **47**, 674 (1975).

167. C. II. Hartmann, U.S. Patent No. 3,607,096 (Sept. 21, 1971),

168. J. A. Lubkowitz, B. P. Semonian, J. Galobardes, and L. B. Rogers, *Anal. Chem.* **50**, 672 (1978).

169. P. L. Patterson and R. L. Howe, *J. Chromatogr. Sci.* **16**, 275 (1978).

170. R. Greenhalz, J. Müler, and W. A. Aue, *J. Chromatogr. Sci.* **16**, 8 (1978).

171. B. Kolb and J. Bischoff, *J. Chromatogr. Sci.* **12**, 625 (1974).

172. C. H. Hartman, *Bull. Environ. Contam. Toxicol.* **1**, 159 (1966).

173. W. Ebing, *Chromatographia* **1**, 382 (1968).

174. N. Ostojic and Z. Sternberg, *Chromatographia* **7**, 3 (1974).

175. C. M. Klimcak and J. E. Wessel, *Anal. Chem.* **52**, 1233 (1980).

176. J. N. Driscoll and F. F. Spaziani, *Res. and Dev.* **27**, 50 (1976).

177. J. N. Driscoll. *J. Chromatogr.* **134**, 49 (1977).

178. J. F. Jaramillo and J. N. Driscoll, *J. High. Resoln. Chromatogr./Chromatogr. Commun.* **2**, 536 (1979).

179. J. E. Lovelock, *Nature* (*London*) **188**, 401 (1960).

180. J. E. Lovelock, *Anal. Chem.* **33**, 163 (1961).

181. M. Yamane, *J. Chromatogr.* **9**, 162 (1962).

182. M. Yamane, *J. Chromatogr.* **11**, 158 (1963).

183. M. Yamane, *J. Chromatogr.* **14**, 355 (1964).

184. K. Watanabe, *J. Chem. Phys.* **26**, 542 (1957).

185. J. F. Roesler, *Anal. Chem.* **36**, 1900 (1964).

186. A. Karmen, L. Giuffrida, and R. L. Bowman, *Nature* (*London*) **191**, 906 (1961).

187. A. Karmen and R. L. Bowman, *Nature* (*London*) **196**, 62 (1962).

188. D. C. Locke and C. E. Meloan, *Anal. Chem.* **37**, 389 (1965).

189. J. G. W. Price, D. C. Femimore, P. G. Simmonds, and A. Zlatkis, *Anal. Chem.* **40**, 541 (1968).

190. R. R. Freeman and W. E. Wentworth, *Anal. Chem.* **43**, 1987 (1971).

191. S. Kapila and C. R. Vogt, *J. High. Resoln. Chromatogr./Chromatogr. Commun.* **4**, 233 (1981).

192. F. J. Yang, *J. High Resoln. Chromatogr./Chromatogr. Commun.* **4**, 83 (1981).

193. S. L. Grotch, *Anal. Chem.* **42**, 1214 (1970).

194. J. Eyem, *Chromatographia* **8**, 456 (1975).

195. M. Novotny, *Chromatographia* **2**, 350 (1969).

196. B. F. Maume and J. A. Luyten, *J. Chromatogr. Sci.* **11**, 607 (1973).

197. D. Henneberg and G. Schomburg, *Z. Anal. Chem.* **211**, 55 (1964).

198. P. Schulze and K. H. Kaiser, *Chromatographia* **4**, 381 (1971).

199. W. Henderson and G. Steel, *Anal. Chem.* **44**, 2302 (1972).

200. J. C. Giddings, *Anal. Chem.* **34**, 314 (1962).

201. N. Sellier and G. Guiochon, *J. Chromatogr. Sci.* **8**, 147 (1970).

202. F. Vangaever, P. Sandra, and M. Verzele, *Chromatographia* **12**, 153 (1979).

203. J. C. Giddings, S. L. Seager, L. H. Stucki, and G. H. Stewart, *Anal. Chem.* **32**, 867 (1960).

204. C. A. Cramers, G. J. Scherpenzeel, and P. A. Leclercq, *J. Chromatogr.* **203**, 207 (1981).

205. K. Grob, *Chromatographia* **9**, 509 (1976).

206. H. W. Durbeck, I. Buker, and W. Leymann, *Chromatographia* **11**, 295 (1978).

207. H. W. Durbeck, I. Buker, and W. Leymann, *Chromatographia* **11**, 372 (1978).

208. N. Neuner-Jehle, F. Etzweiler, and G. Zarske, *Chromatographia* **6**, 211 (1973).

209. F. Etzweiler, *J. Chromatogr.* **167**, 133 (1978).

210. F. Rinderknecht and B. Wenger, *J. High Resoln. Chromatogr./Chromatogr. Commun,* **2**, 746 (1979).

211. D. Henneberg, U. Henrichs, and G. Schomburg, *Chromatographia* **8**, 449 (1975).

212. D. Henneberg, U. Henrichs, and G. Schomburg, *J. Chromatogr.* **112**, 343 (1975).

213. F. A. Thome and G. W. Young, *Anal. Chem.* **48**, 1423 (1976).

214. D. Henneberg, U. Henrichs, H. Husmann, and G. Schomburg, *J. Chromatogr.* **167**, 139 (1978).

215. J. F. K. Huber, E. Matisova, and E. Kenndler, *Anal. Chem.* **54**, 1297 (1982).

216. P. P. Schmid, M. D. Muller, and W. Simon, *J. High Resoln. Chromatogr./Chromatogr. Commun.* **2**, 225 (1979).

217. R. B. Hurley, Jr., *J. High Resoln. Chromatogr./Chromatogr. Commun.* **3**, 147 (1980).

218. J. Meili, F. C. Walls, R. McPherron, and A. L. Burlingame, *J. Chromatogr. Sci.* **17**, 29 (1979).

219. N. Neuner-Jehle, F. Etzweiler, and G. Zarske, *Chromatographia* **7**, 323 (1974).

220. W. Blum and W. J. Richter, *J. Chromatogr.* **132**, 249 (1977).

221. R. F. Brady, Jr., *Anal. Chem.* **47**, 1426 (1975).

222. B. Krakow, *Anal. Chem.* **41**, 815 (1969).

223. P. R. Griffiths, *Fourier Transform Infrared Spectroscopy,* Vol. 1. Academic, New York, 1978.

224. L. V. Azarraga and C. A. Polter, *J. High Resoln. Chromatogr./Chromatogr.* Commun. **4**, 60 (1981).

225. J. A. deHaseth and T. L. Isenhour, *Anal. Chem.* **49**, 1977 (1977).

226. D. A. Hanna, G. Hangae, B. A. Hohne, G. W. Small, R. C. Wieboldt, and T. L. Isenhour, *J. Chromatogr. Sci.* **17**, 423 (1979).

227. P. Coffey, D. R. Mattson, and J. C. Wright, *Am. Lab.* **10**, 126 (1978).

228. D. R. Mattson and R. L. Julian, *J. Chromatogr. Sci.* **17**, 416 (1979).

229. M. D. Erickson, D. L. Newton, E. D. Pellizzari, and K. B. Tomer, *J. Chromatogr. Sci.* **17**, 449 (1979).

230. A. Hanna, J. C. Marshall, and T. L. Isenhour, *J. Chromatogr. Sci.* **17**, 434 (1979).

231. D. L. Wall and A. W. Mantz, *Appl. Spectrosc.* **31**, 552 (1977).

232. K. H. Shapes, A. Bjorseth, J. Tabor, and R. J. Jakobsen, *J. High Resoln. Chromatogr./Chromatogr. Commun.* **3**, 87 (1980).

233. O. Piringer and M. Pascalau, *J. Chromatogr.* **8**, 410 (1962).

234. J. W. Dolan and R. C. Hall, *Anal. Chem.* **45**, 2198 (1973).

235. P. Jones and G. Nickless, *J. Chromatogr.* **73**, 19 (1972).

236. J. F. Lawrence and A. H. Moore, *Anal. Chem.* **46**, 755 (1974).

237. R. C. Hall, *J. Chromatogr. Sci.* **12**, 152 (1974).

238. B. E. Pape, D. H. Rodgers, and T. C. Flynn, *J. Chromatogr.* **134**, 1 (1977).

239. R. J. Anderson and R. C. Hall, *Am. Lab.* **12,** 108 (1980).

240. S. O. Farwell, D. R. Gage, and R. A. Kagel, *J. Chromatogr. Sci.* **19,** 358 (1981).

241. D. T. Davis, *Gas Chromatographic Detectors.* John Wiley & Sons, New York, 1974, p. 194.

242. V. Lopez-Avila and R. Northcutt, *J. High. Resoln. Chromatogr./Chromatogr. Commun.* **5,** 67 (1982).

243. M. Novotny, F. J. Schwende, M. J. Hartigan, and J. E. Purcell, *Anal. Chem.* **52,** 736 (1980).

244. H. P. Burchfield, R. J. Wheeler, and J. B. Bernos, *Anal. Chem.* **43,** 1976 (1971).

245. H. P. Burchfield, E. E. Green, R. J. Wheeler, and S. M. Billedeau, *J. Chromatogr.* **99,** 697 (1974).

246. P. J. Freed and L. R. Faulkner, *Anal. Chem.* **44,** 1194 (1972).

247. R. P. Cooney, T. Vo-Dinh, and J. D. Winefordner, *Anal. Chim. Acta* **89,** 9 (1977).

248. R. P. Cooney and J. D. Winefordner, *Anal. Chem.* **49,** 1057 (1977).

249. J. A. Warren, J. M. Hayes, and G. J. Small, *Anal. Chem.* **54,** 138 (1982).

250. J. M. Hayes and G. J. Small, *Anal. Chem.* **54,** 1202 (1982).

251. K. B. Brenner, *J. Chromatogr.* **167,** 365 (1978).

252. B. D. Quimby, P. C. Uden, and R. M. Barnes, *Anal. Chem.* **50,** 2112 (1978).

253. B. D. Quimby, M. F. Delaney, P. C. Uden, and R. M. Barnes, *Anal. Chem.* **51,** 875 (1979).

254. S. A. Estes, P. C. Uden, M. D. Rausch, and R. M. Barnes, *J. High Resoln. Chromatogr./Chromatogr. Commun.* **3,** 471 (1980).

255. B. D. Quimby, M. F. Delaney, P. C. Uden, and R. M. Barnes, *Anal. Chem.* **52,** 259 (1980).

256. R. J. Lloyd, R. M. Barnes, P. C. Uden, W. G. Elliot, *Anal. Chem.* **50,** 2025 (1978).

257. D. S. Treybig and S. R. Ellebracht, *Anal. Chem.* **52,** 1633 (1980).

258. D. L. Windsor and M. B. Denton, *Appl. Spectrosc.* **32,** 366 (1978).

259. D. L. Windsor and M. B. Denton, *J. Chromatogr. Sci.* **17,** 492 (1979).

260. D. G. Sutton, K. R. Westburg, and J. E. Melzer, *Anal. Chem.* **51,** 1399 (1979).

261. J. E. Melzer and D. G. Sutton, *Appl. Spectrosc.* **34,** 434 (1980).

262. G. W. Rice, J. J. Richard, A. P. D'Silva, and V. A. Fassel, *Anal. Chem.* **53,** 1519 (1981).

263. F. Baumann, J. L. Hendrickson, and D. Wallace, *Chromatographia* **7,** 529 (1974).

264. A. C. Brown III, W. J. Ballantyne, D. L. Wallace, and F. Baumann, "Advanced Instrumentation for Chromatography Automation," Varian Instrument Division Technical Bulletin, No. 81-105, 1981.

265. D. S. Wallace, J. M. Marino, and S. W. Kung, "Design of An Advanced Terminal for Chromatographic Data Handling and Automation," Varian Instrument Division Technical Bulletin, No. 81-102, 1981.

266. D. R. Deans. *J. Chromatogr.* **18,** 477 (1965).

267. D. R. Deans, *Chromatographia* **1,** 18 (1968).

268. D. R. Deans, M. T. Huckle, and R. M. Peterson, *Chromatographia* **4,** 279 (1971).

269. D. R. Deans, *J. Chromatogr.* **203,** 19 (1981).

270. G. Schomburg, H. Kötter, and F. Hack, *Anal. Chem.* **45,** 1236 (1973).

271. G. Schomburg, H. Husmann, and F. Weeke, *J. Chromatogr.* **112,** 205 (1975).

272. G. Schomburg, R. Dielmann, H. Husmann, and F. Weeke, *J. Chromatogr.* **122,** 55 (1976).

273. J. A. Rijks and C. A. Cramers, *Chromatographia* **7,** 99 (1974).

274. J. A. Rijks, J. H. M. VanDenBerg, and J. P. Diependaal, *J. Chromatogr.* **91,** 603 (1974).

275. D. E. Willis, *Anal. Chem.* **50**, 827 (1978).

276. W. G. Jennings, S. G. Wyllie, and S. Alves, *Chromatographia* **10**, 426 (1977).

277. R. J. Miller, S. D. Stearns, and R. R. Freeman, *J. High Resoln. Chromatogr./Chromatogr. Commun.* **2**, 55 (1979).

278. E. L. Anderson, M. M. Thomason, H. T. Mayfield, and W. Bertsch, *J. High Resoln. Chromatogr./Chromatogr. Commun.* **2**, 335 (1979).

279. W. Bertsch, in *Recent Advances in Capillary Gas Chromatography,* W. Bertsch, W. G. Jennings, and R. E. Kaiser, editors, Hüthig, Heidelberg, 1981, p. 1.

280. W. Bertsch, E. Anderson, and G. Holzer, *Chromatographia* **10**, 449 (1977).

281. J. J. C. Schefter, A. Koedam, M. T. I. W. Shuesier, and S. A. Baerheim, *Chromatographia* **10**, 669 (1977).

282. J. C. Liao and R. R. Browner, *Anal. Chem.* **50**, 1683 (1978).

283. R. A. Chapman and C. R. Harris, *J. Chromatogr.* **166**, 513 (1978).

284. J. Krupcik, J. Kriz, D. Prusova, P. Sucharek, and Z. Cervanka, *J. Chromatogr.* **142**, 797 (1977).

285. H. L. Christ, . L. Harless, R. F. Moseman, and M. H. Callis, *Bull. Environ. Contam. Toxicol.* **24**, 231 (1980).

286. F. P. Disanzo, P. C. Uden, and A. S. Siggia, *Anal. Chem.* **52**, 906 (1980).

287. K. T. Joseph and R. F. Browner, *Anal. Chem.* **52**, 1083 (1980).

288. I. Akharan, R. E. Wrolstad, and D. G. Richardson, *J. Chromatogr.* **190**, 452 (1980).

289. S. P. Cram, A. C. Brown III, E. Freitas, and R. E. Majors, "A Coupled HPLC/GC System—Instrumentation and Automation," Pittsburgh Conference on Analytical Chemistry and Applied Spectroscopy, paper No. 155. Cleveland, Ohio, 1979.

290. R. E. Majors, E. L. Johnson, S. P. Cram, and A. C. Brown III, "A Coupled HPLC/GC System—Applications," Pittsburgh Conference on Analytical Chemistry and Applied Spectroscopy, paper No. 116. Cleveland, Ohio, 1979.

291. J. A. Apffel, T. V. Alfredson, and R. E. Majors, 13th International Symposium on Chromatography, Cannes, France, June 1980.

292. R. E. Majors, *J. Chromatogr. Sci.* **18**, 571 (1980).

FIVE

PRACTICE AND TECHNIQUES

5.1 INTRODUCTION

State-of-the-art open tubular column GC provides excellent separation efficiency, peak capacity, chemical inertness, thermal stability, and minimum adsorption and catalytic effects. Complex samples at picogram levels can be eluted from an open tubular column and detected with sensitive GC detectors. However, because of the very low sample capacity and carrier gas flow rates, and the limitations in sampling techniques, difficulties have been encountered in the use of open tubular column GC for routine analysis, particularly for the quantitative analysis of trace components in wide volatility range samples.

In the practice of open tubular column GC, it is important to recognize that good chromatographic results depend on selecting the right chromatographic equipment and conditions. This chapter focuses on the practice and techniques of open tubular column GC. Emphasis will be placed on sample considerations, column selection, carrier gas and flow rate, syringe, septum, column temperature and programming, sampling mode selection and techniques, and qualitative and quantitative analysis.

5.2 SAMPLE CONSIDERATIONS

One of the major problems in open tubular column GC is the presence of interfering substances in the sample. Good laboratory practice in sample preparation is very important. For example, nonvolatile material and particulate matter should not be present; the solvent should be compatible with the column and detector; and the sample concentration should be within the applicable range for the selected sampling mode and detector. Table 5.1 shows the applicable effective sample volume, sample concentration, and sample boiling point ranges for four selected sampling modes using FID and ECD detectors with the assumption that the maximum sample capacity for the column is 50 ng.

The applicable sample mass range when using an FID is between the minimum detector detection limit for the FID (6 pg) and the maximum sample capacity for the column (50 ng). For detectors with a limited linear range, sample mass limits may not correspond to the minimum and the maximum detection limits of the detector. For example, for an ECD with a linear range of four decades used with an open tubular column having a sample capacity of 50 ng, the applicable sample mass limits are between 5 fg and 50 pg; 50 ng is well above the linear range for the ECD.

The range of effective sample volume is determined by the volumes that can be effectively transferred into the column. For split sampling, the sample volume range for the injector is normally between 0.1 and 1 μL in order to ensure good split reproducibility and to minimize injector band spreading. This range gives an effective sample volume transferred into the column of 0.1 and 200 nL assuming a split ratio of between 10:1 and 500:1. For splitless sampling, the effective sample volume is approximately 70–90% of the injected sample volume, depending on the volume injected, the splitless time (purge-off time), and the speed of sample injection. The effective sample volume for splitless injection is thus approximately between 0.1 and 7 μL. For direct and on-column sampling, the effective sample volume is the injected sample volume. The lower limit depends on the minimum volume that

TABLE 5.1. Applicable effective sample volume, concentration, and boiling point range in open tubular column GC with FID and ECD detectors

Detectors	Sampling modes	Ranges		
		Volume(μL)	Concentration (ppm)	Boiling point range
FID	Split	10^{-4} to 2×10^{-2}	3×10^{-1} to 5×10^5	Narrow[a]
	Splitless	10^{-1} to 7	9×10^{-4} to 5×10^2	$<n\text{-}C_{28}$
	Direct	10^{-1} to 10	6×10^{-4} to 5×10^2	$<n\text{-}C_{34}$
	On-column	$\geq 2 \times 10^{-1}$	$\leq 2.5 \times 10^2$	$\sim n\text{-}C_{60}$
ECD	Split	10^{-4} to 2×10^{-2}	5×10^{-4} to 5×10^2	Narrow[a]
	Splitless	10^{-1} to 7	7×10^{-7} to 5×10^{-1}	$<n\text{-}C_{28}$
	Direct	10^{-1} to 10	5×10^{-7} to 5×10^{-1}	$<n\text{-}C_{34}$
	On-column	$\geq 2 \times 10^{-1}$	$\leq 2.5 \times 10^{-1}$	$\sim n\text{-}C_{60}$

[a] Within 10 carbon numbers.

the sampling device can reproducibly inject (e.g., 0.1 μL for a 1 μL syringe and 0.2 μL for an on-column injector syringe). The upper limit depends on the sample capacities of the injector and the column and on the effects of solute focusing techniques. Using a direct injector with a 0.3 mm i.d. glass column and the solvent effect technique, the maximum sample volume can be as high as 10 μL with no significant deterioration of peak shape. For an on-column injector with a 0.3 mm i.d. fused-silica column, the maximum sample volume may exceed 10 μL if the solute focusing technique [1,2] is utilized.

The concentration range for open tubular column GC can be calculated by dividing the allowable mass range by the effective volume range for the system. As shown in Table 5.1, an open tubular column with all available sampling modes covers a concentration range as high as nine decades. Split sampling allows the analysis of concentrated samples and major components. Direct, on-column, and splitless sampling allow trace analysis approaching a level as low as sub-ppt.

In practice, the selection of sampling mode depends on the concentration of the sample to be analyzed. For example, sample concentrations below 3×10^{-1} ppm should not be used with split injection and FID detection. Sample concentrations higher than 5×10^2 ppm may not be suitable for direct, on-column, or splitless injection and FID detection.

5.3   SYRINGE CONSIDERATIONS

Syringe handling technique is very important in open tubular column GC sampling. Details will be discussed in connection with each sampling technique in Section 5.11. Sampling errors normally overlooked are nonreproducible losses through leaky syringes, selective vaporization of the sample from the syringe, and syringe contamination. A brief discussion with examples of syringe effects is given below.

Syringe Leakage Effect

Poor sealing between the syringe plunger and barrel or the presence of a large compressible air gap in the syringe barrel allows sample backflow during injection. The effect of a leaky syringe on the quantitation of a wide volatility range sample is shown in Table 5.2. An average relative response factor (RF) near 1.0 with respect to n-C_{16} internal standard for n-alkanes from C_{12} to C_{32} was observed when a leak-tight plunger-in-needle 1 μL syringe was used. Under identical analytical conditions, but with a broken syringe seal, the syringe gives increasing discrimination and poor precision for the low volatility n-alkanes. In routine practice, the syringe should be frequently checked for leaks, especially when small sample sizes are used. A simple method is to insert the empty syringe into an injector that is pres-

TABLE 5.2. **Effect of syringe leakage on linear sample recovery and reproducibility.**[a]

n-alkane	Concentration (μg μL^{-1})	Leaky syringe RF[b] $\pm$ % RSD[c]	Leak-tight syringe RF[b] $\pm$ % RSD[c]
n-C_{12}	0.597	1.03 $\pm$ 1.45	1.03 $\pm$ 2.11
n-C_{16}	1.188	1.00	1.00
n-C_{22}	0.948	1.14 $\pm$ 2.55	1.03 $\pm$ 0.33
n-C_{32}	0.805	1.58 $\pm$ 3.60	1.08 $\pm$ 1.78

[a] Split sampling at 400:1 split ratio, autosampler injection of 0.65 μL sample volume with 1 μL plunger-in-needle syringe.

[b] Average relative response factor for n-alkanes with respect to hexadecane internal standard.

[c] Percent relative standard deviation.

surized to the column inlet pressure and then check for bubbling at the top of the syringe barrel with a drop of isopropanol or other volatile organic solvent. Plunger-in-needle syringes and syringes with Teflon tip plungers are normally found to have fewer leakage problems.

Selective Vaporization Effect

Selective vaporization in a heated injector is the other major source of quantitative error. This results from incomplete vaporization of low volatile components from the syringe in the heated injector during injection. Selective vaporization can be minimized by using a plunger-in-needle syringe with slow injection and/or long syringe needle residence time in the heated injector: both methods allow the syringe needle to be heated, therefore giving more reproducible vaporization of the sample. Figure 5.1 compares the recovery of a wide volatility range n-alkane sample using plunger-in-needle and plunger-in-barrel gas-tight syringes with a direct injector and using the solvent effect sampling technique. When a plunger-in-needle syringe is used, linear sample recovery for a mixture of n-alkanes from n-C_{12} to n-C_{34} is observed (Figure 5.1A). There is discrimination against n-C_{22} and higher number n-alkanes (Figure 5.1B) when a 5 μL plunger-in-barrel syringe is used.

Syringe Contamination

Syringe contamination or sample carry-over can produce poor quantitative reproducibility and accuracy. The syringe needle and plunger should be washed with solvent to reduce contamination. The efficiency of the solvent wash is affected by the solubility and viscosity of the wash solvent, and the inner diameter of the syringe needle. Syringe needles with i.d.s less than 100 μm are difficult to clean by solvent washing because of the high resist-

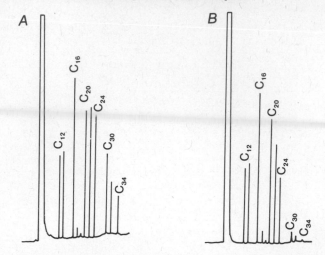

FIGURE 5.1. Chromatograms of *n*-alkanes using (A) a plunger-in-needle syringe and (B) a plunger-in-barrel syringe. Note: 20 m × 0.25 mm i.d. SE-30 column; 2 μL sample volume, 2 min sampling time, 300°C injector and detector temperature, temperature program from 35 to 270°C at 5°C min^{-1} after an initial 2 min isothermal period, He carrier gas at 2 cm^3 min^{-1}, direct injection. Sample was mixture of C_{12} (36.05 ppm), C_{13} (44.02 ppm), C_{16} (71.82 ppm), C_{20} (56.40 ppm), C_{22} (56.71 ppm), C_{24} (53.41 ppm), C_{30} (47.95 ppm), C_{32} (53.06 ppm), and C_{34} (55.95 ppm) in isooctane.

ance to solvent flow and the persistence of air bubbles in the syringe. In such cases, satisfactory results may be obtained by using special syringe cleaning devices that allow heating and evacuation of the syringe after solvent washing.

For plunger-in-needle syringes, an easy and effective way of eliminating sample carry-over in the syringe is to insert the syringe needle into a heated GC injector after two or three solvent washes. Solvent washing prior to heat treatment is particularly important when viscous, high boiling point, or thermally polymerizable samples are used. It should also be noted that the syringe needle should not be left in the heated injector for long without frequently moving the syringe plunger.

5.4 SEPTUM CONSIDERATIONS

Septum leakage allows sample vapor backflow during sample injection and withdrawal processes, especially when high column inlet pressures are used with narrow-bore columns. This effect can be minimized by periodically replacing the septum and by allowing sufficient sampling time for the complete transfer of sample to the column prior to withdrawing the syringe needle from the injector.

The effect of septum bleed should also be considered. Ghost peaks originating from septum bleed can coelute with the components of interest,

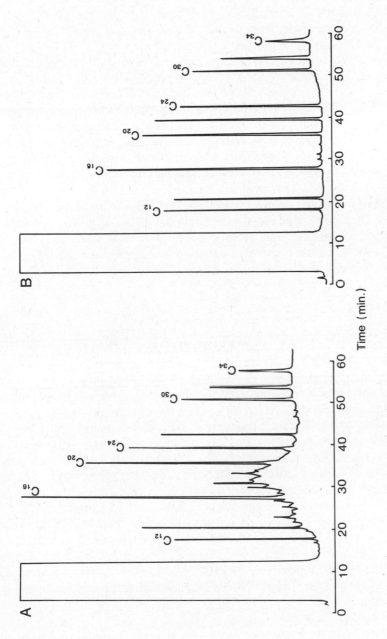

FIGURE 5.2. Chromatograms of *n*-alkanes showing septum contamination using (A) F-532 Microsep and (B) Thermogreen septa. Note: sample composition and analytical conditions were the same as those for Figure 5.1, except that the sample size was 1 μL, the injector temperature was 360°C, and the sampling time was 10 min.

thereby introducing errors. Figure 5.2 shows the effect of septum bleed on quantitation. The Microsep F-532 (Canton Bio-Medical Products, Inc.) is clearly unsatisfactory at a 360°C injector temperature due to interferences in solute peak area measurements (Figure 5.2A). The thermogreen TM LB-1 septum (Supelco, Inc.) shows very low bleed (Figure 5.2B), except for three peaks eluting between C_{16} and C_{20}, under the same chromatographic conditions as those for Figure 5.2A.

Ghost peaks can also originate from the release of adsorbed contaminants in the injector or from septum fragments deposited in the vaporization chamber by coring of the septum by the syringe. This is a particular problem when the column temperature is low enough to allow trapping of the contaminants such as in temperature programming, solvent effect, and cold trapping operations. The effect is minimized by using a properly designed syringe needle tip. Syringe needles of less than 0.007 in. i.d. do not significantly core the septum under normal conditions of septum compression and temperature.

The other effect of septum bleed is associated with the desorption of water which is adsorbed during manufacturing. The desorbed water may cause degradation of the column and loss of water sensitive samples.[3] It is important to precondition the septum at an elevated temperature prior to use, for example, in a GC oven at 100–200°C for 2–3 h.

5.5   SAMPLING TIME

Short sampling times should be used for small sample volumes (e.g., <0.5 μL), heated splitter injectors, and isothermal analysis. Slow injection (longer sampling times) should be used for large samples and temperature-programming operation (e.g., on-column sampling with solute focusing and direct sampling with the solvent effect). For split sampling and on-column injection of sample volumes less than 1.0 μL, the sample is normally rapidly discharged from the syringe into the vaporization chamber. The syringe needle residence time in the heated injector is also short to ensure constant split linearity and minimum loss of resolution. The effect of sampling time on peak broadening, expressed in terms of the number of theoretical plates per unit column length for samples with partition ratios between 1 and 15, is shown in Figure 5.3. In this example, sampling times longer than 3 s should not be used for components with $k < 10$.

The sampling time, Δt, for a 10% loss in resolution can be calculated from Equation (5.1):

$$\Delta t = \frac{1.5 r^2 (1 + k)(hL)^{1/2}}{F} \tag{5.1}$$

Here r, L, and F are column radius, column length, and carrier gas flow rate, respectively. As an example, for solutes with k between 1 and 15, the

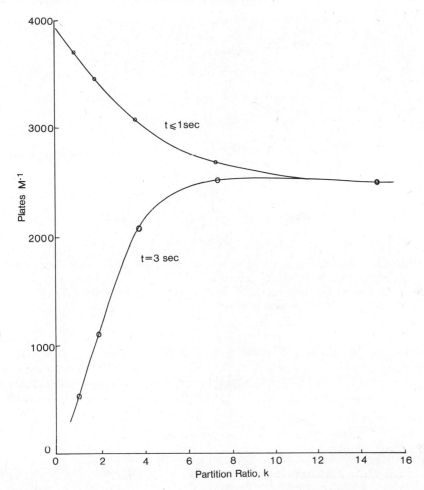

FIGURE 5.3.   Effect of split sampling time on column efficiency. Data were obtained using the Varian Model 8000 autosampler with normal 3 s and modified 1 s sampling times. Note: 20 m × 0.32 mm i.d. SF-96 column, 80°C isothermal column temperature, 250°C injector temperature, 0.1 μL sample size, 250:1 split ratio. Sample was *n*-alkane mixture from C_6 to C_{12}.

calculated sampling times, Δt, lie between 0.18 and 1.71 s for a 25 m × 0.25 mm i.d. column with a flow rate of 1 cm³ min⁻¹.

For the analysis of large sample volumes of wide volatility range samples using a heated injector (e.g., direct or splitless) or an on-column injector, the slow injection and solute focusing techniques allow good quantitation and peak shapes. The effect of sampling time on the elution peak width at half height for the *n*-alkanes is shown in Figure 5.4. Data were obtained for 1 μL direct injections of *n*-alkanes at 0.1 μL s⁻¹ injection rate; a 0.7 mm i.d. vaporization chamber at 300°C and a 0.66 mm o.d. needle plunger-in-needle syringe were used.

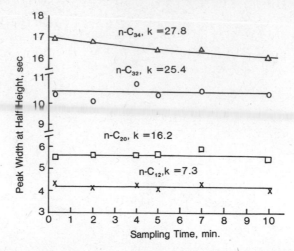

FIGURE 5.4. Effect of sampling time on peak width at half height during direct sampling. Manual syringe injection at a rate of appxoximately 1 μL s^{-1}, 1 μL sample, 5 μL plunger-in-needle syringe with a 0.66 mm o.d. needle. Sample was C_{12} (36.05 ppm), C_{20} (56.40 ppm), C_{32} (47.95 ppm), and C_{34} (55.95 ppm) in isooctane.

By applying the solute focusing technique, the sampling time does not significantly affect the elution peak width for up to 10 min injection and residence time. This implies that long injection and syringe needle residence times in the heated injector (direct or splitless with solvent effect sampling) could be utilized for minimizing sampling errors, selective vaporization, and

TABLE 5.3. Sampling time effect on sample recovery for wide volatility range n-alkanes.[a]

Sampling time[b] (min)	Relative response factor, RF			
	C_{12}	C_{20}	C_{30}	C_{34}
0.25	1.035	1.000	1.050	1.143
2	1.054	1.000	1.010	0.968
4	1.062	1.000	0.997	0.966
5	1.079	1.000	0.957	0.952
7	1.041	1.000	1.014	0.968
10	1.058	1.000	1.034	0.978
Mean, $\overline{RF}$	1.054	1.000	1.010	0.996
± % RSD	1.36	—	2.91	6.65

[a] Direct sampling with 0.1 μL s^{-1} injection speed, 1 μL sample size, and n-eicosane internal standard.

[b] Includes the syringe needle residence time inside the injector.

septum leakage. In addition, a long sampling time can also allow the use of a low injector temperature for the analysis of thermally labile compounds.

The effect of sampling time on sample recovery for the n-alkanes (Table 5.3) indicates excellent sample recovery for wide volatility range samples with a sampling time between 2 and 10 min. The variation in $\overline{RF}$ ranges from 1.1 to 2.4% RSD, excluding the RF values for the 0.25 min sampling time. The RF value for C_{34} with a 0.25 min sampling time is high due to the slow vaporization of C_{34}.

In general, a 0.5–1 μL s^{-1} injection speed and a short syringe needle residence time inside the injector are recommended for split, direct, or on-column injection of small samples (<1 μL).

For large samples (1 μL) where the solvent effect or solute focusing techniques are applied to direct, splitless, or on-column sampling, slow injection at 0.1–0.2 μL s^{-1}, and long syringe needle residence time are required.

5.6 INJECTOR TEMPERATURE

The injector temperature should be high for split injection. For direct and splitless sampling where the solvent effect or cold trapping is applied, a low injector temperature can be used. For thermally unstable compounds, direct injection with a low injector temperature or cold on-column sampling should be used.

Discrimination against high molecular weight compounds from low injector temperature and short sampling time is illustrated in Figure 5.5. Data

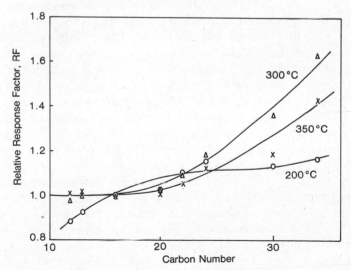

FIGURE 5.5. Effect of injector temperature on the linearity of sample recovery. Syringe and analytical conditions used were the same as those for Figure 5.1.

were obtained for a direct injector with a 1.02 mm i.d. vaporization chamber using autosampler injection (3 s sampling time). The sample size was 2.6 μL. It appears that injector temperature alone affects the range of discrimination of the low volatility components. For an injector temperature of 300°C, discrimination was observed for n-alkanes beyond n-C_{20}. Some improvement in the recovery of n-C_{22} was obtained by increasing the injector temperature to 350°C. By lowering the injector temperature to 200°C, discrimination against the n-C_{16} internal standard was indicated by the low RF values for n-C_{12} and n-C_{13}.

5.7 COLUMN SELECTION

Generally speaking, the desired column for open tubular column GC is one that is efficient, well-deactivated, and thermally stable. Other factors that are important for specific analyses include stationary phase selectivity, film thickness, column length, and internal diameter. These qualities can vary to some degree depending on the particular analysis of interest, and on the specific level of performance desired. In this section, particular attention is given to the relevant considerations in selecting an open tubular column for a specific chromatographic application.

Column Material and Surface Activity

As discussed in Chapter 3, a number of materials have been used for the fabrication of open tubular columns. These include plastics, nickel, copper, stainless steel, glass, and purified silica. At present, fused silica and purified quartz have largely replaced the other materials because of their low catalytic activity, and the ability to modify the silica surface easily, both physically and chemically, to accomodate better selected stationary phases and to produce surfaces that are inactive toward trace organic compounds. The inherent flexibility of commercially available fused-silica and quartz open tubular columns is another significant advantage of these materials.

While the preceding remarks are generally true, there are situations in which other materials may be used without serious hindrance to the separation. In fact, in some cases other materials may be preferred. For example, the analysis of low molecular weight aliphatic hydrocarbons has been carried out for many years using coated stainless steel columns. Stainless steel is more robust than glass or silica, and its somewhat catalytic surface is not a problem for the relatively unreactive aliphatic hydrocarbons.

There are some cases in which glass may be preferred over quartz or fused silica. Alkali glasses are often used to make columns for separating basic compounds such as the amines. In addition, thick-wall glass columns can be leached, etched, roughened, and so on, in order to provide increased wettability or modification for polar stationary phases. Similar treatment of fused-silica columns is limited because of the thin column walls and the resistance of pure silica to some of these treatments.

The proper deactivation of the column wall surface is extremely important in open tubular column GC and has been discussed in detail in Chapter 3. Proper deactivation is not only important from the standpoint of adsorption or reaction with sensitive analyte compounds, but also in determining column stability and lifetime. For instance, a column that has not been properly deactivated can catalyze the decomposition of the stationary phase at elevated temperatures leading to low thermal stability. This, in turn, can lead to nonuniform films and, hence, poor efficiency. On the other hand, a well-deactivated surface can also be produced which is not properly wetted by the stationary phase. This also leads to low thermal stability and poor efficiency.

The fused-silica column material is very inert and requires little or no deactivation for many types of analyses. Except for strong bases, nondeactivated fused-silica columns can be used to chromatograph almost all types of compounds at levels of approximately 10 ng or greater. Figure 7.23 (Chapter 7) shows a chromatogram of basic nitrogen-containing polycyclic aromatic compounds on an undeactivated fused-silica column coated with SE-52. Full-scale response was approximately 10 ng.

The critical surface tension of untreated fused silica is high, and almost all liquid phases can be easily coated. Unfortunately, surface deactivation usually reduces the critical surface tension, thereby limiting effective coating to the more nonpolar phases.[4] If deactivation is required for trace analysis, high-temperature silylation,[5-7] cyclic siloxane reaction,[8,9] and polymer decomposition[10,11] are the most effective methods. These have been described in Chapter 3. Other methods generally appear to be incomplete or only temporary.

The decomposition of Carbowax 20M was initially used for the deactivation of fused-silica columns.[12] This treatment produced a critical surface tension ($\sim$45 dyne cm^{-1}) that was sufficiently high for the deactivated surface to be wettable by many nonpolar and medium polarity phases. Unfortunately, its thermostability was limited to temperatures below 250°C, and the polar undercoating produced undesirable selectivity effects when compounds were chromatographed using nonpolar phases. However, this procedure does provide the ideal base for columns coated with various polyethylene glycol phases.

The ultimate goal in deactivation is to produce a completely inert or neutral surface to all solutes that are to be chromatographed. In practice, this is very difficult to obtain but can be accomplished at the nanogram level. Analyses at lower levels usually require additional selected deactivation steps. Such columns, therefore, are tailor-made for specific analyses such as amines, alcohols, carboxylic acids, and so on, but cannot be used as general purpose columns for all types of compounds.

Stationary Phases

The selection of a stationary phase ultimately depends on the sample to be analyzed and the desired resolution of mixture components. It was generally assumed in the past that nonpolar stationary phases should be used for nonpolar compounds, and polar phases for polar compounds. This was especially true before adequate surface deactivation methods were developed. Polar phases tend to interact and block to some degree the polar active sites on an active surface, thereby providing less adsorptive interaction with polar solutes.

With proper surface deactivation, polar solutes can be adequately analyzed on nonpolar stationary phases. In most cases, the resolution of complex mixtures of polar compounds is better than on polar phase columns. This is because the nonpolar methylpolysiloxane gum phases (e.g., OV-1, SE-30, SE-52, and SE-54) provide the most uniform and stable stationary phase films available, which results in the highest column efficiencies. These stationary phases can be used as all-purpose phases for almost all sample types.

If it appears necessary to resolve one or several important pairs of similar compounds, which are unresolved on the methylpolysiloxane phases, more polar phases can be utilized. Unfortunately, more polar phases generally are less efficient and have lower thermostabilities. The lower useable upper temperature limits of these phases are even more of a disadvantage when one considers that solute retention is greater. Of the medium to slightly polar phases available, the methylphenylpolysiloxanes,[13,14] polyethylene glycols (Carbowax 20M and Superox 20M), and cyanopropylpolysiloxanes[15-17] are the most useful.

Attempts to characterize and tabulate the various stationary phases according to polarity have resulted in several useful classifications. Rohrschneider[18] selected five standard compounds of different chemical structure (benzene, ethanol, 2-butanone, nitromethane, and pyridine) and calculated the difference between their Kovats retention indices on selected stationary phases and the nonpolar hydrocarbon phase, squalane.

$$\Delta I = I_{\text{polar phase}} - I_{\text{squalane}} \tag{5.2}$$

The Rohrschneider constant is given by

$$\frac{\Delta I}{100} = \text{Rohrschneider constant} \tag{5.3}$$

This constant is designated as x, y, z, u, or s depending on which of the five standard compounds was measured. A liquid phase with a large y constant would retain alcohols.

McReynolds constants,[19] which are slightly modified from the Rohrschneider method, are widely used today to compare the polarities of different liquid phases. Tabulations are found in the literature[19] and in catalogs from chromatographic supplies distributors.

There are two additional types of stationary phases which are used for specific separations not based on polarity. These are the chiral and liquid crystal phases. Alcohol, amine, carboxylic acid, and amino acid enantiomers can be resolved on various chiral phases (see Chapter 6). Liquid crystal phases, on the other hand, are very useful for the separation of structural isomers such as the polycyclic aromatic compounds (see Section 6.2). Separations using these phases are based on molecular geometry (size and shape).

All stationary phases have specific temperature operating ranges. The lower temperature is determined by the melting point or viscosity of the phase. At temperatures below the melting point or at high viscosities, solutes are not readily dissolved in the stationary phase, and mass transfer is slow. Use at these temperatures results in broad and sometimes tailing chromatographic peaks.

The upper temperature limit, in some cases, is determined by the volatility of the stationary phase, but is most often limited by its thermal stability. The upper temperature limit can be readily recognized during chromatography as the temperature at which there is a marked increase in the column bleed.

Table 3.8 lists the more commonly used stationary phases in capillary column chromatography along with their temperature ranges and McReynolds constants. The upper temperature limits given are in many cases quite liberal.

Practical experience has demonstrated that gum phases give columns of consistently higher quality. These phases coat more evenly and are thermally more stable. Recent cross-linking of coated phases using free-radical initiators stems from efforts to increase the viscosity and stability of these films. These so-called "bonded" phases are resistant to solvent washing and are stable to higher temperatures.

Selecting the Column

Assuming that most analyses will be performed using fused-silica or vitreous-silica open tubular columns, the remaining parameters that must be selected for a specific separation include: (a) the stationary phase (cross-linked or coated); (b) the stationary phase film thickness; (c) the column length; and (d) the column internal diameter.

In choosing a stationary phase, factors such as performance and stability should usually be more important than selectivity. This is due to the fact that the high efficiencies of open tubular columns usually more than compensate for the selectivities of specific phases, and most separations can be carried out on relatively few stationary phases. On the other hand, as open tubular columns are increasingly applied to routine analysis, with emphasis on productivity, there is an increased need for selective phases.

Stark et al.[20] discussed the considerations for designing and choosing selective phases for open tubular column GC. They defined selectivity as the interplay of the various types of interactions, such as dispersion, dipole, and base–acid, between solute and stationary phase. In contrast to selectivity, polarity refers to phases having a significant concentration of substituent groups with large permanent dipoles. Using these definitions, all stationary phases are selective for different types of compounds, but not all phases are polar. It is important to differentiate selectivity and polarity when selecting the stationary phase for a particular application.

It would be extremely desirable to have a short list of "standard" or "preferred" stationary phases covering a range of selectivities applicable to the majority of separations desired. This would greatly simplify the selection of a stationary phase for a given analysis. Using McReynolds data and the principal component analysis by McCloskey and Hawkes,[21] Stark et al.[20] recommended the use of trifluoropropyl-, methyl-, and phenylpolysiloxanes for dispersion-selective substituents, with trifluoropropylpolysiloxanes being the least polarizable and phenylpolysiloxanes being the most polarizable. For polar phases, cyanopropyl- has the largest permanent dipole and should be the most useful. It also exhibits strong dispersion as well as a modest level of basicity (see Table 5.4). The polyethylene glycols are also polar, and they demonstrate some dispersion and basicity. There are presently no thermally stable acidic phases for open tubular column GC.

Based on the above considerations, the following polysiloxane and polyethylene glycol gum phases were selected[20] as providing a sufficiently broad selectivity range: (a) methylpolysiloxane; (b) 50–70% phenyl methylpolysiloxane; (c) cyanopropyl methylpolysiloxanes of medium (25–50%) and high (70–90%) cyanopropyl incorporation; (d) trifluoropropyl methylpolysiloxane, and (e) polyethylene glycol or a suitable polysiloxane substitute (e.g., 40–45% cyanopropyl, 20–25% phenyl methylpolysiloxane). Future development of selective phases might include amino-, hydroxy-, alkoxy-, and acidic functional groups.

TABLE 5.4. Interaction indices[a] for selected stationary phases.

Stationary phase	Dispersion	Polarity	Base	Acid
Dimethyl silicone	9.0	0	0	0
Phenylmethyl silicone	11.6	0	0	0
Polyethylene glycol	8.6	8[b]	4[b]	0
3-Cyanopropyl silicone	10.4	11	3	0
3,3,3-Trifluoropropyl silicone	8.6	3	0	1
1,2,3-Triscyanoethoxypropane	8.4	11	3[b]	0
Trimer acid	8.5	3	6	3
Diglycerol	8.8	9	9	4

[a] Taken from ref. 20.

[b] Estimated.

For the optimization of specific separations, it has been observed that intermediate polarities can be obtained by mixing nonpolar and polar phases. Several methods have been proposed to determine the best mixture for separating any set of compounds.[22-25] One of the most successful methods is called the *window diagram method*.[24,25] Here, a parameter is selected which can be derived from chromatographic data, the relationship of which with the composition of the mixed phase can be precisely described with a model. This parameter is plotted against the weight composition of the phase, and the window diagram is constructed. An example of the application of this technique to the separation of twelve 4-ring polycyclic aromatic thiophene isomers[26] is given for illustration.

In this case, the capacity factors (k) of the solutes were determined and plotted as a function of the fraction of BBBT (liquid crystal) in BBBT/SE-52 stationary phase mixtures. Figure 5.6 shows the resultant plot. This re-

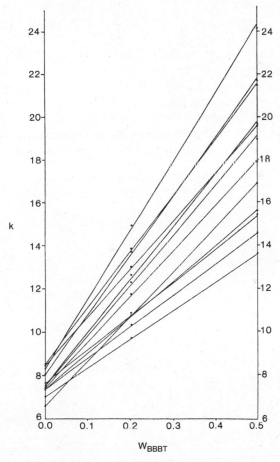

FIGURE 5.6. Plot of capacity factor (k) vs. W_{BBBT} for 4-ring polycyclic aromatic thiophene isomers. (Reproduced with permission from ref. 26. Copyright Preston Publications, Inc.)

lationship can be described simply by the equation:

$$k_M = k_{BBBT} W_{BBBT} + k_{SE-52} W_{SE-52} \tag{5.4}$$

where k_M, k_{BBBT}, and k_{SE-52} represent the capacity factors of solutes on the mixed-phase, 50% BBBT, and pure SE-52 columns, respectively, and W_{BBBT} and W_{SE-52} are the weight fractions of each phase in the mixed phase. The window diagram is constructed by plotting the relative retention (α), which is calculated from the k values of all pairs, versus the percent composition of the phase. After all lines are plotted, only the areas underneath the lines are used to give the window diagram, as shown in Figure 5.7. From this diagram, the best stationary phase composition should be around 16% BBBT in SE-52 with the relative retention (α) of the least separated pair equal to about 1.011. The required plates (N_{req}) for baseline resolution ($R = 1.5$) of all 4-ring thiophene isomers can be calculated from the following equation:

$$N_{req} = 24 \left(\frac{\alpha}{\alpha - 1} \right)^2 \left(\frac{k + 1}{k} \right)^2 \tag{5.5}$$

For a capacity factor of 10, a total of 245,795 plates is needed to resolve all twelve isomers. This would require using a column of greater than 90 m in length. The results obtained from a 23.3 m column coated with 16% BBBT in SE-52 are shown in Figure 5.8A. For comparison, the same mixture separated on pure SE-52 is shown in Figure 5.8B.

The length of open tubular column is the next parameter to consider. The high separation efficiencies of open tubular columns in GC are generally

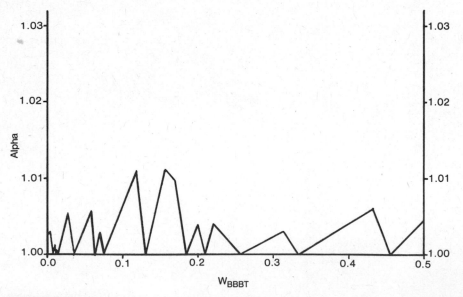

FIGURE 5.7. Window diagram constructed from data in Figure 5.6. (Reproduced with permission from ref. 26. Copyright Preston Publications, Inc.)

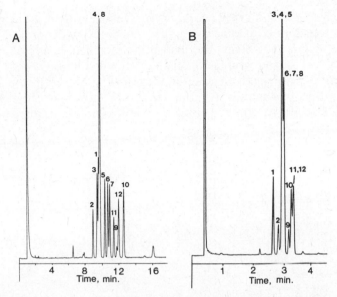

FIGURE 5.8. Chromatograms of 4-ring polycyclic aromatic thiophene isomers on open tubular columns coated with (A) 16% BBBT in SE-52 and (B) pure SE-52. Note: (A) 23.3 m fused-silica column, (B) 19.6 m fused-silica column, 140°C isothermal column temperature, H₂ carrier gas at 40 cm s⁻¹. Selected peak assignments: (2) benzo[*b*]naphtho[1,2-*d*]thiophene; (4) phenanthro[4,3-*b*]thiophene; (8) phenanthro[3,4-*b*]thiophene; (10) phenanthro[2,1-*b*]thiophene. (Reproduced with permission from ref. 26. Copyright Preston Publications, Inc.)

attributed to the relatively long column lengths that can be employed. Columns of over 100 m in length can be used because of their high permeabilities. Columns of such lengths can generate extremely large numbers of theoretical plates of up to 500,000.

There are, however, several undesirable characteristics of open tubular column GC with long columns. Analysis times are long; higher temperatures

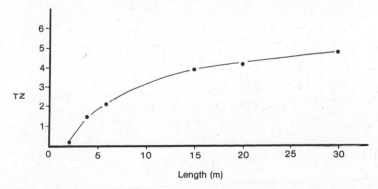

FIGURE 5.9. Plot of TZ values for the isomer pair benzo[*e*]pyrene and perylene vs. column length. (Reproduced with permission from ref. 28. Copyright Dr. Alfred Huethig Publishers.)

are required for elution of sample components; higher column inlet pressures are required to maintain suitable carrier gas velocities; and more active sites are available within the column for adsorption of sensitive compounds.

According to chromatographic theory, column resolution is proportional to the square root of length. In addition, Jennings[27] showed that as the column temperature is programmed, the separation power of the column is reduced. This seems to offset to some degree the gain in efficiency due to

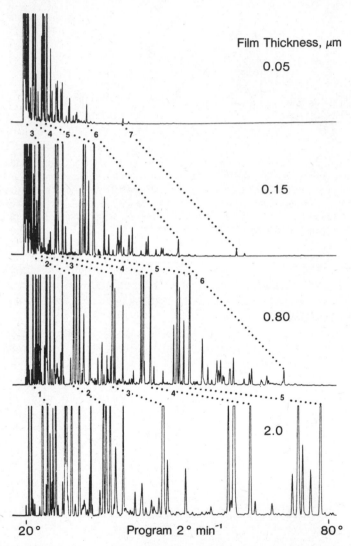

FIGURE 5.10. Chromatograms of gasoline on open tubular columns with varying film thickness. Note: 15 m SE-52 glass column. (Reproduced with permission from ref. 30. Copyright Dr. Alfred Huethig Publishers.)

increased length. Figure 5.9 shows a plot of the Trennzahl or separation number (TZ) for the isomer pair benzo[e]pyrene and perylene as a function of column length.[28] As can be seen, little resolution is lost in going from a 30 m column to a 15 m column. Of course, the real decision concerning column length is dependent on the complexity of the sample to be analyzed and the resolution requirements for close-eluting components of the sample.

The stationary phase film thickness, d_f, and the column internal diameter (or radius, r) are more important parameters than the column length. Both parameters can be related to the phase ratio, β (the ratio of the gas volume in the column and the volume of the stationary phase), as given in Equation (2.19) (Chapter 2).

Typical β values for open tubular columns are 50–500.[29] Changes in either the film thickness or the column internal diameter strongly influence the phase ratio and, hence, the solute retention. An increase in film thickness leads to a decrease in the phase ratio and an increase in retention. In addition, a decrease in column efficiency is obtained (see Section 2.3). Grob and Grob[30] illustrated this as shown in Figure 5.10. Generally, thick-film columns (>0.5 μm) are used for the analysis of highly volatile compounds, and thin-film columns (<0.5 μm) are used for the analysis of higher molecular weight compounds. A typical average film thickness is about 0.25 μm.

An alternative to using thick-film columns for the analysis of highly volatile compounds is to use low temperatures. The capacity ratio, k, is inversely proportional to the column temperature and, therefore, the solute spends more time in the liquid stationary phase. It has recently been shown[31] that cross-linking of conventional stationary phases leads to lower glass transition temperatures, which extends the lower temperature ranges of these phases.

Smaller diameter columns have a lower sample capacity, and sample introduction is more difficult. On the other hand, as was discussed in Chapter 2, faster speeds of analysis and higher efficiencies are possible with the proper adjustment of carrier gas velocity. A typical average column diameter is about 0.25 mm.

5.8 CARRIER GAS SELECTION AND MEASUREMENT

The selection of carrier gas for open tubular column GC depends on three important factors: (a) separation efficiency in terms of resolution and speed; (b) detector compatibility and sensitivity; and (c) physico-chemical properties of the carrier gases. Practical constraints, such as column pressure drop, gas impurity effects, and explosion hazard must also be considered. The inert gases, such as He and N_2, purified so as to contain low concentrations of impurities such as oxygen and water, are normally used in open tubular column GC. Hydrogen is often suggested as carrier gas for high-speed GC. However, its effects on detector stability and sensitivity, as well as the explosion hazard and the reactions of H_2 with metallic oxides, should also be taken into account.

The purity of the carrier gas and the nature of the impurities should be monitored. The impurities that can react with stationary liquid phases and sample solutes, and affect detector stability and sample response, should be removed. The carrier gas purity is critical in temperature-programming operation. The stability of the baseline is affected, and ghost peaks appear if carrier gas impurities are trapped at the inlet of the column (at low column temperature) and eluted at higher programmed temperatures. Table 5.5 gives the purities of some commonly used carrier gases in open tubular column GC.

In cases where speed of analysis outweighs the necessity for high resolution, a carrier gas which has a high optimum linear velocity, u_{opt}, and for which the dependence of h on u is small, is preferred. As demonstrated by Equation (2.33), the optimum carrier gas velocity is linearly proportional to the parameters D_G° and j. The diffusivity of the solute molecules in the carrier gas, D_G°, is approximately inversely proportional to the square root of the molecular weight or the density of the carrier gas. Therefore, a carrier gas of low density such as H_2 and He allows a greater analysis speed. The j factor is approximately inversely proportional to the viscosity of the carrier gas. For a narrow-bore column, for which a high pressure gradient across the column is obtained, H_2 which has approximately half the viscosity of He or N_2 allows approximately twice the analysis speed than can be obtained with either He or N_2.

As discussed earlier [Equation (2.30)], the minimum theoretical plate height, h_{min}, is proportional to the square root of both j and D_G°; the smaller j and D_G°, the smaller h_{min}. For difficult separations in which maximum resolving power is required, carrier gases with low j and D_G° values are important.

Many devices are provided for carrier gas flow-rate measurement. In practice, a soap bubble flow meter gives better than 1% precision over a 10 cm^3 volume.

If the measured flow rate, F, is used for accurate retention volume calculations, corrections for water vapor pressure over the detergent solution

TABLE 5.5.   Levels of contaminants in some commonly used high-purity
carrier gases[a] in open tubular column GC.

Gases	Purity (%)	Maximum contaminant concentration (ppm)								
		N_2	He	O_2	Ar	CO_2	Ne	H_2O	CH_4	H_2
He	99.995	14	—	1	1	1	14	12	1	1
H_2	99.999	3	50	1	—	1	—	5	0.8	—
N_2	99.998	—	—	10	—	1	—	10	0.1	—

[a] Matheson Co., East Rutherford, New Jersey.

and temperature difference must be made. The flow rate corrected to column temperature, T, and outlet pressure, P_o, is given by

$$F_c = F \left(\frac{T}{T_w}\right) \frac{(P_o - P_w)}{P_o} \tag{5.6}$$

where T_w is the temperature measured at the flow meter and P_w is the vapor pressure of water at T_w; temperatures are in Kelvin and F_c is the corrected flow rate.

The corrected flow rate can also be calculated by using the retention time of a nonretained or a retained test component using Equation (5.7) or (5.8):

$$F_c = \frac{\pi r^2 L}{j t_0} \tag{5.7}$$

$$F_c = \frac{\pi r^2 L (1 + k)}{j t_R} \tag{5.8}$$

These relationships are of great practical importance because they also provide the best means of measuring the mean column gas phase volume, V_G, and the mean column radius, r.

5.9 CARRIER GAS FLOW PROGRAMMING

Carrier gas flow programming[32-40] is a technique that allows the carrier gas flow rate in the column to be gradually increased to reduce the analysis time and to increase the detectability of highly retained samples. It has the following advantages over constant flow GC: (a) It allows rapid elution of compounds with high boiling points at relatively low column temperatures. (b) It allows better baseline stability and extends column lifetime in comparison to temperature programming. The evaporation rate of the stationary liquid phase changes linearly with flow rate and exponentially with temperature. (c) Low-temperature GC is particularly favorable for the analysis of thermally labile compounds. (d) It is particularly favorable for open tubular column GC because the H vs. $\bar{u}$ plot is relatively flat. (e) It demands no special apparatus. It can be effectively applied with adjustment of the column inlet pressure. (f) The initial carrier gas flow rate can be restored very rapidly and, thus, more chromatographic runs can be made in a given time.

In practice, a low carrier gas flow rate is used initially to elute components of high volatility, followed by an increase of flow rate for the more strongly retained components. Carrier gas flow-rate programming can be changed either stepwise or continuously.[32,38-40] The average carrier gas velocity in the column increases linearly with an increase in the column inlet pressure. If the column inlet pressure increases exponentially with time, the retention times of the solutes decrease in a similar manner as in linear temperature-

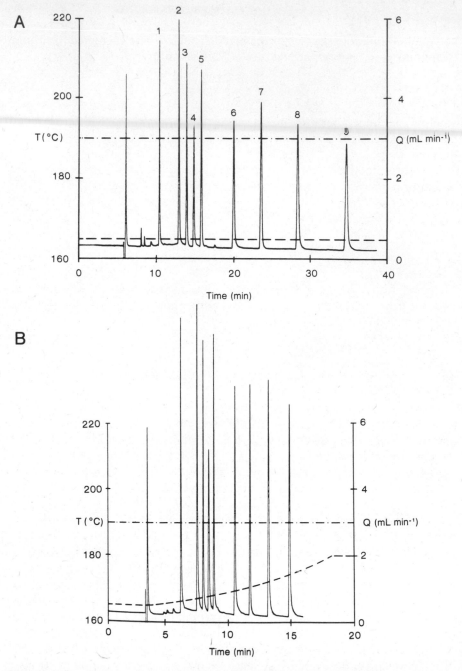

FIGURE 5.11. Chromatograms of a mixture of chlorinated hydrocarbons using a carrier gas flow rate (A) isorheic at 0.5 cm³ min⁻¹ and (B) programmed exponentially from 0.5 to 2.0 cm³ min⁻¹. Note: 25 m SE-30 glass column, 190°C isothermal column temperature, split ratio of 1:100, N₂ carrier gas, ECD detection, temperature curve (–··–), flow rate curve (– – –). Selected peak assignments: (1) pentachlorobenzene; (2) technazene; (6) pyrazon; (7) heptachlor; (8) aldrin; (9) heptachlorepoxide. (Reproduced with permission from ref. 39. Copyright Elsevier Scientific Publishing Company.)

programming GC. Figure 5.11 compares the separation speed and peak shapes resulting from chromatography of a mixture of chlorinated hydrocarbons using a constant flow rate and a programmed flow rate at the same column temperature.[39] The analysis time for the chlorinated hydrocarbons was reduced from 35 to 15 min with an exponentially increasing flow program from 0.5 to 2 cm^3 min^{-1}.

Flow programming can also be combined with temperature programming. Figure 5.12 shows how a further reduction in analysis time can be achieved by simultaneous programming of both the column temperature and the carrier gas flow rate.

Flow programming requires the use of a high inlet pressure control valve and associated pneumatics, and a compatible detector. The stability and sensitivity of the detector must be independent of the variation in carrier gas flow rate. When hydrogen carrier is used with detectors such as the FID, FPD, and TSD where detector response changes with hydrogen gas flow rates, a special arrangement for maintaining a constant hydrogen flow to the detector is required. Splitting the flow or adjusting the hydrogen fuel gas to compensate for the column flow-rate change are useful when the detector gas flow rates need to be maintained constant.

Flow programming combined with the use of short columns can greatly reduce the analysis times for simple mixtures. Figure 5.13 shows a 45 s

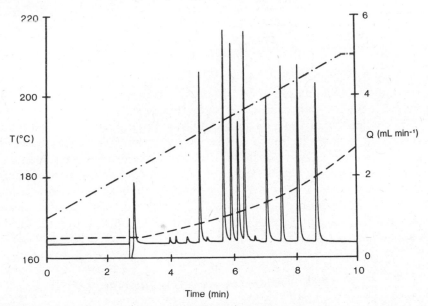

FIGURE 5.12. Chromatogram of a mixture of chlorinated hydrocarbons (as in Figure 5.11) with both temperature and flow programming. Note: temperature program from 170 to 210°C at 4°C min^{-1}, otherwise same conditions as in Figure 5.11B. (Reproduced with permission from ref. 39. Copyright Elsevier Scientific Publishing Company.)

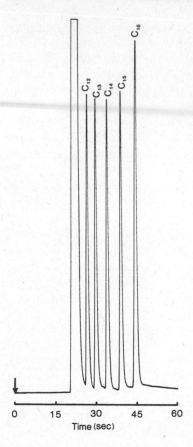

FIGURE 5.13. Chromatogram of a mixture of *n*-alkanes under flow-programming conditions. Note: 5 m column, isothermal column temperature, flow program from 0.5 to 25 cm^3 min^{-1}. (Reproduced with permission from ref. 40. Copyright Dr. Alfred Huethig Publishers.)

chromatogram of a mixture of *n*-alkanes, C_{12} to C_{18}, with flow programming from 0.5 to 25 cm^3 min^{-1} on a 5 m open tubular column.[40]

The advantages of short columns and flow programming are that a wide programmable flow-rate range can be easily obtained without excessively high inlet pressures, and the flow rate can be rapidly restored, leading to short intervals between runs. However, a short column has limited sample and peak capacities. In practice, it is preferrable to start with a long column to separate the desired components with greater resolution than is necessary at a given flow rate and temperature. A carrier gas flow program can then be applied so that peak resolution is just adequate.

5.10   COLUMN TEMPERATURE PROGRAMMING

The advantages of column temperature programming in open tubular column GC are the following: (a) Analysis times are reduced for samples with a wide boiling point range. (b) The solvent-effect, cold-trapping, or solute focusing

TABLE 5.6.   Comparison of programmed temperature and isothermal GC.

Parameters	Isothermal	Temperature programming
Sample boiling point range	<100°C	≥100°C
Sample volume	<1 μL	≤10 μL
Sampling speed	Fast ($t \le 0.05 \, W_h$ for the first peak or for the peak of interest)	Can be as slow as 0.1 μL s^{-1} if solvent effect or solute focusing techniques are applied
Carrier gas purity	Not as critical	High purity required
Peak capacity	≤10 Components	>10 Components
Stationary phase selection	Wide choice	Restricted to low bleed and thermally stable phases
Peak detection	Poor for long retained components	Improved by increasing the temperature-programming rate
Analysis speed	Slow	Fast
Sampling mode compatibility	Split, direct, and on-column	Split, direct, on-column, splitless, multidimensional, column switching, headspace, and pyrolysis GC

techniques can be utilized to reduce the effects of slow sample injection of large sample volumes. (c) Better retention reproducibility is possible. (d) Sample detection is improved. In addition, temperature programming at a high rate (e.g., 10°C min^{-1}) for a wide temperature range (e.g., from 40°C to the maximum allowable programmable temperature for the column) provides the most useful approach for the analysis of an unknown sample. The chromatogram from such a single analysis gives a very useful indication of the range of compounds present and allows selection of analysis conditions for an optimized separation. Temperature-programmed and isothermal GC are compared in Table 5.6.

For programmed temperature GC, a number of operational parameters may need to be optimized for achieving the desired speed and resolution. These are discussed below.

Initial Temperature

For unknown samples, the initial temperature is normally chosen to be at ambient. In general, the selection of the initial column temperature in temperature-programming operation is based on the following three factors: (a) The volatilities of the sample components: the initial temperature is normally chosen to be lower than the elution temperature of the most volatile component in the sample. (b) the viscosity of the stationary phase: the initial column temperature should be above the freezing temperature of the stationary phase, or be restricted to a temperature just higher than that which leads to low column efficiency and sample capacity due to increased stationary phase viscosity at low temperature. (c) The need for the effective application of the solvent effect, cold-trapping, or the solute focusing techniques: For the solvent effect technique, the initial column temperature should be about 20°C below the solvent boiling point to allow reconcentration at the inlet of the column. The initial column temperature can be as low as −50°C for cold-trapping and 10°C above the solvent boiling point for on-column injection with the solute focusing technique. In such cases, the initial column inlet is normally isolated from the rest of the column so that the heating rate of the column inlet is not limited to the column-programming rate. The column inlet is then rapidly heated to vaporize the trapped sample components so that the initial sample band spreading is minimized.

Final Temperature

The selection of the final temperature for temperature-programmed analysis is normally based on two factors: (a) the temperature stability of the stationary liquid phase and the sample components and (b) the boiling point of the least volatile component in the sample.

The final temperature for open tubular columns is normally suggested by column manufacturers. It should be noted that a new column tends to have higher bleed which limits the actual operational temperature to much less than that suggested. A new column should be programmed gently to prevent the stationary liquid phase from forming droplets.

Heating Rate and Flow Rate

Heating rate and flow rate affect chromatographic resolution and speed of analysis. The interplay of these two parameters is complex. For example, for weakly retained components, an increase in the temperature-programming rate will not significantly reduce the analysis time, if the carrier gas flow rate is constant. On the other hand, an increase in flow rate reduces proportionally the elution temperature and the analysis time, if the temperature-programming rate is held constant. The most effective way to improve the speed of analysis of weakly retained samples is to increase the flow rate. For strongly retained components, the speed of analysis depends on the ratio of heating rate to flow rate. At small heating-rate to flow-rate ratios, a change of heating rate with constant flow rate will not significantly reduce the analysis time. However, at high heating-rate to flow-rate ratios, an increase in heating rate gives a proportional reduction in analysis time; whereas an increase in flow rate gives no significant change in the speed of analysis.

In open tubular column temperature-programmed GC, the heating-rate to flow-rate ratio is normally high because of the low flow rates used. This implies that the heating rate will significantly affect the retention times of the strongly retained components, and that the heating rate should be kept constant if retention time reproducibility is important. For weakly retained components, however, retention time reproducibility is mainly affected by the flow-rate stability and is less dependent on the stability of the heating rate.

In terms of resolution, it is generally true that the smaller the heating-rate to flow-rate ratio, the better the resolution. In open tubular column GC, since the flow rate is small, small heating rates are required to obtain optimum resolution. The heating rates are normally below 5°C min^{-1} for the separation of closely spaced components. Heating rates higher than 10°C min^{-1} are rarely chosen except in exploratory runs, in high-speed analyses, and with the solvent effect or cold trapping in which fast initial temperature programming is used.

In temperature-programmed GC, the heating rate can affect the column lifetime. Polar phase columns form droplets relatively easily when programmed at high heating rates. Low bleed columns such as SE-30 and chemically bonded columns may be used with high heating rates, however.

5.11 SAMPLING MODE SELECTION AND TECHNIQUES

Proper selection of the sampling mode and the practice of the sampling technique are crucial to the success of open tubular column GC for qualitative and quantitative analysis.

Split Sampling

Split sampling is used in the analysis of mixtures containing uniformly distributed components in high concentration. It is very useful in method development and in optimizing separations. The advantages of split sampling are the following: (a) Initial solute molecular zone spreading is negligible at high split ratio. (b) The split ratio can be easily adjusted for the sample volume and concentration used. (c) Excellent retention time reproducibility and good quantitative precision can be obtained with autosampler injection. (d) The procedure is compatible with both isothermal and temperature programming operations.

The disadvantages of split sampling are the following: (a) Split sampling is discriminatory (i.e., it does not have the same split ratio for samples with wide concentration and volatility ranges, and poor sample recovery and precision are obtained for low volatility components). (b) Trace analysis is difficult. (c) There is a large consumption of carrier gas, especially when a high split ratio is used. (d) The carrier gas flow rate is not constant during temperature programming if the flow rate is regulated by a pressure regulator. (e) Quantitative precision and accuracy are heavily dependent on injection repeatability and technique.

Operational parameters such as split ratio, injector temperature, sample–carrier gas mixing, injection sample volume and concentration, and injection repeatability need to be considered in split sampling. The split ratio is generally determined by the effective volume or mass of the sample that the open tubular column allows. The effective volume allowable is the volume that may not contribute more than 10% to solute zone spreading in the column, while the effective mass allowable is a function of the column sample capacity.

The splitter injector temperature is determined by the volatility of the sample and the heat capacity of the carrier gas. Maximum injector temperature should be employed when there is a high split ratio and high flow rate in the injector, a large sample volume, a high boiling point sample, and a low heat capacity carrier gas such as H_2 or He. However, a high injector temperature cannot be used if thermally labile compounds are injected or high bleed septa with no septum purge are used.

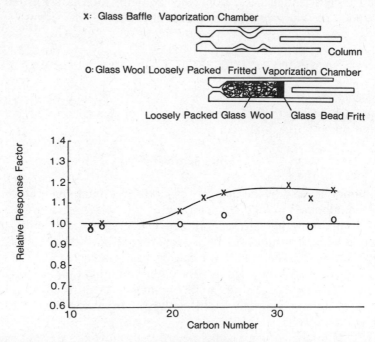

FIGURE 5.14. Measurement of split sampling linearity for C_{12} (0.97 ng), C_{13} (1.18 ng), C_{16} (1.93 ng), C_{20} (1.52 ng), C_{22} (1.54 ng), C_{24} (1.44 ng), C_{30} (1.29 ng), C_{32} (1.31 ng), and C_{34} (1.50 ng). Note: 20 m × 0.25 mm i.d. SE-30 glass column, temperature program from 35 to 270°C at 4°C min^{-1} after a 3 min isothermal period, 400:1 split ratio, He carrier gas, 300°C injector temperature, 0.65 µL injection volume using an autosampler with a 1 µL plunger-in-needle syringe.

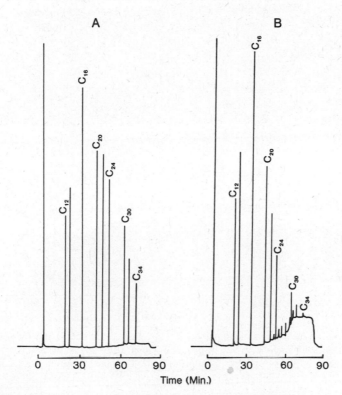

FIGURE 5.15. Chromatograms showing the sample size effect on the recovery of wide volatility range samples for (A) 0.65 µL and (B) 0.1 µL injection volumes. Note: 400:1 split ratio; other conditions were the same as those for Figure 5.14.

The sample can be effectively mixed with the carrier gas by using a splitter vaporization chamber designed to create a local vortex. Loosely packed glasswool inside the vaporization chamber increases both heat capacity and mixing efficiency and improves the linearity and reproducibility of split sampling. Split linearity for an n-alkane mixture of C_{12} to C_{34} in isooctane using two vaporization chamber designs is compared in Figure 5.14. Marked loss of nonvolatile components ($\geq C_{20}$) was observed for the simple baffle-type chamber which has a low heat capacity and mixing effect. Excellent split linearity was obtained, however, for the same mixture when a loosely packed glasswool vaporization chamber was used.

Figure 5.15 shows the effect of injection sample volume on splitter linearity: severe loss of n-alkanes above C_{20} was observed for a 0.1 µL sample volume, and relative response factors for n-alkanes ranging from 0.72 to 15.4 with respect to C_{20} internal standard were measured. For a 0.65 µL injection sample volume, however (Figure 5.15A), split linearity was excellent. For a small sample size, the volatiles lost in the syringe comprise a significant portion of the total volume so that there is severe nonlinearity in sample

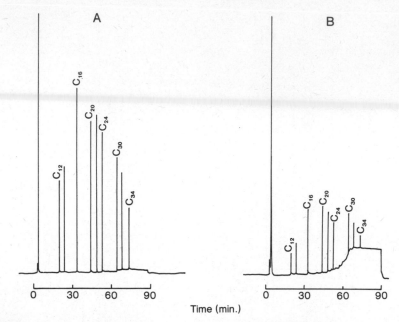

FIGURE 5.16. Chromatograms showing the concentration effect on the recovery of wide volatility range samples for (A) ~20 ng per component (S/N ≅ 1000:1) and (B) ~1.0 ng per component (S/N ≅ 50:1). Analytical conditions were the same as those given for Figure 5.14.

recovery. A small sample volume (≤0.2 μL) should not be used for quantitative analysis of wide volatility range samples with split sampling. This is particularly relevant for plunger-in-barrel type syringes which have needle sample volumes of about 0.8 μL.

The range of sample concentrations appropriate for the split sampling mode was given in Table 5.1. Examples of quantitative errors originating from inappropriate sample concentration are shown in Figure 5.16. Linear sample recovery was observed for all *n*-alkanes (between 0.97 and 1.93 ng each) in a 0.65 μL injection sample volume (Figure 5.16A). Figure 5.16B shows that a 20-fold dilution of the same sample gave significantly different relative response factors and sample recoveries.

Quantitative precision in split sampling depends mainly on the repeatability of the injection processes. Thus, automation allows reproducible injection and provides better than ±1% precision in split sampling of *n*-alkanes with concentrations between 0.5 and 2 ng μL^{-1}.

Splitless Sampling

The splitless sampling technique of Grob and Grob[41–43] is effective in extending the usefulness of a splitter injector for trace analysis because large sample volumes and wide ranges of sample concentration can be employed (see Table 5.1). The advantages of splitless sampling are the following: (a) Analysis of very dilute samples without preconcentration is possible. Sample concentrations as low as 9×10^{-4} ppm can be analyzed with FID detection. (b) Sample components eluted near the solvent can be resolved from the solvent tail and have narrow peak widths due to the solvent reconcentration effect. (c) Large sample volumes can be used with slow injection. The initial sample bandwidth is minimized by the solvent effect. (d) Thermally labile compounds can be analyzed because rapid vaporization of the sample is not required and, consequently, relatively low injection port temperatures can be used to minimize sample degradation. (e) The technique is easily automated.

The disadvantages of splitless sampling are the following: (a) The quantitative results depend upon the injection technique. (b) A column temperature below the boiling point of the solvent is required for the solvent effect and, consequently, solvent selection is important for splitless sampling. (c) Column efficiency can be rapidly lost. (d)The retention time shift[44] for solute components eluted near the solvent tail is a function of solvent boiling point, polarity, and volume injected. Precise reproducibility of sample volume is necessary from run to run if reproducibility of retention time is required. (e) Sample recovery for splitless sampling depends on sampling time, and therefore optimization and automation are required for quantitative splitless sampling. (f) Solutes eluted ahead of the solvent will not be separated efficiently and quantitatively because of peak shape distortion due to displacement chromatography. (g) The applicable solute boiling point range is normally below $n\text{-}C_{28}$.

The practical aspects of the splitless sampling technique have been discussed in detail by Yang et al.[44] Operational considerations such as solvent selection, sample volume, and sampling time are discussed below.

In splitless sampling, a large sample volume and slow injection are required to ensure proper sample vaporization and transfer into the column. It is particularly important to utilize the "solvent effect" to overcome the effects of the slow sample injection and large initial sample bandwidth. The solvent selected affects the peak shape and resolution of the weakly retained peaks. Figure 5.17 illustrates the effect of solvent boiling point on peak shapes and retention times.[44] A good solvent effect is obtained using heptane as solvent (i.e., 10,900 theoretical plates were measured for the C_8 peak and baseline separation from the solvent peak was obtained). With the use of pentane as solvent, as shown in Figure 5.17A, a solvent peak tail was observed, and only 2700 theoretical plates were measured for the C_8 peak. Figure 5.17 also shows that the higher the boiling point of the solvent, the longer are the retention times of those solutes eluted near the solvent. The relationship between the retention time and the boiling point of the solvent in splitless sampling is explored further in Figure 5.18. The retention time of $n\text{-}C_{10}$ correlates well with the boiling point of the solvents on an OV-101 column at 30°C. All of the solvents, with the exception of toluene, showed

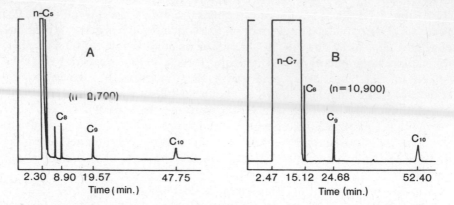

FIGURE 5.17. Chromatograms showing the effect of the solvent type and column temperature in splitless sampling on the retention time, separation efficiency, and resolution. (A) 6.0 μL injection of *n*-alkanes diluted $10^5:1$ in *n*-C_5; (B) 6.0 μL injection of the same *n*-alkanes diluted $10^5:1$ in *n*-C_7 on the same column and temperature at 32°C. (Reproduced with permission from ref. 44. Copyright Elsevier Scientific Publishing Company.)

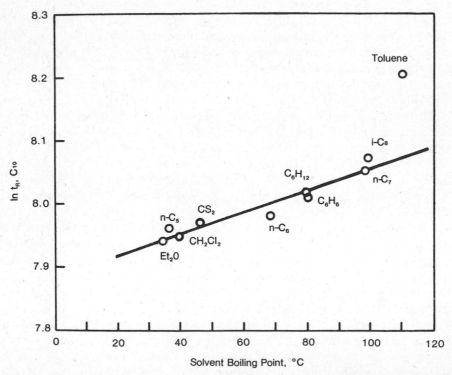

FIGURE 5.18. Relationships between the retention time shift for splitless sampling and the boiling point of the solvent. Note: OV-101 column, 30°C isothermal column temperature, *n*-decane test solute. (Reproduced with permission from ref. 44. Copyright Elsevier Scientific Publishing Company.)

206

an excellent fit to a straight line regardless of their chemical type. Toluene, which has a high boiling point relative to the column temperature (30°C), is effective in inducing very large retention time shifts.

The effects of the sample volume are shown in Figure 5.19. The most obvious of these is the dependence of the retention time and the separation efficiency on the sample volume. A shift in retention time from 9.43 to 18.85 min was measured for the C_8 peak when the injection volume was increased from 1 to 9.5 μL.

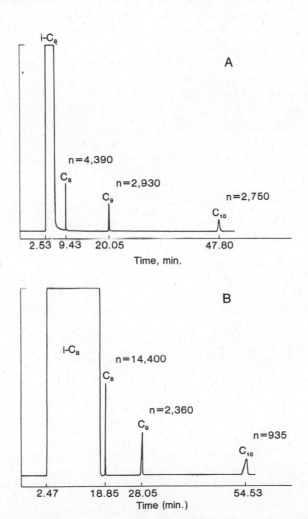

FIGURE 5.19.   Chromatograms showing the effect of sample size in the splitless mode on the retention time, separation efficiency, and resolution for (A) 1.0 μL injection of *n*-alkanes diluted $10^4 : 1$ in isooctane and (B) 9.5 μL injection of the same sample. Note: 25 m × 0.25 mm i.d. OV-101 column, 34°C isothermal column temperature. (Reproduced with permission from ref. 44. Copryright Elsevier Scientific Publishing Company.)

The solvent effect is thus effective for an injection sample volume of 1 μL or larger, and the effectiveness increases with increasing volume. The limitation here may be the sample capacity of the column. However, small sample volumes (e.g., ≤2 μL) have the advantage of minimizing stationary liquid phase stripping and thus increasing the lifetime of the column. A reproducible sample size is required from run to run if retention times are to be reproducible in splitless sampling.

A maximum injection rate of 0.1 μL s^{-1} was found to give good quantitative results. A slow rate of sample introduction is advantageous because the solvent effect sharpens the broad injection profile, and overloading of the injector vaporization chamber is avoided. Figure 5.20 is a graph of the relative peak height as a function of the sampling time for peaks with both large and small k. The relative peak heights are seen to converge for sampling times longer than 20 s; that is, the loss of low-k components is greater when sampling times are less than 20 s.

The delay time in activating the injector purge flow is the other important operational parameter. For short delay times, the injector is purged too soon and thus a large fraction of the total sample is vented from the injector. Sample recovery is low and the least volatile components are preferentially

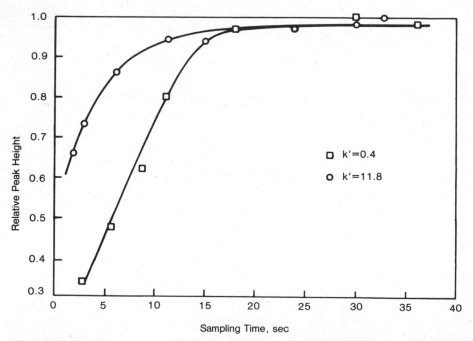

FIGURE 5.20. Effect of splitless sampling time (Δt_1) on the sample recovery and sample discrimination on the basis of k. Note: 25 m × 0.25 mm i.d. OV-101 column; sample contained 2 μL of n-C$_8$ ($k = 0.4$) and n-C$_{10}$ ($k = 11.8$) diluted 10^5:1 in carbon disulfide. (Reproduced with permission from ref. 44. Copyright Elsevier Scientific Publishing Company.)

sampled onto the column. With a long purge delay time, back diffusion of the sample in the injector onto the column begins to occur. The result is that the column efficiency is reduced and the solvent peak profile begins to tail appreciably. A purge delay time corresponding to a purge flow of more than six times the injector volume is recommended for optimum sample recovery.

General guidelines for splitless sampling are the following: (a) Sample volumes between 1 and 2 μL are preferred. (b) A sampling time longer than 20 s and a sampling rate of about 0.1 μL s^{-1} are recommended. (c) A purge delay time of longer than 40 s or a total purge flow of more than six injector volumes is required. (d) A sample dilution higher than 10^5 to 1 is preferred. (e) The injector temperature should be as low as possible. (f) The initial column temperature should be more than 20°C below the boiling point of the solvent. (g) Solutes eluted ahead of the solvent will not be separated efficiently or quantitatively unless there is a large variation in k between the solutes and the solvent.

Direct Sampling

Direct sampling allows the sample injected into the injector to be totally transferred onto the column. The advantages of direct sampling are the following: (a) The range of sample concentration that can be analyzed is more than one decade wider than that for splitless sampling. (b) The solvent effect allows slow direct sampling and analysis of a large volume of very dilute sample without preconcentration. (c) Thermally labile compounds can be analyzed because rapid vaporization of the sample is not required and, consequently, relatively low injector temperatures can be used. (d) No splitting of carrier gas and sample occurs, so that a minimum amount of carrier gas is consumed and very small sample sizes can be analyzed. (e) The method can be easily automated.

The disadvantages of direct sampling are the following: (a) Low bleed septa are necessary for temperature-programming operation if the initial column temperature is lower than 100°C and no septum purge is employed. (b) A heated injector is used and, therefore, samples containing low volatility components ($> C_{34}$) are difficult to analyze quantitatively. (c) Differences between the i.d. of the vaporization chamber and the o.d. of the syringe needle must be less than 0.05 mm so that a positive pressure gradient is obtained.

Direct sampling can be used with both isothermal and temperature-programming operations. In the former, a small sample volume (<0.5 μL) should be used to minimize the loss in column efficiency due to the initial sample bandwidth. For the injection of a large volume (>0.5 μL), temperature programming should be utilized to allow on-column focusing. Figure 5.21 shows a chromatogram resulting from direct sampling of 9 μL of a solution of trace compounds in isooctane. The large sample volume and slow

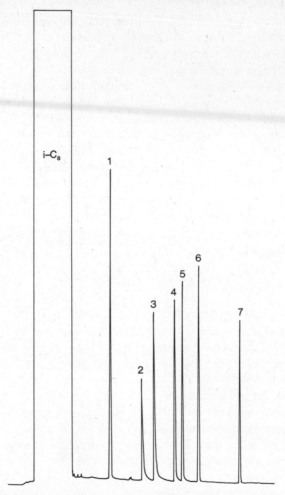

FIGURE 5.21. Chromatogram showing direct sampling with the solvent effect for trace analysis. Note: temperature program from 60 to 150°C at 3°C min⁻¹, 200°C injector temperature, He carrier gas at 2 cm³ min⁻¹ flow rate; FID detector at 512 × 10⁻¹² AFS, 9 μL sample size, 0.7 mm i.d. glass-lined stainless steel vaporization chamber, 10 μL plunger-in-needle syringe with a 0.66 mm o.d. needle. Peak assignments: (1) 2-octanone; (2) 1-octanol; (3) naphthalene; (4) 2,6-dimethylphenol; (5) 2,4-dimethylaniline; (6) dodecane; (7) tridecane.

injection rate of 0.1 μL s⁻¹ using a plunger-in-needle syringe have no effect on peak broadening.

Cold On-Column Sampling

Cold on-column sampling allows the injection of a liquid sample onto the column, and thus produces the best reproducibility and accuracy for wide volatility range samples. The additional advantages of the cold on-column sampling technique are the following: (a) Thermally labile compounds can be analyzed. (b) There is no splitting and waste of carrier gas and sample. (c) No septum is used. (d) Quantitative analysis of relatively nonvolatile samples is possible.

The disadvantages of the cold on-column sampling technique are the following: (a) Sample clean-up is extremely important, particulates and nonvolatiles should be removed before sampling. (b) A column i.d. smaller than 0.2 mm is difficult to use because of constraints in the construction of the syringe and the injector. (c) A liquid solvent may strip the stationary liquid phase coated on the inlet end of the column and result in droplet formation and rapid loss of column efficiency.

The control of the initial sample dispersion at the inlet of the column is very important in on-column sampling. A poor injection sample profile arising from droplet formation due to flooding and poor wettability of the inner wall of the column by the solvent can result in multiple peak splitting and thus poor quantitation. Figure 5.22 illustrates peak splitting from a cold on-column injection. The nonuniform flooding of a 0.25 mm i.d. fused-silica column by injection of a 3 μL sample at 35°C is evident. Peak splitting may be minimized by reducing the sample volume (e.g., to 1 μL or less) and by applying the solute focusing technique.[1,2]

In the on-column solute focusing method, the injector temperature is chosen to be below the solvent boiling point for liquid sample injection, and the column oven temperature is set 10–20°C above the solvent boiling point so that the solvent molecules can be vaporized and separated from the solute molecules as soon as they enter the heated column oven zone, thus preventing solvent flooding of the column. The injection step is then followed by temperature programming (normally at a high rate, i.e., 100°C min^{-1}) to minimize initial sample band spreading in the injection zone. However, when thermally labile compounds are analyzed, the injector temperature should be programmed at a lower rate (e.g., below 10–20°C min^{-1}) to minimize thermal degradation. The column temperature-programming rate is determined by the required degree of resolution and the speed of analysis. Faster analysis times can be achieved because the column oven only needs to be cooled down to the temperature determined by the elution temperature and the required resolution of the solute components and not the solvent boiling point.

Constant flow control pneumatics are required for the on-column solute focusing injection of large sample volumes. In a constant pressure control pneumatics system, the build-up pressure of the vaporized solvent molecules inside the column could force sample backup resulting in sample loss and peak shape distortion. A constant pressure pneumatics system also limits the operable column oven temperature to a very narrow range (e.g., about 10–15°C above the solvent boiling point). A column oven temperature too high above the solvent boiling point results in solvent vaporization and high pressure buildup and flow reversal in the injector.

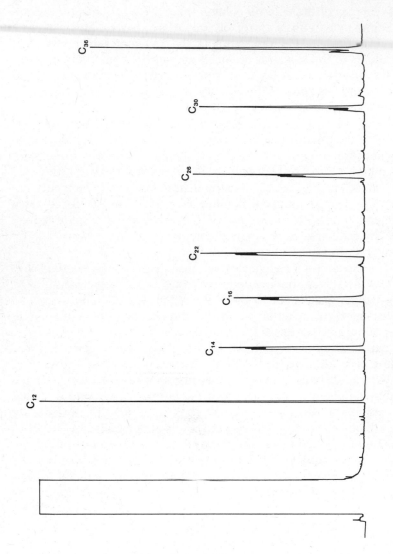

FIGURE 5.22. Chromatogram of *n*-alkanes showing peak splitting from a cold on-column injector. Note: 12 m × 0.25 mm i.d. SE-30 fused-silica column, 3 μL sample size of *n*-alkanes in isooctane, He carrier gas at 2 cm³ min⁻¹, injector temperature program from 70 to 300°C at 10°C min⁻¹, column temperature program from 70 to 300°C at 10°C min⁻¹.

The on-column solute focusing technique with slow injection (e.g., 0.1 $\mu L\ s^{-1}$) allows sample sizes as large as 10 μL to be satisfactorily analyzed.[1,2] Using open tubular columns with immobilized stationary phases, quantitative trace analysis of volatile complex samples is possible. Figure 5.23 shows chromatograms of 3 and 8 μL sample sizes using on-column injections with solute focusing. No peak splitting for high molecular weight n-alkanes was observed.

On-column injection of small sample sizes ($\leq 1\ \mu L$) can be satisfactorily accomplished with or without the application of solute focusing. The column oven temperature can either be above or below the solvent boiling point, depending on the required resolution of the solute components from the solvent peak. The injection speed must be as rapid as possible to minimize liquid sample backup into the space between the column wall and the syringe needle by capillary action and carrier gas flow reversal. The mechanical

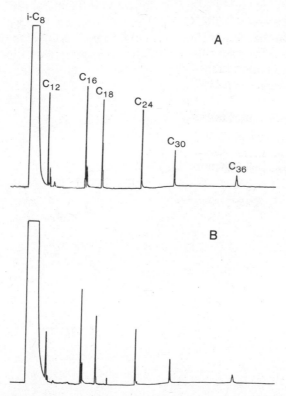

FIGURE 5.23.   Chromatograms showing on-column injections of (A) 3 μL and (B) 8 μL sample volumes using the solute focusing technique. Note: 14 m SE-54 fused-silica column, injection zone and column oven temperatures were at 20 and 110°C, respectively, during injection, injection zone temperature program from 20 to 300°C at 180°C min^{-1}, column temperature program from 110 to 290°C at 15°C min^{-1} after a 2 min isothermal period, He carrier gas at 2 cm^3 min^{-1}. (Reproduced with permission from ref. 1. Copyright Dr. Alfred Huethig Publishers.)

action of the syringe plunger forces the sample to spread away from the tip of the syringe needle and thus minimizes problems of sample loss and peak shape distortion.[45]

The selection of solvent for on-column sampling is not as critical as that required in splitless and solvent effect direct sampling of large sample volumes. A temperature-programmable on-column injector allows low boiling point solvents, such as pentane, to be satisfactorily employed in quantitative analysis. If the on-column injector does not provide temperature control, the selection of the solvent depends on the column temperature.[45] Figure 5.24 shows how there was no loss of dodecane for the solvents, pentane, hexane, and heptane, when the column temperature was 40°C. For increasing column temperatures, a greater loss of dodecane was measured. Above 100°C, more than 75% of the dodecane was lost when pentane was used as solvent.

A leak-tight on-column injection system is essential for the analysis of samples containing highly volatile solutes. Gas backflow during injection of a sample containing volatile C_6 to C_{10} n-alkanes on a leaky on-column injector results in the loss of those volatiles from the injector (Figure 5.25A). Both absolute and relative losses of the n-alkanes were measured. Figure 5.25B is a chromatogram obtained by using a leak-tight on-column injector

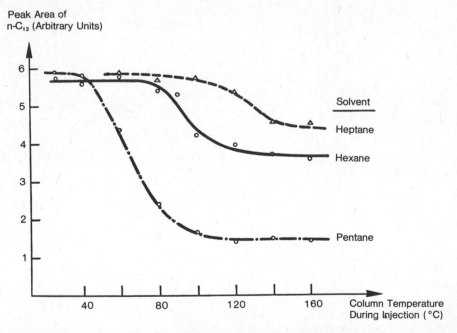

FIGURE 5.24. Sample losses at high injection temperatures. Peak area of n-dodecane in three different solvents. Note: 20 m × 0.30 mm i.d. SE-52 (0.6 μm) column, 1 μL injection volume, 2.5 cm³ min⁻¹ carrier gas flow rate. (Reproduced with permission from ref. 45. Copyright Elsevier Scientific Publishing Company.)

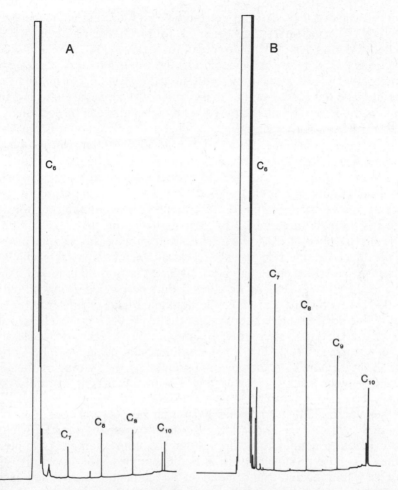

FIGURE 5.25. Chromatograms showing the effect of gas leakage during on-column injection: (A) leaky on-column injector and (B) leak-tight on-column injector. Note: 10 m × 0.3 mm i.d. SE-30 fused-silica column, He carrier gas at 2.4 cm^3 min^{-1} flow rate, column temperature program from 35 to 120°C at 5°C min^{-1}, injector temperature program from 35 to 200°C at 180°C min^{-1}, 1 μL sample injected.

for the same sample. It shows not only an increase in total peak area for all components, but also an increase in peak areas of the more volatile components as compared to that of *n*-decane.

Pyrolysis GC

The major factors affecting reproducibility in pyrolysis GC are pyrolyzer design and selection, pyrolysis temperature control, sample size, and pyrolyzer–GC interfacing.

The selection between the pulse-mode and continuous-mode pyrolyzer strongly affects the pyrogram obtained. In most pulse-mode pyrolyzers, the coated thin-film sample is in direct contact with the hot wire and secondary reactions are therefore minimized. On the other hand, the exact pyrolysis temperature is not measurable, although the temperature profile is reproducible, and the coated sample is difficult to weigh accurately. For the continuous-mode pyrolyzer, the pyrolysis temperature can be accurately measured, and quantitative measurement of the sample amount is possible with a high-precision balance. However, secondary reactions of pyrolysis products are more likely to occur due to contact with the hot zone of the furnace.

The reaction temperature should be high enough to ensure adequate degradation, but not so high that secondary reactions occur before the pyrolysis products have been purged to the column. The optimum reaction temperature is normally determined by repeated runs on the same sample at increasing pyrolysis temperature without removing the sample probe. The temperature heating rate is also important for obtaining reproducible pyrograms. Farré-Rius and Guiochon[46] argued that for the practical conditions in which pyrolysis GC can be carried out, polymer decomposition is controlled by the heating rate of the filament and by heat transfer to the sample, not by the equilibrium temperature of the filament. Rapid and uniform heating of the sample to its pyrolysis temperature is highly desirable. Heating rates as high as $20,000°C \ s^{-1}$ are commercially available. The Curie-point pyrolyzer heating rate, however, depends on the output of the high-frequency power supply and normally requires hundreds of milliseconds to reach the desired pyrolysis temperature.

Sample size affects the final pyrogram because of poor heat transfer. There may be a large temperature gradient across the sample, while a uniform reaction temperature throughout the entire sample is important to achieve good reproducibility. For liquid samples, it is imperative that the sample is thin and uniformly coated on the filament. A sample size as uniform and as small as possible ($<15 \ \mu g$) should be used to ensure rapid heating to a well-defined reaction temperature.

The most important additional requirements in pyrolysis GC are rapid transfer of the pyrolysis products to the GC column with no losses from recondensation or secondary reactions, and avoidance of loss of column efficiency due to a poor sample injection profile. It is particularly desirable to interface directly the uncoated column inlet to pyrolyzer.

5.12   QUALITATIVE ANALYSIS

Retention Measurements

The narrow peaks inherent in open tubular column GC, especially if recorded via a computerized data system, lend themselves well to qualitative analysis from retention data. Correlations between retention times (or temperatures) for an unknown component and a reference compound on two different open tubular columns of different polarity allows identification of the unknown, although extra information from selective or spectroscopic detection may be necessary for positive identification. A number of procedures have therefore been proposed for the comparison of retention data.[47]

Retention relative to a single standard eliminates the effects of variations in carrier gas flow rate and film thickness but is less reliable than a retention index based on retention relative to a homologous series of compounds. The most widely used of these is the Kovats retention index,[48] I, which utilizes the linear relation between log retention time, t'_R (corrected for column dead volume, conventionally measured from an unretained peak—generally that of methane—or by an extrapolation method), and the carbon number of the n-alkane standards. The retention time of compound x [$t'_R(x)$] is interpolated between the values of t'_R for two adjacent n-alkanes with carbon numbers z and $z + 1$ [retention times $t'_R(z)$ and $t'_R(z + 1)$]:

$$I(x) = 100z + 100 \frac{\log t'_R(x) - \log t'_R(z)}{\log t'_R(z + 1) - \log t'_R(z)} \qquad (5.9)$$

A similar expression applies in the case of temperature-programmed GC, except that log t'_R is replaced by the elution temperature, T_R. The temperature-programmed retention index, $I_P(x)$, for a compound eluted between two n-alkanes with carbon numbers z and $z + 1$ is given by[49]

$$I_P(x) = 100z + 100 \frac{T_R(x) - T_R(z)}{T_R(z + 1) - T_R(z)} \qquad (5.10)$$

provided that the program begins at a low enough temperature.

A repeatability of 0.05–0.10 units for values of I measured on open tubular columns has been reported:[50] reproducibility, especially between laboratories is much less precise, however, often of the order of about one unit.[50] This arises mainly from errors in temperature measurement, but also as a result of differences between batches of stationary phase and changes during operation, and between the degrees of column wall deactivation.[51] The latter difficulties are accentuated by polarity differences between sample and reference compounds, which are especially evident in temperature programming. For this reason, a variety of homologous series other than the n-alkanes have been chosen as calibration standards. These include the methyl esters of n-alkanoic acids,[52,53] certain steranes,[54] primary n-alcohols,[55] n-propyl ethers,[54] and 2-alkanones.[56] In general, reproducibility is much improved if the unknown and the reference compounds belong to the same chemical class. For example, a retention index system for polycyclic aromatic compounds based on the standards naphthalene, phenanthrene, chrysene, and picene has been found[51] to be much less dependent on film thickness and column history than indices based on the n-alkanes as standards

TABLE 5.7.   Retention indices from open tubular columns with different
stationary phase film thicknesses.

Compound	Normal retention index (n-alkanes)		PAH index	
	Column A	Column B	Column A	Column B
Naphthalene	1168.74	1166.21	200.00	200.00
Acenaphthylene	1425.03	1413.00	244.65	244.67
Fluorene	1555.87	1549.28	268.14	268.22
Phenanthrene	1744.70	1734.95	300.00	300.00
Anthracene	1754.20	1744.40	301.73	301.76
4H-cyclopenta[def]phenanthrene	1876.18	1864.28	321.95	322.09
Pyrene	2063.99	2048.56	351.13	351.25
Benzo[a]fluorene	2167.27	2153.04	366.64	366.75
Benzo[e]pyrene	2770.65	2751.03	450.66	450.80
Perylene	2812.49	2815.42	456.12	456.23

(Table 5.7). Interlaboratory comparisons also show much better reproducibility.

If the methyl n-alkanoates are chosen as standards,[52,53] retention can be expressed in terms of the equivalent chain length (or chain length) of the methyl alkanoate with the same retention. Other retention index systems closely related to the Kovats procedure include the methylene units of VandenHeuvel.[57] Here the C_{22} and C_{24} n-alkanes (MU = 22 and 24, respectively) are used as reference compounds. The steroid number, however, uses[54] a calibration line drawn between the retention times of the steranes, androstane (SN = 19) and cholestane (SN = 27). Detailed assessment of the contributions of the steroid skeleton and various functional groups have been made and these can be used in identification.[47]

The latter approach has been applied in great depth in the determination of the influence of structure on Kovats retention indices and use of the resulting parameters in qualitative analysis.[47,50] Differences between values of I measured on polar and nonpolar phases, ΔI, are of intrinsic interest, while the homomorph retention increment factor H,

$$H = I(x) - I(n\text{-alkane}) \tag{5.11}$$

where the homomorphous n-alkane has the same number of atoms in its carbon skeleton, also finds application, either directly or as $\Delta(\Delta I)$ which is given by

$$\Delta(\Delta I) = \Delta I(x) - \Delta I(\text{homomorph}) \tag{5.12}$$

The temperature dependence of I has also been proposed[47] as a parameter for qualitative analysis, especially for hydrocarbons.[58] In this way compounds with similar ΔI values may be distinguished.[59]

Selective Detection

In addition to utilizing gas chromatographic retention data for qualitative analysis, selective detectors can provide extremely useful information concerning the structures of the compounds analyzed. The available GC detectors can be divided into two groups: element selective detectors and functional group or structural-feature detectors. These have been discussed in Chapter 4 with applications in Chapters 6 and 7.

Multiple Detection

The use of multiple detectors for the simultaneous detection of column effluent provides increased analytical information. There are significant advantages in the use of a universal detector in combination with one or more selective detector(s). The specificity of the selective detector can be used to simplify the chromatogram and to allow the determination of the chemical nature of the components in a complex mixture. Many applications have been reported concerning the use of the multiple detection technique for qualitative identification of samples. Selective detectors such as FPD,[60-63] TSD,[61,63-75] ECD,[63,66-68] and MS[69] have been combined with a universal detector, FID, for many types of applications in environmental, food and flavor, and biological samples (see Chapter 7).

On-line application of multiple detectors for column effluent detection requires a column effluent splitter for interfacing of the detector(s). A column effluent splitter for high-resolution open tubular column GC should meet many criteria including splitter linearity independent of flow rate, temperature, and molecular weight of the sample; low dead volume band spreading; and good mechanical strength and chemical inertness.

Many types of column effluent splitters[63,65,70,71] have been reported. Among these, T splitters with either fixed or valve adjustable split ratios are often utilized. These generally do not maintain constant split ratios for wide boiling point range samples unless the "Coanda flow switching effect"[72,73] is avoided. This is a flow switching mechanism in which the outgoing stream tends to prefer one splitter channel (and may switch) because of disturbances resulting from minor changes in flow rate, temperature, or the viscosity of the fluid stream.

The application of a combined FPD and FID via a linear column effluent splitter[63] for the analysis of volatile coal hydrogenation fractions is shown in Figure 5.26. The simplicity of the FPD chromatogram (Figure 5.26A) and the excellent detection limits for the sulfur compounds illustrate the advantage of multiple detection.

The calculation of response ratios from two different detectors can be helpful in qualitative analysis. Thus the relative responses of an FID and an ECD allow the differentiation of structural isomers for a large number of polycyclic aromatic compounds.[74] The electron affinities and hence ECD

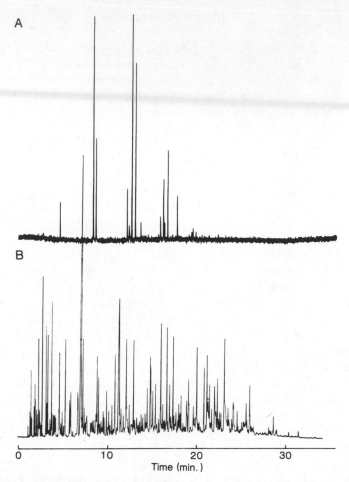

FIGURE 5.26. Chromatograms of a coal hydrogenation fraction splitting the column effluent to a (A) FPD and (B) FID. Note: 25 m OV-101 glass column, temperature program from 34 to 160°C, He carrier gas at 2 cm^3 min^{-1}, effluent split ratio 1.24:1 (FID:FPD). (Reproduced with permission from ref. 63. Copyright Preston Publications, Inc.)

responses of these compounds depend on their structures. Similarly, using the PID in series with an FID allows the classification of hydrocarbon mixture components into aromatic or aliphatic groups.[75]

5.13   QUANTITATIVE ANALYSIS

Open tubular column GC overcomes four major difficulties of conventional packed column GC, namely: surface adsorption and catalytic decomposition; limited column resolving power and peak capacity; peak broadening; and high stationary phase bleed rates. However, because of the low column sample capacity and the complexity of sampling techniques, quantitative analysis using open tubular column GC is not practiced routinely in many laboratories.

Many fundamental problems and practical considerations discussed previously, such as sampling mode selection and technique, column stability, injector reproducibility, column temperature and gas flow stabilities, detector sensitivity, stability and linearity, are all important in open tubular column GC quantitative analysis. In this section, peak area or height measurement and standardization are discussed.

Data Sampling Rate

The data sampling rate or analog-to-digital converter (ADC) clock rate should be faster than 1/10 of the peak width at half height for accurate peak area measurement. For example, a 1 s peak width at half height requires at least 10 data points (i.e., a 10 Hz ADC clock rate) for accurate peak area measurement.

Peak Height Versus Peak Area

Peak height is the simplest measure of peak size. It can give good quantitative results if the following conditions can be met: (a) accurate location of the baseline; (b) constant carrier gas flow rates and peak elution times (a 1% retention time variation can cause a 1% change in peak height); (c) constant column temperature (the peak height can change as much as 3% per 1°C column temperature variation); (d) absence of peak shape distortion due to peak splitting, adsorption, or overloading (adsorption is particularly critical at trace levels); (e) reproducibility of sampling time and splitter in split sampling; (f) effectiveness of the solvent effect and reproducibility in sampling conditions in solvent-effect temperature programming.[44]

When a mass flow-rate-dependent detector is used, peak area measurement is preferred if chromatographic conditions such as flow rate and temperature cannot be maintained constant throughout the analysis.

Several techniques[76-81] of peak area measurement, such as electronic or computer integration, planimeter, cut and weigh, triangulation, and so on, are used. Among these, computer integration, which can accurately handle drifting baseline, fused peak deconvolution, and noise spike filtering and smoothing is preferred. The fundamental principles[82-87] of data acquisition and handling were briefly discussed in Chapter 4.

Peak Size Standardization

The four commonly used techniques of peak size standization are area normalization, internal standard, external standard, and standard addition. The composition of an unknown mixture may be simply determined by calculating the percent area of each component in the chromatogram. However, if detector responses are not equal, a response factor is required for normalization. Percent area normalization may be used if all of the sample components are separated and detected, the detector responds linearly to all sample components (even if one represents 99% of the sample), and there is no loss of sample due to injector discrimination or column adsorption and chemical interaction. The method cannot be used with a nonlinear splitter injector or any of the selective detectors such as FPD, ECD, and TSD.

The advantage of percent area normalization is that the amount of mixture injected into the column need not be known exactly, thus eliminating the major error source in quantitative analysis by GC. It is often used in conjunction with an FID, which is relatively insensitive to flow and temperature change, and has a response to a variety of classes of compounds which is relatively constant.

The response factor, f, with respect to a reference compound can be calculated from

$$f_i = \frac{(area)_{ref}\,(amount)_i}{(area)_i\,(amount)_{ref}} \tag{5.13}$$

Table 5.8 shows how the response factors for five components (A, B, C, D, and E) in a simple mixture relative to A are calculated.

By use of the response factor, the normalized area percent for each component in the sample matrix can be calculated from Equation (5.14):

$$\text{Normalized area \% of } i\text{th component} = \frac{(area \times f)_i \times 100\%}{\sum\limits_{n=1}^{n} (area \times f)_n} \tag{5.14}$$

The solvent area may be excluded from the sum of the peak areas if there is complete resolution of solvent and sample peaks.

The internal standard method does not require complete resolution, elution, and detection of all sample components, only that there is complete resolution of the peak(s) of interest and the internal standard, which must meet a number of requirements: it must not be originally present in the

TABLE 5.8. Response factors, area percent, and normalized area percent for B, C, D, and E with respect to A.

Components	Mass (ng)	Peak area (μV s)	Area %	Response factors	Normalized area %
A	33	13,200	19.82	1.00	13.14
B	28	9,700	14.56	1.15	11.10
C	45	12,500	18.77	1.44	17.92
D	60	17,200	25.83	1.40	24.97
E	85	14,000	21.02	2.43	33.87

unknown sample; it must be completely resolved from all sample components; it should be eluted as close as possible to the components of interest; it must have similar concentration, physico-chemical properties, and detector response to these component(s) of interest; it must be stable, nonreactive, and highly pure; and it must be sufficiently nonvolatile to allow long storage time and minimum loss in sampling processes.

The response factors for the components of interest relative to the known internal standard are calculated according to Equation (5.13) using a standard mixture.

It is best first to construct a calibration curve using standard solutions containing the component(s) of interest at several concentration levels, but all containing the same concentration of internal standard: a plot of mass ratio vs. area ratio is then made (Figure 5.27). This plot should be linear, and the response factor is calculated from the slope.

The concentration of internal standard is calculated so as to yield about the same peak area as the analytes, so that errors in measuring peak areas cancel. The sample containing the internal standard may then be carried through the sample pretreatment step(s) and is then chromatographed. The mass and mass percent of the component(s) of interest can then be calculated from Equations (5.15) and (5.16) and the area ratio of analyte to internal standard:

$$(Mass)_i = \frac{(area)_i \, (mass)_{int\ std} \times f}{(area)_{int\ std}} \tag{5.15}$$

$$(Mass\ \%)_i = \frac{(area)_i \, (mass)_{int\ std} \times f \times 100\%}{(area)_{int\ std} \, (mass)_{unknown}} \tag{5.16}$$

An example of the internal standardization method is shown in Table 5.9.

The technique is particularly useful in quantitative analysis because it has all of the advantages of percent area normalization. It minimizes quantitative errors due to sample preparation and injection, allows the quantitation of one or more than one component in the sample matrix, and requires that

TABLE 5.9. Internal standardization calculations.

Components	Peak area (μV s)	Response factor	Mass (ng)	Mass %
A	13,200	1.10	53.78	13.32
B	9,700	1.30	46.70	11.56
I[a]	13,500	1.00	50	12.38
C	12,500	1.40	64.8	16.05
D	17,200	1.25	79.63	19.72
E	14,000	2.10	108.89	26.97

[a] Assumed to be the internal standard of known mass, 50 ng. The total sample mass is 403.81 ng.

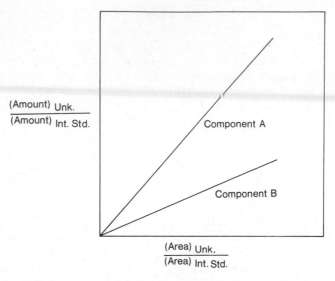

$$\frac{\text{(Amount) Unk.}}{\text{(Amount) Int. Std.}}$$

Component A

Component B

$$\frac{\text{(Area) Unk.}}{\text{(Area) Int. Std.}}$$

FIGURE 5.27. Calibration curve for the internal standard technique. The slopes of the curves are the response factors for components A and B.

chromatographic resolution only be optimized for the separation of the component(s) of interest and the internal standard.

In external standardization, peak areas or peak heights of unknowns are compared with those of a series of known (or external) standards. The compounds of interest present in the unknowns are also present in the external standards, but in known quantities. Standard solutions at the same concentration levels as the component(s) of interest in the unknown samples are prepared and chromatographed, and calibration curves (e.g., Figure 5.28) are plotted for all components of interest. The same volume of unknown sample is then chromatographed. The areas of the peaks from unknowns are then interpolated on the standard calibration curves. Extrapolation should not be used. The external standardization technique requires that the external standards have the same concentration levels as components of interest; that chromatographic conditions for both unknowns and standards are identical; that the concentration ranges of the standards cover the entire range expected in the unknown; and that calibration curves are obtained by using the same volume of standards.

External standardization is normally applied when reproducible sample sizes can be injected onto the column. In open tubular column GC, it is difficult to obtain a high degree of precision by this method because of the small sample sizes and injector discrimination, but these difficulties may be partly overcome by on-column injection.

The standard addition method involves the use of both external and internal standardization. The unknown sample is first chromatographed.

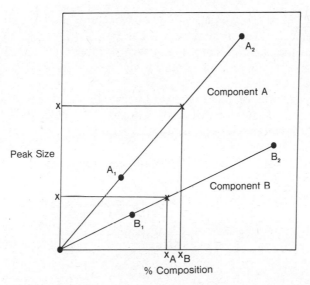

FIGURE 5.28.   Calibration curve for the external standard technique. A_1 and A_2 are calibration standard compositions for components A at two levels. B_1 and B_2 are calibration standard compositions for component B at two levels. X_A and X_B are the calculated % compositions of A and B, respectively.

Known amounts of the pure components of interest are then added to a known amount of the unknown. The solution is then chromatographed under the same conditions.

This procedure may use one of the components in the original sample as the reference peak or a known amount of pure compound may be added as

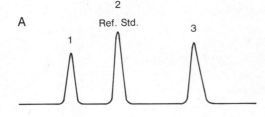

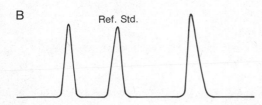

FIGURE 5.29. Chromatograms illustrating the standard addition technique. Chromatogram A is the unknown sample where peaks 1 and 3 are of interest. Chromatogram B is obtained after adding known amounts of components 1 and 3 into the sample.

a reference. The method is advantageous because matrix effects are cancelled. Either peak area or peak height can be used for the calculation. The use of one of the sample components as a reference is particularly advantageous for a complex mixture where a high-purity internal standard is difficult to obtain.

The standard addition technique is illustrated by Figure 5.29 where peaks 1 and 3 are components of interest and peak 2 is selected as a reference. Addition of known amounts of compounds 1 and 3 to the unknown sample yields the chromatogram shown in Figure 5.29B. By using peak 2 as the reference, the amount of peaks 1 and 3 in the unknown sample can be calculated from the following formulas:

$$\text{Amount of 1 in unknown} = \frac{(\text{mass 1 added}) \left(\dfrac{\text{peak size of 1}}{\text{peak size of 2}}\right)_A}{\left(\dfrac{\text{peak size of 1}}{\text{peak size of 2}}\right)_B - \left(\dfrac{\text{peak size of 1}}{\text{peak size of 2}}\right)_A} \quad (5.17)$$

$$\text{Amount of 3 in unknown} = \frac{(\text{mass 3 added}) \left(\dfrac{\text{peak size of 3}}{\text{peak size of 2}}\right)_A}{\left(\dfrac{\text{peak size of 3}}{\text{peak size of 2}}\right)_B - \left(\dfrac{\text{peak size of 3}}{\text{peak size of 2}}\right)_A} \quad (5.18)$$

REFERENCES

1. F. J. Yang, *J. High Resoln. Chromatogr./Chromatogr. Commun.* **6**, 448 (1983).
2. J. V. Hinshaw, Jr. and F. J. Yang, *J. High Resoln. Chromatogr./Chromatogr. Commun.* **6** (1983).
3. E. D. Smith, K. E. Sorrells, and R. G. Swinea, *J. Chromatogr. Sci.* **12**, 101 (1974).
4. K. D. Bartle, B. W. Wright, and M. L. Lee, *Chromatographia* **14**, 387 (1981).
5. Th. Welsch, W. Engewald, and Ch. Klaucke, *Chromatographia* **10**, 22 (1977).
6. K. Grob, G. Grob, and K. Grob, Jr., *J. High Resoln. Chromatogr./Chromatogr. Commun.* **2**, 31 (1979).
7. B. W. Wright, M. L. Lee, S. W. Graham, L. V. Phillips, and D. M. Hercules, *J. Chromatogr.* **199**, 355 (1980).
8. T. J. Stark, R. D. Dandeneau, and L. Mering, *1980 Pittsburgh Conference*, Atlantic City, New Jersey, March 10, 1980, abstract 002.
9. L. Blomberg, K. Markides, and T. Wännman, *J. High Resoln. Chromatogr./Chromatogr. Commun.* **3**, 527 (1980).
10. D. A. Cronin, *J. Chromatogr.* **97**, 263 (1974).
11. G. Schomburg, H. Husmann, and H. Borwitzky, *Chromatographia* **12**, 651 (1979).
12. R. D. Dandeneau and E. H. Zerenner, *J. High Resoln. Chromatogr./Chromatogr. Commun.* **2**, 351 (1979).
13. J. Buijten, L. Blomberg, K. Markides, and T. Wännman, *J. Chromatogr.* **237**, 465 (1982).

14. P. A. Peaden, B. W. Wright, and M. L. Lee, *Chromatographia* **15**, 335 (1982).

15. K. Markides, L. Blomberg, J. Buijten, and T. Wännman, *J. Chromatogr.*, **254**, 53 (1983).

16. K. Markides, L. Blomberg, J. Buijten, and T. Wännman, *J. Chromatogr.*, in press.

17. B. E. Richter, J. C. Kuei, L. W. Castle, B. A. Jones, J. S. Bradshaw, and M. L. Lee, *Chromatographia*, in press.

18. L. Rohrschneider, *J. Chromatogr.* **22**, 6 (1966).

19. W. O. McReynolds, *J. Chromatogr. Sci.* **8**, 685 (1970).

20. T. J. Stark, P. A. Larson, and R. D. Dandeneau, *J. Chromatogr.*, in press.

21. D. H. McCloskey and S. J. Hawkes, *J. Chromatogr. Sci.* **13**, 1 (1975).

22. P. H. Weiner and J. F. Parcher, *J. Chromatogr. Sci.* **10**, 612 (1972).

23. S. D. West and R. C. Hall, *J. Chromatogr. Sci.* **14**, 339 (1976).

24. R. J. Laub and J. H. Purnell, *Anal. Chem.* **48**, 799 (1976).

25. R. J. Laub and J. H. Purnell, *Anal. Chem.* **48**, 1720 (1976).

26. R. C. Kong, M. L. Lee, Y. Tominaga, R. Pratap, M. Iwao, R. N. Castle, and S. A. Wise, *J. Chromatogr. Sci.* **20**, 502 (1982).

27. W. G. Jennings, *Gas Chromatography with Glass Capillary Columns*, 2nd. ed. Academic Press, New York, 1980, p. 112.

28. B. W. Wright and M. L. Lee, *J. High Resoln. Chromatogr./Chromatogr. Commun.* **3**, 352 (1980).

29. R. R. Freeman, *High Resolution Gas Chromatography*. Hewlett-Packard, Avondale, Pennsylvania, 1981, p. 7.

30. K. Grob and G. Grob, *J. High Resoln. Chromatogr./Chromatogr. Commun.* **2**, 109 (1979).

31. T. J. Stark and P. A. Larson, *J. Chromatogr. Sci.* **20**, 341 (1982).

32. S. R. Lipsky, R. A. Landowne, and J. E. Lovelock, *Anal. Chem.* **31**, 852 (1959).

33. A. Zlatkis, D. Fenimore, L. S. Ettre, and J. E. Purcell, *J. Gas Chromatogr.* **3**, 75 (1965).

34. C. Costa Neto, J. T. Koffer, and J. W. deAlencar, *J. Chromatogr.* **15**, 301 (1964).

35. L. Mázor and J. Takács, *J. Gas Chromatogr.* **4**, 322 (1966).

36. R. P. W. Scott, in *Gas Chromatography 1964*, A. Goldup, editor. Institute of Petroleum, London, 1965, p. 25.

37. W. G. Jennings and S. Adam, *Anal. Biochem.* **69**, 61 (1975).

38. S. Nygren and P. E. Mattsson, *J. Chromatogr.* **123**, 101 (1976).

39. S. Nygren, *J. Chromatogr.* **142**, 109 (1977).

40. S. Nygren, *J. High Resoln. Chromatogr./Chromatogr. Commun.* **2**, 319 (1979).

41. K. Grob and G. Grob, *J. Chromatogr. Sci.* **7**, 584 (1969).

42. K. Grob and G. Grob, *J. Chromatogr. Sci.* **7**, 587 (1969).

43. K. Grob and K. Grob, Jr., *J. Chromatogr.* **94**, 53 (1974).

44. F. J. Yang, A. C. Brown III, and S. P. Cram, *J. Chromatogr.* **158**, 91 (1978).

45. K. Grob, Jr. and H. P. Neukom, *J. Chromatogr.* **189**, 109 (1980).

46. F. Farré-Rius and G. Guiochon, *Anal. Chem.* **40**, 998 (1968).

47. J. K. Haken, *Adv. Chromatogr.* **14**, 367 (1976).

48. E. Kovats, *Helv. Chim. Acta,* **41**, 1915 (1958).

49. H. Vanden Dool and P. D. Kratz, *J. Chromatogr.* **11**, 463 (1963).

50. G. Schomburg and G. Dielmann, *J. Chromatogr. Sci.* **11**, 151 (1973).

51. M. L. Lee, D. L. Vassilaros, C. M. White, and M. Novotny, *Anal. Chem.* **51**, 768 (1979).

52. F. P. Woodford and C. M. van Gent, *J. Lipid Res.* **1**, 188 (1960).

53. T. K. Miwa, K. L. Mikoljczak, F. R. Earle, and I. A. Wolff, *Anal. Chem.* **32**, 1739 (1960).

54. W. J. A. VandenHeuvel and E. C. Horning, *Biochim. Biophys. Acta* **64**, 416 (1962).

55. S. J. Hawkes, *J. Chromatogr. Sci.* **10**, 535 (1972).

56. R. G. Ackman, *J. Chromatogr. Sci.* **10**, 536 (1972).

57. W. J. A. VandenHeuvel, W. L. Gardiner, and E. C. Horning, *Anal. Chem.* **36**, 1550 (1964).

58. J. A. Rijks and C. A. Cramers, *Chromatographia* **7**, 99 (1974).

59. C. A. Cramers, J. A. Rijks, V. Pacakova, and I. Ribeiro le Andrade, *J. Chromatogr.* **51**, 13 (1970).

60. W. Bertsch, F. Shunbo, R. C. Chang, and A. Zlatkis, *Chromatographia* **7**, 128 (1974).

61. M. Hrivnac, W. Frischknecht, and L. Cechova, *Anal. Chem.* **48**, 937 (1976).

62. G. Goretti and M. Possanzini, *J. Chromatogr.* **77**, 317 (1973).

63. F. J. Yang, *J. Chromatogr. Sci.* **19**, 523 (1981).

64. M. J. Hartigan, J. E. Purcell, M. Novotny, M. L. McConnell, and M. L. Lee, *J. Chromatogr.* **99**, 339 (1974).

65. D. W. Later, B. W. Wright, and M. L. Lee, *J. High Resoln. Chromatogr./Chromatogr. Commun.* **4**, 406 (1981).

66. K. Grob, *Chromatographia* **8**, 423 (1975).

67. W. Giger, M. Reinhard, C. Schaffner, and F. Zürcher, in *Identification and Analysis of Organic Pollutants*, L. H. Keith, editor. Ann Arbor Science Publishers, Ann Arbor, Michigan, 1976, p. 433.

68. B. Brechbühler, L. Gay, and H. Jäger, *Chromatographia* **10**, 478 (1977).

69. N. Neuner-Jehle, F. Etzweiler, and G. Zarske, *Chromatographia* **7**, 323 (1974).

70. E. L. Anderson and W. Bertsch, *J. High Resoln. Chromatogr./Chromatogr. Commun.* **1**, 13 (1978).

71. F. Etzweiler and N. Neuner-Jehle, *Chromatographia* **6**, 503 (1973).

72. C. J. Miller, *Electron World* **77**, 23 (1967).

73. S. P. Chavez and C. G. Richards, *Fluid* **Q2**, 40, (1969).

74. A. Bjorseth and G. Eklund, *J. High Resoln. Chromatogr./Chromatogr. Commun.* **2**, 22 (1979).

75. S. Kapila and C. R. Vogt, *J. High Resoln. Chromatogr./Chromatogr. Commun.* **4**, 233 (1981).

76. D. L. Ball, W. E. Harris, and H. W. Habgood, *J. Gas Chromatogr.* **5**, 613 (1967).

77. D. L. Ball, W. E. Harris, and H. W. Habgood, *Anal. Chem.* **40**, 129 (1968).

78. H. M. McNair and E. J. Bonelli, *Basic Gas Chromatography*. Varian Aerograph, Walnut Creek, California, 1968, p. 156.

79. F. Bauman and F. Tao, *J. Gas Chromatogr.* **5**, 621 (1967).

80. J. D. Hettinger, J. R. Hubbard, J. M. Gill, and L. A. Miller, *J. Chromatogr. Sci.* **9**, 710 (1971).

81. F. J. Debbrecht, in *Modern Practice of Gas Chromatography*, R. L. Grob, editor. John Wiley & Sons, New York, 1977, p. 166.

82. S. N. Chesler and S. P. Cram, *Anal. Chem.* **43**, 1922 (1971).

83. P. Bocek, J. Novak, and J. Janak, *J. Chromatogr.* **42**, 1 (1969).

84. F. Baumann, E. Herlicska, and A. C. Brown, *J. Chromatogr. Sci.* **7**, 680 (1969).

85. D. Derge, *Chromatographia* **5**, 415 (1972).

86. D. W. Grant and A. Clarke, *J. Chromatogr.* **92**, 257 (1974).

87. W. Kipiniak, *J. Chromatogr. Sci.* **19**, 332 (1981).

SIX

APPLICATIONS: CHEMICAL CLASS

6.1 INTRODUCTION

Much of the development of gas chromatographic instrumentation, column technology, and methodology is a direct result of the properties of the chemical classes which are to be analyzed and the information about these chemicals which is hoped to be obtained from the analysis. Some of these properties include, polarity, acidity, basicity, chirality, thermal stability, volatility, reactivity, and complexity. The objective of this chapter is to outline the important aspects of open tubular column GC of the major classes of compounds analyzed using this technique. This chapter is not intended to be inclusive of all chemical classes analyzed or all published material pertaining to each subject. Instead, problems and approaches to their solutions are outlined with references to previously published work. One major class of chemicals, the steroids, is not treated in this chapter, but is included in Chapter 7. The examples described in this chapter provide an overview of the types of analytical problems that can be solved using this powerful technique.

6.2 HYDROCARBONS

The hydrocarbons are relatively nonpolar and thermally stable when compared to most other organic compounds. This leads to their excellent chromatographic properties of minimal adsorption and peak tailing even on columns with less than complete column wall deactivation. For this reason and the fact that the hydrocarbons are one of the most important classes of organic chemicals, most of the early applications and much of today's use of open tubular chromatographic columns are centered around hydrocarbon analysis.

Since adsorption is generally not a problem, the main problems encountered during the chromatography of hydrocarbons are the extreme complexities and wide molecular weight ranges of various hydrocarbon mixtures. This places high demands on the separation efficiencies and thermal stabilities of available chromatographic columns, and on the performance of sample introduction devices, detectors, and electrometers. Open tubular column GC has certainly emerged as the most effective tool for the analytical resolution of these complex mixtures.

In this section, the application of open tubular column GC to the analysis of hydrocarbons will be discussed. The section is divided into discussions of the various classes of hydrocarbons: alkanes (acyclic hydrocarbons having no double or triple bonds), alkenes and alkynes (unsaturated acyclic hydrocarbons), cycloalkanes and cycloalkenes, aromatic hydrocarbons (hydrocarbons containing a benzene ring), and polycyclic aromatic hydrocarbons (hydrocarbons containing two or more fused benzene rings).

Alkanes

In GC, compound retention is determined by the nature of the intermolecular interactions between the compounds analyzed and the stationary phase. Since the polarity of the stationary phase has only a minor effect on the retention of alkanes, the order of retention corresponds to that of their boiling points. Matukuma[1] discussed the relationship between the boiling points of alkanes and their retention values and proposed the boiling point index, I_B, which can be calculated from the equation

$$I_B = 10^{(0.00134052\ T_b + 2.558916)} - 440.5 \qquad (6.1)$$

where T_b is the boiling point in K at 760 mmHg. Sultanov and Arustamova[2] calculated the boiling points of a large number of C_9 and C_{10} isoalkanes from I_B values obtained on a 150 m $\times$ 0.25 mm i.d. open tubular column coated with BM-4 vacuum oil. The relationship between I_B and the Kovats retention index, I, has been described.[1]

Over the years, much effort has been made to standardize retention parameters for the precise identification and characterization of peaks in complex chromatograms. As discussed in Section 5.7, the Kovats retention index system has been universally accepted for the standardization of retention data.

The Kovats retention indices of a large number of hydrocarbons have been measured on polar and nonpolar capillary columns under precisely controlled conditions.[3–5] Excellent precision (standard deviation of 0.03

TABLE 6.1. Retention indices (I) and ΔI Values of
C_6-hydrocarbons on squalane and polypropylene
glycol at 100°C.[a]

Compound	I_{100}^{sq} [b]	I_{100}^{ppg} [c]	$\Delta I_{100}^{sq\text{-}ppg}$
n-Hexane	600.00[d]	600.00[d]	—
2-Methylpentane	571.3	568.4	2.9
1-Hexene	586.9	610.5	23.6
Cyclohexane	675.5	699.2	23.5
Benzene	660.7	767.9	107.2

[a] Data taken from ref. 6.

[b] Index measured on a 100 m × 0.25 mm i.d. open tubular column
coated with squalane.

[c] Index measured on a 100 m × 0.25 mm i.d. open tubular column
coated with polypropylene glycol.

[d] By definition.

index units) was obtained for all types of hydrocarbons. Accurate control
of temperature, carrier flow, and retention measurement must be maintained
for this kind of reproducibility. Column-to-column reproducibility of I values
depends on the reproducibility of the column wall activity and the polarity
of the stationary phase. These parameters have been discussed in detail by
Schomburg and Dielmann.[6] Generally speaking, with the proper skill and
equipment, a repeatability of 0.05 index units of hydrocarbons on nonpolar
stationary phases, and 0.1 index units on polar stationary phases, should be
obtained.

The use of the difference in retention indices on polar and nonpolar
phases, ΔI, was also introduced by Kovats[7] as an aid to recognizing different
types of compounds. Values of I and ΔI resulting from retention measurements on squalane and polypropylene glycol for C_6-hydrocarbons of various
types are given in Table 6.1.[6] The alkanes, which are nonpolar and also not
significantly polarizable, show similar retention on both phases and, hence,
minimal ΔI values. Hydrocarbons containing various degrees of unsaturation
are polarizable and interact with polar stationary phases to yield larger ΔI
values in order of their degree of polarization.[8]

It is also well known that the temperature variation of a retention index
is a sensitive function of the chemical nature of the compound. The temperature coefficient of the retention index, dI/dT, has been used as an identification parameter for characterizing different types of hydrocarbons.[8] Values of dI/dT for a number of C_6-hydrocarbons on nonpolar (squalane) and
polar (acetyltributylcitrate) open tubular columns are given in Table 6.2.[5]
Using dI/dT values, branched alkanes can be subclassified into mono-, di-,
tri-, and tetra-substituted isomers;[9] naphthenes into monocyclic, bicyclic,
tricyclic, and polycyclic; and aromatics into phenyl and biphenyl.[8,10] It has

TABLE 6.2.   Retention indices (I) of C_6-hydrocarbons on squalane and acetyltributylcitrate at 50 and 70°C and their temperature coefficients (dI/dT).[a]

Compound	$I_{50}^{sq\,b}$	$I_{70}^{sq\,b}$	$\dfrac{dI^{sq}}{dT_{50\text{-}70}}$	$I_{50}^{atc\,c}$	$I_{70}^{atc\,c}$	$\dfrac{dI^{atc}}{dT_{50\text{-}70}}$
n-Hexane	600.00[d]	600.00[d]	—	600.00[d]	600.00[d]	—
2-Methylpentane	569.7	570.0	0.017	569.6	570.0	0.020
2,2-Dimethylbutane	536.8	538.5	0.082	536.9	538.6	0.084
1-Hexene	582.3	582.7	0.024	616.9	617.1	0.009
Cyclohexane	662.7	667.2	0.222	677.0	681.6	0.234
Benzene	637.2	641.8	0.232	779.6	779.4	0.190

[a] Data taken from ref. 5.

[b] Index measured on a 100 m × 0.25 mm i.d. open tubular column coated with squalane.

[c] Index measured on a 50–75 m × 0.25 mm i.d. open tubular column coated with acetyltributylcitrate.

[d] By definition.

been shown that the number, nature, and arrangement of rings for naphthenes and aromatics exert a stronger influence on dI/dT values than do alkyl substituents on the rings. There are, however, considerable limitations to the use of dI/dT values for classifying hydrocarbons into types. For example, highly branched alkanes and cyclic alkenes, cyclic alkanes and aromatics, and alkenes and monoalkyl alkanes could not be distinguished by their dI/dT values.[8]

As discussed previously, there are limitations in the use of either ΔI or dI/dT values individually for the classification of hydrocarbons into types. When both these extensions to the retention index system (dependence on temperature and polarity) are considered together, much more definitive information is obtained as follows:[8]

1. Linear and branched alkanes can be distinguished from other types of hydrocarbons by using ΔI values, but further classification into mono-, di-, tri-, and tetra-substituted alkanes must be done using dI/dT studies.

2. Cyclic alkanes, cyclic alkenes, aromatics, and highly branched alkanes can be classified into their respective groups by using the ΔI system.

3. dI/dT studies are very useful in distinguishing between cyclic alkanes and monoalkenes, and cyclic alkenes and bicyclic compounds.

4. Monoalkenes, dialkenes, and acetylenes can be characterized by ΔI values.

5. The dI/dT system is useful in differentiating highly branched and non-branched alkenes.

The general approach to hydrocarbon classification using the extension of the retention index system was outlined by Mitra et al.[8] as follows:

1. When I values of peaks on phases of different polarity are approximately the same, they represent linear and branched alkanes.
2. If condition 1 is satisfied, the study of dI/dT values on a nonpolar phase reflects the distribution between nonbranched and highly branched alkanes.
3. When I values of peaks on phases of opposite polarity are not the same, the peaks represent compounds other than linear and branched alkanes, such as alkenes, cyclic alkanes, and aromatics.
4. If condition 3 is observed, dI/dT values on a nonpolar phase are applicable in distinguishing alkenes, cyclic alkanes, and aromatics.

In addition to efforts in the past directed toward the standardization of retention data for hydrocarbons, much work has been done to correlate retention with molecular structure, with the hope of being able to predict the retention of compounds for which no standards or data are available. Although exact correlations have not been discovered to date, some useful predictions can be made. Schomburg and Dielmann[6] have discussed the prediction of retention for unknown compounds using calculated homomorph retention increments, H and $\Delta(\Delta I)$, defined by the equations

$$H = I(\text{compound}) - I(\text{homomorphous compound}) \qquad (6.2)$$
$$\Delta(\Delta I) = \Delta I(\text{compound}) - \Delta I(\text{homomorphous compound}) \qquad (6.3)$$

where I is the retention index, and the homomorph for the hydrocarbon of interest is the n-alkane of the same carbon number or a compound with the same carbon skeleton. Table 6.3 illustrates the calculation of retention indices from retention data obtained on a squalane column. In one case H^{sq} values are calculated from I^{sq} data for 2-methylundecane and 3-methylun-

TABLE 6.3. Calculation of retention indices (I) from homomorph retention increments (H).[a]

Compound	I^{sqb}	H^{sq}	$I^{sq}_{(calc.)}$
2-Methylundecane	1164.0	−36.0	—
3-Methylundecane	1169.6	−30.4	—
2,3-Dimethylundecane	1251.4	—	1233.6
2,9-Dimethylundecane	1232.6	—	1233.6
2-Octene	802.5	+2.5	—
4-Methylheptane	767.0	−33.0	—
4-Methyl-2-heptene	746.2	—	769.5

[a] Data taken from ref. 6.

[b] Index measured on squalane.

TABLE 6.4. Calculation of ΔI values from homomorph retention increments $(H).^{a}$

Compound	$I_{80}^{sq\,b}$	H^{sq}	$I_{80}^{ppg\,c}$	H^{ppg}	ΔI	$\Delta I_{(calc.)}$
n-Hexylcyclopropane	913.0	+13.0	943.8	+43.8	—	—
1-Octene	702.7	+17.3	810.3	+10.3	—	—
1,7-Octadiene	765.0	—	821.0	—	56.0	55.2
5-Hexenylcyclopropane	895.3	—	954.7	—	59.4	58.4

[a] Data taken from ref. 6.

[b] Index measured on a 100 m × 0.25 mm i.d. open tubular column coated with squalane.

[c] Index measured on a 50 m × 0.25 mm i.d. open tubular column coated with polypropylene glycol.

decane and the defined retention for the homomorphous compound, dodecane ($I = 1200$). The sum of these two H^{sq} values is the predicted retention increment for both 2,3-dimethylundecane and 2,9-dimethylundecane. Note that the calculated retention index is extremely close for the 2,9-isomer and 17.8 units off for the 2,3-isomer. In the second example, the increments resulting from a double bond and a methyl substituent are summed to predict the retention for 4-methyl-2-heptene.

Table 6.4 illustrates the calculation of ΔI values from retention data and homomorph retention increments on two columns of different polarity. The calculated ΔI values are within 1 index unit for the two examples shown.

Although the prediction of retention indices on the basis of the homomorph principle is not often exact, the qualitative prediction of elution orders of compounds of related structure can be extremely valuable. This is especially true if standards of the specific compounds of interest are not available.

The selection of an appropriate stationary phase for open tubular column GC of alkanes is determined predominantly by the volatility range of the hydrocarbon sample. Squalane (2,6,10,15,19,23-hexamethyltetracosane) has been extensively used to analyze hydrocarbons with boiling points up to 150°C. It was found that squalane was oxidized to some extent when stored for long periods of time, and, therefore, other nonpolar phases with higher working temperature ranges were desirable.[8]

The most widely used stationary phases for hydrocarbon analysis today are the methylsilicone liquids (e.g., SF-96, OV-101, and SP-2100) and gums (e.g., SE-30 and OV-1). The gum phases coat more efficiently on open tubular columns and are stable to temperatures around 300°C. Figure 6.1 shows a gas chromatogram of the aliphatic hydrocarbon fraction of a coal liquid on a fused-silica open tubular column coated with SE-52. Alkanes and alkenes from C_7 to C_{31} were eluted during temperature programming to the upper limit of 280°C.

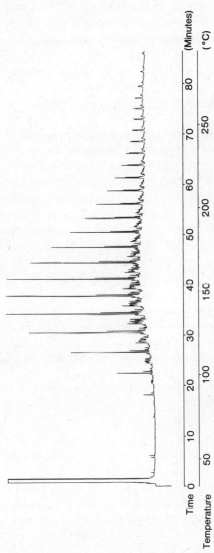

FIGURE 6.1. Chromatogram of the aliphatic hydrocarbon fraction of a coal liquid. Note: 25 m SE-52 fused-silica column, H₂ carrier gas.

235

Johansen and Ettre[11] have measured the retention indices of 43 hydrocarbons on OV-101 and compared these values to those obtained on squalane. They found that squalane index values could be correlated to analyses carried out on methylsilicone columns, except for compounds having a high temperature dependence.

There are numerous reports in the literature describing the application of open tubular column GC to the analysis of alkanes. Recent studies include the analysis of petroleum products,[8,12–15] oil spills,[16,17] air pollution,[17,18] mussels,[19] shale oil,[20–22] coal liquids,[23] tobacco,[24] and tobacco smoke condensate.[25] The analysis of alkanes is generally straightforward for compounds between C_6 and C_{30}. The major problem is the differentiation between the alkanes and numerous other hydrocarbon types that can be present in the samples. In addition to retention index studies as described earlier, other approaches to this problem have been reported. DiSanzo et al.[20,21] described the use of a precolumn vapor phase reactor containing sulfuric acid on diatomite to subtract alkenes, aromatics, and heterocycles from complex shale oil samples. Furthermore, on-line molecular sieve precolumns were used to remove n-alkanes, leaving branched chain and cyclic alkanes for comparison. Mitra et al.[8] used a postsubtraction column of mercury perchlorate to subtract quantitatively alkenes and aromatics after the analytical column, but before the GC detector. Gallegos et al.[14] recently described the use of mass chromatography with open tubular columns for compound-type analysis of petroleum naphthas. Fragments of m/e of 77, 71, 69, 67, and 65 were monitored as being characteristic of vinyl- or alkylbenzenes, linear or branched alkanes, monocyclics or monoolefins, cyclic olefins or diolefins, and cyclic diolefins or triolefins, respectively.

The analysis of low molecular mass alkanes (C_6 and smaller) and high molecular weight alkanes (C_{30} and larger) present specific problems that

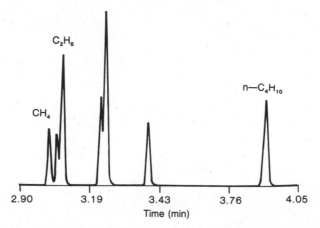

FIGURE 6.2. Chromatogram of C_1–C_4 alkanes. Note: 60 m SP-2100 column, 0°C isothermal temperature, helium carrier gas. (Reproduced with permission from ref. 26. Copyright Dr. Alfred Huethig Publishers.)

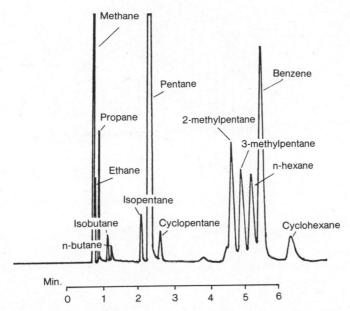

FIGURE 6.3. Chromatogram of short-chain alkanes. Note: 25 ft stainless steel column containing a styrene-divinylbenzene cross-linked polymer phase, 150°C isothermal temperature. (Reproduced with permission from ref. 27. Copyright Preston Publications, Inc.)

must be addressed. The low capacity factors of C_1 to C_6 alkanes on open tubular columns lead to the difficulty in their analysis by this method. However, Glajch and Schunn[26] used a 58 m × 0.25 mm i.d. open tubular column coated with SP-2100 for the rapid analysis of low molecular weight hydrocarbons which were injected as gases. The column was temperature programmed from 0 to 220°C at 25°C/min. A heated (200°C) sample-loop split injection system was used to introduce the sample into the column. Figure 6.2 shows a chromatogram of C_1 to C_4 alkanes which was obtained in less than 4 min. Particular caution was taken to provide the proper split ratio and sample size in order to obtain the observed excellent efficiency.

In another study,[27] a styrene-divinylbenzene polymer was synthesized and cross-linked on the wall of a capillary column and used for the separation of low molecular mass alkanes. Figure 6.3 shows an example of a chromatogram obtained on such a column.

In order to avoid sample discrimination during the analysis of high molecular mass (low volatility) hydrocarbons, several precautions should be taken with regard to sample introduction. As discussed in Chapter 4, aerosols may form during split injection, which leads to discrimination with respect to volatility. Aerosol formation can be avoided by using packing materials in the vaporizer[28] and by selecting a suitable solvent and injector temperature. The injector temperature (as well as all transfer lines and detector temperatures) should be set slightly higher than the maximum column tem-

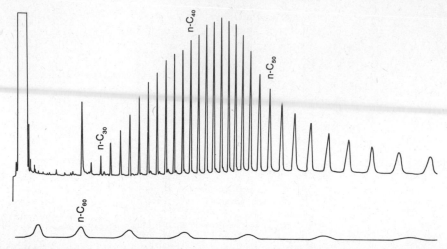

FIGURE 6.4. Chromatogram of a paraffin wax. Note: 7 m SE-54 fused-silica column; temperature program from 200 to 340°C at 15°C min^{-1}; He carrier gas at 80 cm s^{-1}. (Reproduced with permission from ref. 33. Copyright Hewlett-Packard Company.)

perature for both split and splitless injection. Too high temperatures magnify problems of poor solute transfer from the syringe (the solvent is vaporized too fast, leaving the sample in the syringe), explosive vaporization, aerosol formation, condensation on the septum, and catalytic decomposition.

The volatility of the solvent also influences discrimination during analysis of mixtures with a wide range of molecular mass in both split and splitless sampling. Grob and Grob[29] found that optimal resolution could be achieved if the boiling point of the solvent is slightly higher than the initial column temperature. If the solvent is too volatile, explosive vaporization can lead to the partial distribution of the sample into the carrier gas inlet lines which would lead to certain adsorption of the less volatile components.

Although sample discrimination can be avoided with careful control of the parameters during split and splitless injection, on-column injection appears to be superior in this respect.[28-30] Direct introduction of liquid samples into the open tubular column without any prevaporization avoids all of the previously mentioned problems.

Additional suggestions for the analysis of high molecular mass hydrocarbons (or any other high molecular mass species) include the use of hydrogen as carrier gas and the use of thin-film, thermally stable, and short open tubular columns.[31-33] These three suggestions led[33] to the elution of hydrocarbons up to C_{65} in a parafin wax (Figure 6.4).

Alkenes and Alkynes

Analytical interest in the alkenes and alkynes is stimulated by their importance as petrochemical raw materials and intermediates in the production of polymers, detergents, and other products. The similarities in the properties of isomeric alkenes and alkynes and the complexities of their mixtures lead to open tubular column GC as the most promising method for their total analysis. The mass spectra of these compounds can also be quite similar, leaving retention index studies as the only alternative.

As discussed in the previous section, retention data and calculated retention increments are useful in differentiating alkenes and alkynes from other classes of hydrocarbons. When the samples to be analyzed contain only alkenes or alkynes, additional information can be obtained from this data concerning the structure of individual components. Detailed studies of the C_2–C_{14} n-alkenes,[3–6,8–10,34–40] the C_{15}–C_{18} n-alkenes,[41–43] and the C_6–C_{14} n-alkynes[44,45] have been published. Correlations of I and its increments with molecular parameters such as carbon number, position of the double bond, and molecular configuration have been made. General observations are as follows:

1. Retention indices of n-alkenes depend on the stationary phase polarity (including the column wall surface activity[46]), column temperature, carbon number, position of the double bond, and on the geometric configuration of the molecule.

2. In a homologous series of isomers with identical geometry and double bond position, the I values increase linearly with increasing chain length in accordance with the equation

$$I = a + bn \tag{6.4}$$

where a and b are constants, and n is the number of carbon atoms.

3. The b value, which corresponds to the mean contribution of one methylene group to the retention index value, in a homologous series, decreases as the double bond shifts toward the center of the molecule.

4. The elution order of corresponding trans- and cis-isomers remains invariable on polar phases (trans-alkenes are eluted before cis-alkenes), while there are some inconsistencies on nonpolar phases.

5. Internal n-alkenes of the same geometric configuration elute in the order in accordance with the double bond shift from the center to the end of the molecule. With lower alkenes (up to C_9), 1-alkenes are eluted before the other internal isomers on nonpolar phases.

6. On nonpolar phases, the temperature coefficient dI/dT increases as the double bond shifts toward the center of the molecule. There are some inconsistencies on polar phases.

7. Cis-alkenes have larger temperature coefficients than do trans-alkenes.

8. The homomorph retention increments, H, decrease with an increase

in the number of carbon atoms in the molecule, and the change is more pronounced for the *cis*-isomers.

9. The *H* values of internal *n*-alkenes decrease as the double bond shifts to the center of the molecule.

10. The Δ*I* values decrease with the carbon number in homologous series of *n*-alkenes.

11. Among positional isomers with the same carbon number, the *cis*-2-alkenes have the largest Δ*I* values. The Δ*I* values decrease with a shift of the double bond toward the center of the molecule up to the 4-isomers and then they remain practically constant.

As with the alkanes, the most widely used stationary phase for retention index measurements has been squalane. Table 6.5 compares the retention index intervals for the *n*-decenes (the most difficult to separate *n*-alkenes on squalane) on other more polar phases.[41] It can be seen that the retention interval increases with an increase in the polarity of the stationary phase. In order to study the C_{15}–C_{18} linear alkenes, Sojak et al.[42,43] used the more thermally stable C_{87} hydrocarbon and Carbowax 20M as nonpolar and polar stationary phases, respectively. Meltzow et al.[47] reported the use of a strongly selective stationary phase Arneel SD (an aliphatic nitrile) for the separation of all the isomers of *n*-hexene, *n*-hexadiene, and *n*-hexyne (see Figure 6.5).

The complexing properties of transition metals have been used in the past for selective retention of alkenes. Open tubular columns coated with stationary phases doped with silver nitrate[48] and dicarbonyl-rhodium-3-trifluoroacetyl-*d*-camphorate[49] have been reported. Figure 6.6 shows a chromatogram of gaseous alkenes, including C_2H_4 and C_2D_4, on a 100 m × 0.25 mm i.d. open tubular column coated with the rhodium complex in squalane.[49]

TABLE 6.5. Range of retention indices of *n*-decenes on different stationary phases at 50°C.[a]

Stationary phase	Retention interval, *I*	δ*I*
Squalane	979.2–999.2	20.0
Apiezon L	983.0–1005.7	22.7
Bis-2-ethylhexyl tetrachlorophthalate	993.2–1023.3	30.1
7,8-Benzoquinoline	998–1028	30
Di-*n*-butyl tetrachlorophthalate	998.0–1029.8	31.8
Polyethylene glycol 4000	1035.7–1068.9	33.2
Polyphenyl ether[b]	1014.6–1041.0[c]	26.4

[a] Taken from ref. 41.

[b] Taken from ref. 40.

[c] Measured at 60°C.

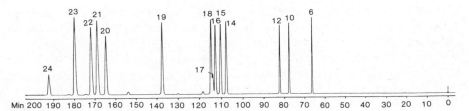

FIGURE 6.5. Chromatogram of isomeric *n*-hexadienes and *n*-hexynes. Note: 200 m Arneel SD stainless steel column, 22°C isothermal temperature, helium carrier gas. Selected peak assignments: (6) 1,5-hexadiene; (12) 1-*cis*-4-hexadiene; (18) 1-*trans*-3-hexadiene; (19) 3-hexyne; (21) 2-hexyne; (22) *cis*-2-*trans*-4-hexadiene. (Reproduced with permission from ref. 47. Copyright Friedr. Vieweg and Sohn.)

In addition to retention index studies on the alkenes themselves, numerous other methods have been used to differentiate between alkenes having different positions of the double bond.[50,51] One of the most successful approaches is to use a simple derivatization procedure such as oxyselenation[52] or hydrogenation[6] followed by gas chromatography–mass spectrometry (GC-MS). The manner in which the derivative is obtained may vary greatly, but the net result is addition across the double bond. Addition of X-Y to an unsymmetrical alkene produces two isomeric substituted alkanes, unless X and Y are identical. The Kovats indices and ΔI values of the obtained derivatives can be used for the characterization and identification of the initial species. For instance, double bonds can be converted stereospecifically into the corresponding substituted cyclopropanes, the geometric isomers of which exhibit more characteristic retention than do the initial alkenes.[6]

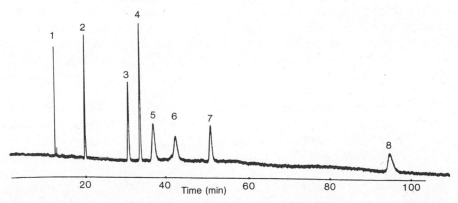

FIGURE 6.6. Chromatogram of gaseous alkenes. Note: 100 m nickel column coated with dicarbonyl-rhodium-3-trifluoroacetyl-*d*-camphorate in squalane, nitrogen carrier gas. Selected peak assignments: (1) methane; (2) isobutene; (3) *trans*-2-butene; (4) *n*-pentane; (5) ethene; (6) tetradeuteroethene; (7) *cis*-2-butene; (8) 1-butene. (Reproduced with permission from ref. 49. Copyright Friedr. Vieweg and Sohn.)

Retention index studies of C_6–C_{14} n-alkynes have been made on open tubular columns coated with squalane, Apiezon L, polyphenylether, polyethylene glycol 4000, Ucon LB 550 X, Ucon 50 HB 280 X, and Carbowax 1000.[53,54] The best resolution of positional isomers was obtained on squalane. The retention indices of n-alkynes are governed by their π-electron system and its polarization, and by steric factors due to the loss of free rotation at the triple bond. The retention rules for the n-alkynes were found to be similar to those discussed previously for the n-alkenes. The lowest I values were obtained on nonpolar phases while polar phases increased the retention. The internal isomers with the same number of carbon atoms were generally eluted in sequence of the shift of the triple bond from the middle to the end of the molecule.

On nonpolar phases (squalane and Apiezon L) the elution sequence was nearly in accordance with the boiling points of the positional isomers, and 1-alkynes were eluted first. On the more polar phases of polyethylene glycol, the 1-alkynes were retained more strongly due to the formation of hydrogen bonds.

As was found with n-alkenes, I exhibited a linear dependence on the number of carbon atoms in the molecule, and the increments of I per CH_2 group were dependent on the number of carbon atoms and the position of the triple bond.

Cycloalkanes and Cycloalkenes

Data on cycloalkanes and cycloalkenes are of considerable interest in organic and geochemistry. Studies have been reported concerning the open tubular column GC of cyclopropanes;[5,6,55] cyclopentanes and cyclopentenes;[5,38,56,57] cyclohexanes and cyclohexenes;[5,38,57–59] dicyclopentadienes;[60,61] adamantanes;[62] and steranes, terpanes, and terpenes.[63–69]

Schomburg and Dielmann[55] recommended converting olefins into the corresponding cyclopropanes followed by retention studies for supporting the identifications of the original olefin isomers. They found that the H^{sq} values of trans- and cis-isomers of substituted cyclopropanes exhibit a significant difference, whereas the cis- and trans-isomers of the corresponding olefins can be separated and characterized only with special liquid phases at high resolution. Furthermore, the difference in retention indices of cis- and trans-isomers of internal olefins are small and decrease as the double bond migrates to the center of the carbon chain. For comparison, the cis–trans differences of the corresponding cyclopropanes are larger and independent of the position in the chain.

The following general conclusions concerning the retention behavior of cyclopropanes and cyclopropenes have been drawn:[55]

1. The H^{sq} values of the trans- and cis-isomers of disubstituted cyclopropanes are characteristic and dependent on the type of substitution and the chain length of the alkyl groups.

2. The differences in the H^{sq} values of *trans-* and *cis-*isomers of the cyclopropanes are constant and independent of the chain length of the substituting alkyl groups. They are not influenced by branchings, rings, or double bonds within the substituting alkyl groups except when the double bonds are in the α and β positions to the ring.

3. The ΔI values of *trans-* and *cis-*isomers are characteristic but similar to those of the corresponding olefins.

4. Diastereoisomeric cyclopropanes can be separated in the cases of α-, β-, or γ-methyl-branched alkyl cyclopropanes and cyclopropyl–cyclopropanes.

5. The application of the retention increments for the different types of double bonds to cyclopropane hydrocarbons presents no difficulties as long as the double bond in the alkyl group is not in conjugation to the ring. For α-unsaturated cyclopropanes, the influence of conjugation of the π electrons with the cyclopropane ring on retention behavior leads to a significant increase in the I^{sq} value.

6. The measured H^{sq} for systems in which two cyclopropane rings are conjugated (α, γ) is lower than that of the cyclopropane double-bond interaction in the case of *trans-*isomers and is nonexistent for the *cis-*isomers.

7. Similar H^{sq} values could be derived for dicyclopropane hydrocarbons having the cyclopropane rings close together inside the carbon chain (α, γ, or δ position).

The retention of alkyl-substituted cyclopentanes, cyclopentenes, cyclohexanes, and cyclohexenes is determined by the ring size, position of the side chain, degree of double-bond substitution, and length and structure of the side chain. Several regularities exist in the variation of I and incremental I values with molecular structure for the cyclohydrocarbons:[57]

1. The H values (obtained on nonpolar phases) increase as the degree of alkyl substitution at the double bond of cycloalkenes increases. This tendency was found for ethyl-, *n*-propyl-, and *n*-butyl-derivatives of cyclopentenes and cyclohexenes. This regularity is not retained for higher *n*-alkylcyclohexenes, due to the combination of complicated steric and electronic effects with intermolecular interaction.

2. The retention indices of *n*-alkylcyclopentenes and cyclohexenes have a linear dependence on the carbon number and an almost linear dependence on column temperature, except for methyl and ethyl derivatives.

3. The variation of I with temperature is more pronounced for hydrocarbons with six-membered rings than for the corresponding ones with five-membered rings.

4. The temperature increments of the retention index of cyclanes and cyclenes are markedly greater than those of *n*-alkenes and *n*-alkynes with similar empirical formulas, respectively.

5. In the *n*-alkylcyclopentene series, the 3-isomers are always eluted before the 1-isomers. The elution order of isomeric *n*-alkylcyclohexenes depends on the length of the side chain and the column temperature. Lower homologs elute in the order 3-, 4-, and 1-isomers. When the length of the side chain approaches 6–8 carbons, the 1-isomers elute first.

6. Isomers with branched alkyl substituents have lower *I* values than those with unbranched side chains, due to steric factors.

7. For *n*-alkylcyclohexenes, the *H* increments are greater than the corresponding *n*-alkylcyclopentenes with equal *n*.

8. *n*-Alkyl substitution in the ring and increasing length of the *n*-alkyl side chain (beginning from the ethyl group, except 1-*n*-alkyl-1-cyclopentenes and 1-*n*-alkyl-1-cyclohexenes) leads to a decrease in *H*.

9. The range of *H* values between 1- and 3-*n*-alkyl-1-cyclopentenes is wide and they are well separated: for 1- and 3-*n*-alkyl-1-cyclohexenes it is much narrower. With lengthening of the *n*-alkyl substituent, the differences in *H* values increase and the separation of 1- and 3-isomers is improved.

10. Branching in a side chain leads to a decrease in *H*.

11. Of the *n*-alkyl-substituted cyclopentenes, cyclohexenes, and corresponding cycloalkanes, the largest ΔI values are found for cyclohexenes, the smallest for cyclopentanes and cyclohexanes. The latter two have similar ΔI values.

12. The ΔI values decrease with lengthening of the *n*-alkyl side chain, and remain practically constant for hydrocarbons $\geq C_{10}$–C_{12}.

13. In most cases, the ΔI values of cyclopentene and cyclohexene derivatives with branched substituents are smaller than corresponding homologs with unbranched side chains. In the cyclohexene series, 3-isoalkyl-substituted compounds have larger ΔI values than 1-isomers.

In addition to the well studied mono- and di-substituted C_6–C_{15} alkylcyclohexanes and alkylcyclohexenes described above, Albaiges and Guardino[59] undertook a systematic study of a series of monoterpene-like hydrocarbons, structurally related to *p*-menthane. Retention index studies on squalane, SE-30, Apiezon L, OV-225, and PEG 20M in conjunction with gas chromatography–mass spectrometry resulted in the positive identification of 17 (10 for the first time) compounds in a pyrolysis naphtha. Figure 6.7 shows one of the chromatograms that they obtained. Saeed et al.[69] published a number of retention indices for monoterpenes on open tubular columns coated with SE-30 and Carbowax 20M.

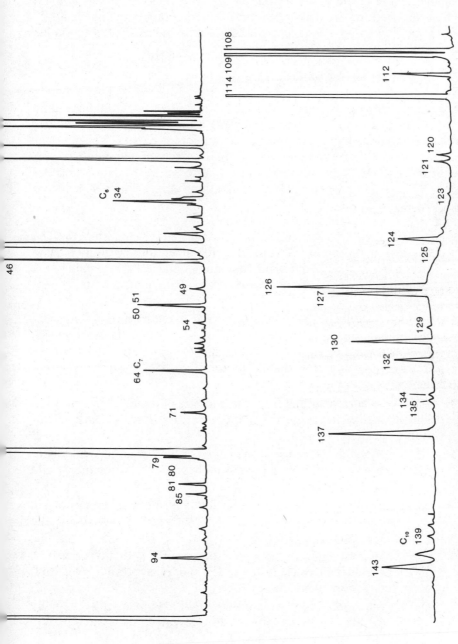

FIGURE 6.7. Chromatogram of pyrolysis naphtha. Note: 100 m squalane stainless steel column; temperature program from 20 to 90°C at 0.75°C min⁻¹; helium carrier gas. Selected peak assignments: (42) hexadiene; (44) 1-methylcyclopentene; (78) toluene; (105) ethylbenzene; (109) *m*-xylene; (114) *o*-xylene; (126) 1-methyl-3-ethylbenzene; (137) 1,2,4-trimethylbenzene. (Reproduced with permission from ref. 59. Copyright Friedr. Vieweg and Sohn.)

The identification of terpanes, steranes, and phenylcycloalkanes in fossil fuels is important in understanding the processes of their formation from plant material and in the fingerprinting of pollutants. Open tubular column GC has been used for the separation of these compounds in shale oil[64–66] and crude oil.[68,69] In combination with MS, a large number of compounds were identified.

Aromatic Hydrocarbons

Aromatic hydrocarbons have been the subject of many gas chromatographic studies in the past, and the measurement of retention data and structure retention correlations have been reported in numerous papers. Recent studies using a high-precision gas chromatographic system in combination with open tubular columns has allowed further fine-structure retention correlations.[70–73] The more general structure retention correlations for aromatic hydrocarbons are as follows:

1. I values for n-alkylbenzenes increase with alkyl chain length. The I_{CH_2} increment increases from 80 to 100 index units with increasing carbon number on both polar and nonpolar phases. A constant value around 100 index units is obtained for n-hexylbenzene and higher homologs.

2. For the dialkylbenzenes, the I_{CH_2} values depend on the mutual positions and lengths of the alkyl chains. The longer the alkyl chain, the smaller is the I_{CH_2} increment on the introduction of a methylene group into the shorter chain. On the other hand, substitution into the longer chain increases the increment.

3. The I_{CH_2} values resulting from the addition of a methylene group to dialkylbenzenes increase in the order *ortho* < *meta* < *para* substitution.

4. The dI/dT values of the first members of the series of n-alkylbenzenes increase slightly with increasing number of carbon atoms and then become constant.

5. For identical structural types of dialkylbenzenes, differing only in the positions of the alkyl groups, the dI/dT values increase in the order *meta* < *para* < *ortho* substitution.

6. For the corresponding positional isomers of C_{10} dialkylbenzenes, the dI/dT values increase in the order: methylisopropyl- < diethyl- < methyl-n-propylbenzene.

7. The largest dI/dT values were observed for asymmetrically vicinally substituted alkylbenzenes and the lowest values for symmetrically substituted 1,3,5-alkylbenzenes.

8. The effect of the position of the alkyl chain on the magnitude of dI/dT increases in the order 1,3,5- < 1,2,4- < 1,2,3-trialkylbenzenes.

9. For trialkylbenzenes, the replacement of a smaller alkyl group with a larger one will result in a decrease in the dI/dT value, and this effect is more significant with symmetrical molecules. The dI/dT value for 1,3,5-triisopropylbenzene approaches that of an *n*-alkane.

10. A marked increase in dI/dT values for alkylbenzenes are observed on polar phases when compared to data obtained on nonpolar phases.

11. The retention of the isoalkylbenzenes is lower than that of the corresponding *n*-alkylbenzenes.

12. In the series of C_{10} alkylbenzenes, the order of retention is methyl-propyl- < dimethylethyl- < tetramethylbenzene, that is, from more bulky to less bulky groups.

13. Retention increases from the more symmetrical 1,3-, 1,3,5-, and 1,2,4,5-alkylbenzenes to the more asymmetrical vicinally substituted 1,2-, 1,2,3-, and 1,2,3,4-alkylbenzenes.

14. The ΔI values decrease with increasing number of carbon atoms in *n*-alkylbenzenes.

15. For the trimethylbenzenes, the ΔI values increase in the order 1,3,5- < 1,2,4- < 1,2,3-trimethylbenzene.

For more specific structure retention correlations, the reader should refer to the published literature.[70–75]

A specific class of aromatic hydrocarbons of commercial interest is the phenylalkanes. These compounds are important intermediates in the production of detergents. Lesko et al.[75] have presented a detailed discussion

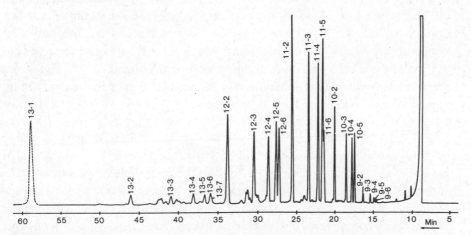

FIGURE 6.8. Chromatogram of a commercial mixture of phenylalkanes. Note: 50 m OV-101 glass column, 200°C isothermal temperature, helium carrier gas. Peak assignments: the first numeral indicates the number of carbon atoms in the linear alkyl chain and the second numeral the position of the phenyl group on the chain. Peak 13-1 is an internal standard. (Reproduced with permission from ref. 75. Copyright Elsevier Scientific Publishing Company.)

of the use of open tubular column GC for their analysis. Figure 6.8 shows a chromatogram of the isomeric phenylalkanes in a commercial mixture obtained on an open tubular column coated with OV-101. It was found that the H increments decreased with the shift of the phenyl group to the middle of the alkyl chain, and for a given location in the alkyl chain, they increased as the number of carbon atoms increased.

Squalane has been used most widely as a stationary phase for aromatic hydrocarbons, again because of the extensive collection of retention data that is available. Other phases that have been used cover the whole range of polarities. A tandem 1,2,3-tris-2-cyanoethoxypropane/DC-550 column was used to separate the aromatic hydrocarbons in crude oil.[76] Kuchhal et al.[77] found Carbowax 1540 to give the best resolution of the isomeric diethylbenzenes. After trying several different stationary phases (SF-96, UCON 50 HB 5100, and Igepal CO-880) all six isomers of the mono ring-substituted dimethyldiphenylmethanes were resolved[78] on a 180 m long open tubular column coated with Emulphor O (polyethylene glycol etherified with octadecanol).

It is possible to achieve extraordinary separations of structural isomers such as the substituted benzenes by using liquid crystalline stationary phases.[79–81] Grushka and Solsky[79] studied the properties of the liquid crystal 4,4'-dimethoxyazoxybenzene coated on a capillary column. It was found that at a given temperature in the nematic region (118°C and higher) of the liquid crystal, the retention time of a selected component such as o-xylene can have two different values depending on whether the column is being heated or cooled. These results indicate that cooling produces changes in the macrostructure of the mesophase due to either supercooling or surface orientation effects. The liquid crystalline structure can be maintained at temperatures below the melting point, which allowed almost complete separation of the hard-to-separate compounds m- and p-xylene.[79–81] Figure 6.9 shows chromatograms of substituted benzenes on a 6 m open tubular column coated with a liquid crystal p-n-heptyloxyphenyl-4-(trans-4-n-propylcyclohexanecarbonyloxy)-2-methylbenzoate.[81] Chromatograms A, B, and C represent results obtained in the crystalline range (48.5°C), nematic range (66.5°C), and supercooled range (48.5°C), respectively. The separation of m- and p-xylene is complete after cooling the phase back down to 48.5°C.

In addition to retention measurements, derivatization reactions and spectroscopic analysis can give extremely helpful information for elucidating the structures of components in complex mixtures. Tirgan and Sharifi-Sandjani[82] catalytically hydrogenated the side chains of alkylstyrenes in a light-oil fraction of a coal distillate without altering the aromatic ring. Open tubular column GC was used to identify and quantify the alkylstyrene peaks which, when hydrogenated, shifted retention.

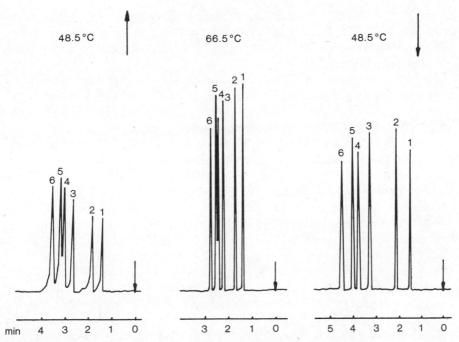

FIGURE 6.9. Chromatograms of substituted benzenes. Note: 6 m glass column coated with *p-n*-heptyloxyphenyl-4-(*trans-n*-propylcyclohexanecarbonyloxy)-2-methylbenzoate liquid crystal. Selected peak assignments: (1) benzene; (2) toluene; (3) ethylbenzene; (4) *m*-xylene; (5) *p*-xylene; (6) *o*-xylene. (Reproduced with permission from ref. 81. Copyright Dr. Alfred Huethig Publishers.)

Polycyclic Aromatic Hydrocarbons

The widespread interest in the analysis of polycyclic aromatic hydrocarbons (PAH) stems from the well-known carcinogenic activity that many of these compounds have demonstrated, and the observation that these compounds are ubiquitous throughout our environment.[83] PAH are usually present in various samples as complex mixtures of isomers, containing hundreds of components of widely ranging concentrations. The identification and quantitation of each component in a PAH mixture is important for understanding the possible health risks of human exposure to these compounds.

Open tubular column GC has emerged as one of the most powerful analytical tools for the detailed characterization of complex PAH mixtures. A review was recently published[84] which discusses in depth the application of open tubular column GC to the analysis of these compounds.

Although a number of different stationary phases have been used in the past for the analysis of PAH,[85] SE-52 (5% phenyl methylsilicone) and SE-54 (1% vinyl, 5% phenyl methylsilicone) are the two most widely used ones. They are chemically very similar, and the McReynolds constants for both are nearly the same. Borwitzky and Schomburg[85] emphasized that the order of elution of PAH was independent of the polarity of the stationary phase, and only minor selectivities could be obtained for selected isomer groups on more polar phases.[86]

Recent attempts have been made to measure and tabulate the Kovats retention indices for the numerous PAH known.[87,88] It was found that the retention indices were influenced by the stationary phase film thickness, the length of column, the temperature programming rate, the carrier gas flow rate, and the injection system (see Table 5.7). Although the column surface effects are much reduced with the fused-silica material, nevertheless, more reliable indices are obtained using the internal standards: naphthalene, phenanthrene, chrysene, and picene.[88,89] The retention indices of over 400 polycyclic aromatic compounds have been measured using this system[88,89] on capillary columns coated with SE-52. The average 95% confidence limits for four measurements on each PAH were ± 0.25 index units.

The elution order of PAH is governed principally by their volatilities[90] which can be approximated by boiling points.[85,87,91] Among components of isomer groups, longer retention is generally associated with more "extended" molecules. Thus, phenanthrene elutes before anthracene. An increase in retention is observed when the PAH are substituted with alkyl groups, and totally unsaturated PAH are usually more retained than partially hydrogenated compounds. Several papers have discussed in detail the structure–retention properties of the alkylbiphenyls and alkylnaphthalenes.[92–95]

Liquid crystal phases have shown quite pronounced selectivities for PAH, but the efficiencies and thermal stabilities of these phases coated on open tubular columns have been quite unsatisfactory.[96] However, mixed phases of liquid crystals with silicone gums have produced columns possessing both high selectivity and moderately high efficiency.[97–99] The synthesis of a silicone gum phase with liquid crystal moieties attached to the silicon atoms was recently reported.[100] Columns coated with this material (PMMS phase) exhibit high efficiencies expected of open tubular columns, excellent ther-

TABLE 6.6. Length-to-breadth
(L/B) ratios for the methylchrysene
isomers.

Isomer	L/B Ratio
2-Methylchrysene	1.85
1-Methylchrysene	1.71
3-Methylchrysene	1.63
4-Methylchrysene	1.51
5-Methylchrysene	1.48
6-Methylchrysene	1.48

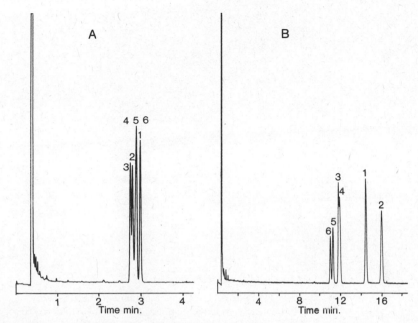

FIGURE 6.10. Chromatograms of the six methylchrysene isomers. Note: (A) 19 m SE-52 fused-silica column, 230°C isothermal temperature, H_2 carrier gas; (B) 19 m PMMS fused-silica column, 230°C isothermal temperature, H_2 carrier gas. Peak assignments: (1) 1-methylchrysene; (2) 2-methylchrysene; (3) 3-methylchrysene; (4) 4-methylchrysene; (5) 5-methylchrysene; (6) 6-methylchrysene. (Reproduced with permission from ref. 101. Copyright American Chemical Society.)

mostability, and a wide nematic range (70–300°C).[101] Temperature programming can easily be utilized with this phase.

The separation of solutes by liquid crystal phases in the nematic state is considered to be a solute molecular ordering based on molecular geometry. Retention is approximately according to the order of the length-to-breadth (L/B) ratio. Figure 6.10 shows chromatograms of the methylchrysene isomers on open tubular columns coated with the liquid crystal silicone gum stationary phase and SE-52. Table 6.6 lists the L/B ratios for the methylchrysenes. As predicted, retention follows the L/B ratios, and the resolution is much better on the liquid crystal phase.

It is observed in Figure 6.10 that the 3- and 4-methylchrysenes are not resolved on the liquid crystal phase, although these two isomers are fully separated on SE-52. Therefore, a blend of the two phases should be superior to either pure material. Using the window diagram method[102] to obtain the optimum phase composition, a 50:50 blend of the two phases was suggested in order to provide baseline (6σ) separation of all solutes provided that the resultant column yielded 100,000 plates (N_{req} at $k' > 10$). A 4σ separation required only 45,000 plates, which should be easily obtained on coupling together two approximately 20 m pure phase columns. Figure 6.11 shows

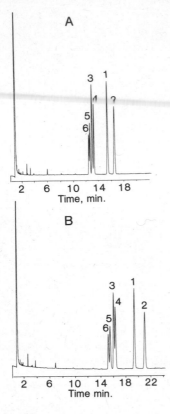

FIGURE 6.11. Chromatograms of the six methylchrysene isomers on tandem-connected columns. Note: (A) SE-52 (19 m): PMMS (19 m) fused-silica column, 230°C isothermal temperature, H_2 carrier gas; (B) PMMS (19 m): SE-52 (19 m) fused-silica column, 230°C isothermal temperature, H_2 carrier gas. Peak assignments same as in Figure 6.10. (Reproduced with permission from ref. 101. Copyright American Chemical Society.)

chromatograms obtained from two 19 m open tubular columns tandem connected in the order SE-52: PMMS (Figure 6.11A) and PMMS: SE-52 (Figure 6.11B). The slight differences in the two chromatograms result from the fact that the pressure drops and, hence, solute residence times in column segments are finite and nonequivalent.

Much effort was recently spent to extend the temperature range of open tubular columns in order to chromatograph higher molecular mass (>300) compounds. Grob[103] chromatographed compounds from coronene ($C_{24}H_{12}$) to rubrene ($C_{42}H_{28}$) on a short (5.5 m) open tubular column coated with OV-101. Compounds of molecular masses up to 376 in a carbon black extract have been chromatographed on a well-deactivated SE-52 column,[84] and Romanowski et al.[104] have reported on the detection of over 100 PAH having molecular masses between 300 and 402 in air particulate matter using a thin-film SE-54 fused-silica open tubular column connected to a mass spectrometer.

More recently,[105,106] cross-linked methylphenylpolysiloxane stationary phases have been prepared that demonstrate extremely high-temperature stabilities and slightly different selectivities than coated phases. Figure 6.12 shows a chromatogram of a carbon black extract that was obtained on a cross-linked 70% phenyl methylphenylpolysiloxane fused-silica open tubular column.[106] This stationary phase is remarkably stable to temperatures as high as 400°C.

The most widely used detector in PAH analysis is the flame ionization detector (FID) because of its excellent response linearity, sensitivity, and day-to-day quantitative reliability. Since the detector is not selective for PAH, much effort must be undertaken to separate and purify PAH fractions before chromatographic analysis. These procedures were recently reviewed.[83] Additional, more recently developed fractionation schemes for isolating clean PAH fractions from lubricating oils,[107] air particulate matter,[108,109] sediments,[109-112] biological material,[112,113] and fossil fuels[114] have been described.

Selective detection of PAH in total hydrocarbon mixtures is desirable because of the reduced effort that would be required for sample preparation. Coupling of open tubular columns to various selective detectors for this purpose include the electron capture detector (ECD),[115-118] photoionization detector (PID),[119] ultraviolet spectrometric detector (UVD),[120] as well as GC-MS.

Bjørseth and Eklund[118] measured the ECD/FID response ratios for 29 PAH and found that many isomers could be differentiated by measurement of these ratios. Similarly, using a PID in series with an FID allowed the classification of eluted peaks into aromatic or aliphatic hydrocarbons based on the normalized relative molar response in the two detectors.[119]

Although infrared spectrometry, itself, has limited potential as a GC detector for identifying components of complex PAH mixtures, the use of

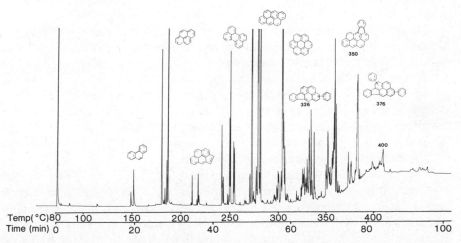

FIGURE 6.12. Chromatogram of a carbon black extract. Note: 12 m 70% phenyl methyl-phenylpolysiloxane fused-silica column, temperature program from 80 to 400°C at 4°C min⁻¹, H₂ carrier gas at 100 cm s⁻¹. (Reproduced with permission from ref. 106. Copyright Friedr. Vieweg and Sohn.)

matrix isolation IR spectrometry provides sufficient resolution to identify specific PAH and their alkylated derivatives.[121] Preliminary results from the analysis of a coal liquid on a SCOT column using matrix isolation IR detection have been reported.[121]

6.3 ALCOHOLS

The alcohols have been one of the most difficult classes of compounds to analyze by gas chromatography. This is a result of their relatively high polarity and ability to interact with silica surface active sites by hydrogen bonding with silanol groups or adsorbing to basic oxide sites ($-O^-$ sites or strained siloxane bridges) where there is relatively high electron density. This means that in cases where a nonpolar stationary phase is to be used, the column wall must not only be neutral (not acidic or basic), but must be relatively free of the previously named active sites. The lower limit of detection of alcohols depends on the degree of deactivation of the column surface. For this reason, alcohols have been popular ingredients in test mixtures designed for evaluation of the degree of deactivation of open tubular columns (see Chapter 3).

Proper leaching, dehydration, dehydroxylation, and blocking of residual silanol groups of glass columns provide sufficient deactivation for chromatography of alcohols at nanogram levels on nonpolar phases.[122] With much less effort, equal or better results can be obtained with fused silica. Figure 6.13 shows a chromatogram of n-alcohols on a fused-silica open tubular column coated with SE-52.[123] Full-scale response represents approximately 1 ng.

Thermally aged nonpolar phase columns can result in increased tailing of alcohols. Injection of propylene glycol at high temperature (250°C) reduced tailing to near new column levels, while silylating agents were found to be ineffective.[124] It is believed that the propylene glycol caps active silanol groups of adjacent stereochemistry to produce silyl alkyl ethers. Symmetrical peaks can also be obtained on incompletely deactivated surfaces if a small amount of a surface active agent or polar phase[125] is added to the nonpolar stationary phase. This latter solution, however, is usually only temporary, and surface activity slowly returns.

Most of the difficulties encountered in the analysis of alcohols on nonpolar stationary phases can be eliminated if they are analyzed in the form of their acetyl esters or trimethylsilyl ethers.[126] Alternatively, polar stationary phases can be used. The retention of alcohols on polar phases is influenced by the degree of polarity of the phase and by the position of the hydroxyl group in the hydrocarbon chain.[126] Polyethylene glycol stationary phases are generally accepted as being the most suitable polar phases for alcohols.[126–128] The elution order of alcohol acetyl esters and trimethylsilyl

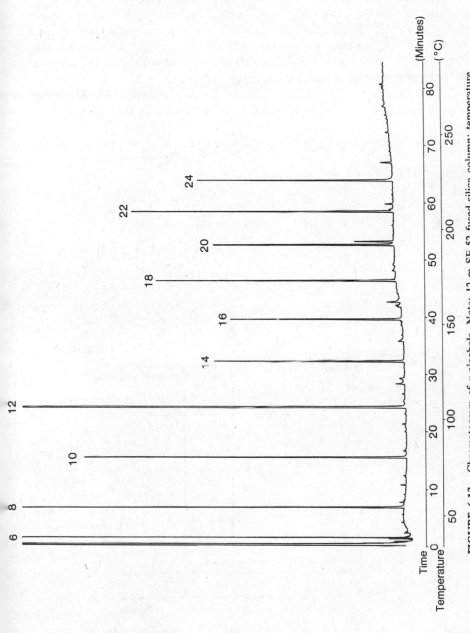

FIGURE 6.13. Chromatogram of *n*-alcohols. Note: 12 m SE-52 fused-silica column; temperature program from 40 to 250°C at 4°C min^{-1} after an initial 2 min isothermal period, H$_2$ carrier gas at 100 cm s^{-1}, full-scale response represents 1 ng of compound injected. Peak assignments: peak numbers refer to carbon chain length for the *n*-alcohols.

ethers on nonpolar stationary phases is usually the same as that of the underivatized alcohols on polyethylene glycols.[126]

Krupcik and co-workers[126,129] studied the retention characteristics of C_9–C_{14} secondary alcohols and their acetyl esters on capillary columns coated with Carbowax 1500 and 20M and OV-101. The best separations were obtained when the alcohols were converted to their acetyl derivatives and chromatographed on OV-101 and Carbowax 1500. Since the rate of increase of k' with film thickness for hydrocarbons was much less than that for alcohols, it was proposed that ethyl esters be used as standards for obtaining retention indices of alcohols when polar columns such as the polyethylene glycols were used.[127]

As can be expected, compounds containing more than one hydroxyl group

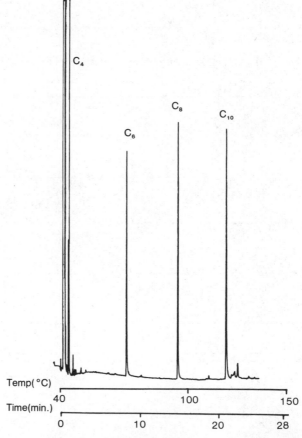

FIGURE 6.14. Chromatogram of diols. Note: 12 m SE-52 fused-silica column; temperature program from 40 to 150°C at 4°C min^{-1} after an initial 2 min isothermal period, H_2 carrier gas at 100 cm s^{-1}, 1 ng of each compound injected. Peak assignments: peak numbers refer to carbon chain length of the terminally substituted diols.

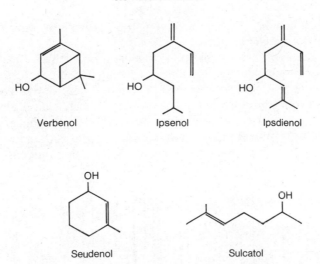

Verbenol   Ipsenol   Ipsdienol

Seudenol   Sulcatol

FIGURE 6.15. Chiral terpene alcohols having pheromone properties for various insect species.

are more difficult to chromatograph than the simple alcohols. Boneva and Dimov[130] used a short (10 m) glass Carbowax 20M column that was etched with 1,2-difluoro-2-chloroethyl ether before coating for the separation of ethylene glycols and ethanolamines without derivatization. Figure 6.14 shows a chromatogram of diols obtained on a properly deactivated fused-silica column.[131] Minimal peak tailing is observed for 1 ng injected of each compound. Schomburg et al.[132] reported a comprehensive study of the gas chromatography of diols as a function of deactivation of both fused-silica and glass open tubular columns. Polar Carbowax 20M and nonpolar (OV-1 and OV-101) phases were studied. Although, excellent results could be obtained on either glass or fused-silica when properly deactivated, fused-silica column preparation was easier to accomplish. Quantitative trace analyses could also be carried out with better reproducibility on the polar Carbowax 20M column, although the tailing behavior of the well-deactivated alkyl-polysiloxane columns was good.

A specific problem in alcohol analysis is the resolution of optically active alcohol enantiomers. Chiral terpene alcohols, such as those shown in Figure 6.15, have been shown to have pheromone properties for different insect species.[133,134]

There are two approaches to the chromatographic separation of enantiomers. Direct separation on optically active stationary phases derived from amino acids or peptides has been mostly applicable to amino acids (see Section 6.6). However, Oi et al.[135,136] obtained partial resolution of some chiral alcohols on optically active copper(II) complexes, and complete resolution (separation factors of 1.012–1.071) on optically active amino acid derivatives, N,N'-[2,4-(6-ethoxy-1,3,5-triazine)diyl]bis(L-valyl-L-valine isopropyl ester), and N,N',N''-[2,4,6-(1,3,5-triazine)triyl]tris(Nα-lauroyl-L-lysine-*tert*-butyl-amide). It was found that the (*S*)-isomers were consistently eluted prior to the (*R*)-isomers.

The second approach to enantiomer separation is the formation of diastereoisomers by introducing a second asymmetric center into the chiral analyte molecule by a simple chemical reaction. These derivatives can then be separated on any conventional stationary phase. Chiral alcohols have been derivatized using L-α-chloroisovaleryl chloride in the presence of a base,[137] or using *N*-trifluoroacetyl-L-alanine in the presence of dicyclohexylcarbodiimide as a condensing agent.[138] The derivatives were subse-

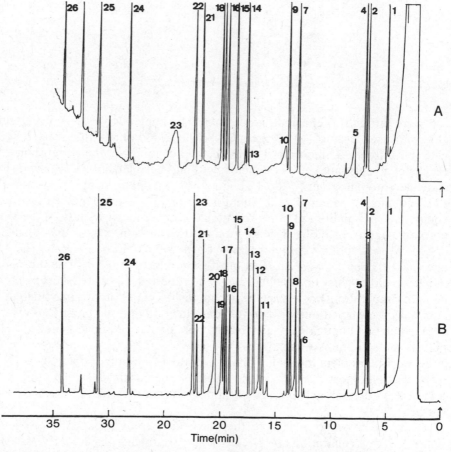

FIGURE 6.16. Chromatograms of a mixture of standard compounds (A) with boric acid subtractor column, and (B) without subtractor column. Note: 40 m OV-101 glass column, temperature program from 40 to 180°C at 4°C min^{-1} after an initial 2 min isothermal period, H$_2$ carrier gas, 10 ng of each compound injected. Selected peak assignments: (2) 2-heptanone; (3) cyclohexanol; (4) heptanal; (8) 1-octanol; (9) nonanal; (10) linalool; (11) borneol; (12) menthol; (13) α-terpineol; (20) 1-decanol; (23) thymol. (Reproduced with permission from ref. 139. Copyright Elsevier Scientific Publishing Company.)

quently separated on glass capillaries coated with OV-17, SE-30, or Emul-phor.

A novel approach to the detection of alcohols in mixtures containing compounds with other functional groups is the use of selective subtraction loops.[139] A 3 m length of 0.35 mm i.d. glass capillary was coated with boric acid and connected to the end of an analytical column (40 m OV-101). The boric acid loop selectively subtracted primary and secondary alcohols within the temperature range of 56–116°C. Tertiary alcohols do not form trialkyl borates but could be differentiated from primary and secondary alcohols by distinct peak broadening caused by partial dehydration. Figure 6.16 illustrates the use of the boric acid subtraction loop.

6.4 PHENOLS

The complete analysis (both qualitative and quantitative) of complex mixtures of phenols by gas chromatography is difficult to achieve. These compounds are highly polar and have low vapor pressures at ambient temperatures.. Furthermore, many isomeric alkyl phenols have similar chemical and physical properties and require high-resolution GC with selective stationary phases for their separation. These phases must be coated on completely inert surfaces in order to prevent peak tailing.

The most selective stationary phases that have been used for open tubular column GC of underivatized phenols include di-n-decylphthalate,[140,141] di(3,3,5-trimethylcyclohexyl)-o-phthalate,[141] tricresylphosphate,[141,142] and tri(2,4-xylenyl)phosphate.[141,143] The best separations were obtained on tri(2,4-xylenyl)phosphate[141] (see Figure 6.17). The elution of phenols on the phthalates was less affected by hydrogen bonding than on phosphate liquid phases. The addition of phosphoric acid minimized adsorption on the metallic column wall and peak tailing.

The major limitation to the use of the previously described phases is their relatively low thermal stabilities: none of them can be used at temperatures above 135°C. Furthermore, peak tailing is still observed at low levels of injected phenols even though several percent phosphoric acid is added to the stationary phase.

Other more thermally stable selective stationary phases that have been used for the analysis of underivatized phenols include diethylene glycol adipate,[145] Pluronic L64,[146] and Superox 20M.[147,148] All six dimethylphenol isomers, the cresols, and phenol were resolved on the Pluronic L64 column, while Superox 20M separates the three methylphenols (cresols) from one another, the six dimethylphenols from one another, and the three ethylphenols from one another, but 2,5-dimethylphenol and 4-methylphenol coelute, as do 3,5-dimethylphenol and 4-ethylphenol.

An advantage to using Superox 20M as stationary phase is the ease at which it can be efficiently coated on fused-silica tubing. Figure 6.18 shows

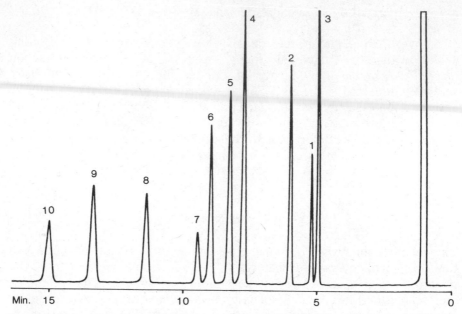

FIGURE 6.17. Chromatogram of phenol and methylphenols. Note: 20 m stainless steel column coated with tri-(2,4-xylenyl)phosphate + orthophosphoric acid modifier, 120°C isothermal temperature. Selected peak assignments: (1) phenol; (2) 2-methylphenol; (3) 2,6-dimethylphenol; (4) 4-methylphenol; (6) 2,4-dimethylphenol; (10) 3,4-dimethylphenol. (Reproduced with permission from ref. 141. Copyright American Chemical Society.)

a chromatogram of over 50 phenolic standard compounds on a fused-silica column coated with Superox 20M.[147] Excellent efficiency was obtained.

White and Li[147] have provided a comprehensive treatment of the retention characteristics of underivatized phenols on Superox 20M. The retention of phenols is a function of the strength of intermolecular hydrogen bonds with the etheric oxygens of the stationary phase, as well as the vapor pressures of the compounds. For alkyl phenols, the order of elution is *ortho*, *para*, and *meta*. Alkyl substituents in the *ortho* position increase steric hindrance of the phenolic hydroxyl group, reducing its ability to form hydrogen bonds with the stationary phase. Thus, 2,6-di-*tert*-butylphenol elutes before phenol and 2-*tert*-butylphenol.

For *para* and *meta* alkyl positional isomers, retention is governed by the ability of the alkyl substituent to influence the electron density around the hydroxyl oxygen atom. The greater the electron density, the weaker the hydrogen bond to the etheric oxygen of the stationary phase. For 4-butyl-phenols, the order of elution is 4-*tert*-butyl-, 4-*sec*-butyl-, and 4-*n*-butyl-phenol. This is explained by the inductive effects of the butyl groups. The *tert*-butyl group has the strongest electron donating inductive effect, and the electron density around the phenolic oxygen atom is highest. Therefore, its intermolecular hydrogen bond is weaker and it elutes before the other 4-butylphenols.

If the substituent changes from alkyl to aryl, as in the case of the phenylphenols, the order of elution becomes *ortho, meta, para*. The reverse in order of the *meta-* and *para-*isomers results because the *para-*phenyl group removes electron density from the hydroxyl atom via a resonance effect, thus increasing the intermolecular hydrogen bonds to the stationary phase.

The retention mechanisms discussed above were substantiated by the observation that alkyl phenols substituted in the 3 and 4 positions displayed a linear free-energy relationship of the Hammett type between the logarithm of the ratio of the activity coefficients of phenol to substituted phenol and the chromatographic substituent constant, σ_c.[147]

The order of elution of various phenols changed with column temperature. This could be due to differences in polarity or selectivity of the stationary phase at different temperatures. For example, 2-*sec*-butylphenol is not completely resolved from 3-ethylphenol and 2,3,5,6-tetramethylphenol at 160°C, while complete resolution is obtained at 150°C.[147]

It was found that the logarithm of the relative retention time as a function of carbon number for the C_1- to C_3-*n*-alkylphenols (substituted in the 2 and 4 positions) was linear. Therefore, the retention times of the 2- and 4-*n*-butylphenols were predicted.[147]

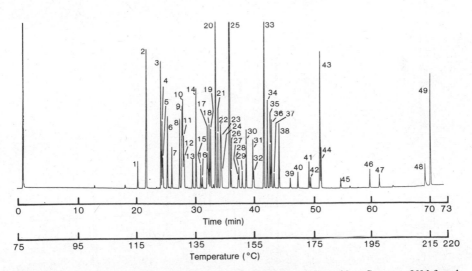

FIGURE 6.18. Chromatogram of standard phenol mixture. Note: 30 m Superox 20M fused-silica column, He carrier gas, 300:1 split ratio. Selected peak assignments: (1) 2,6-dimethylphenol; (2) 2,6-di-*tert*-butylphenol; (3) phenol; (4) 2-methylphenol; (5) 2,4,6-trimethylphenol; (11) 2,4-dimethylphenol; (20) 2-*tert*-butylphenol; (27) 4-*tert*-butylphenol; (29) 4-*sec*-butylphenol; (31) 3,4,5-trimethylphenol; (41) 2-phenylphenol; (46) 1-naphthol; (48) 3-phenylphenol; (49) 4-phenylphenol. (Reproduced with permission from ref. 147. Copyright American Chemical Society.)

The direct analysis of underivatized phenols on open tubular columns has been reported for coal-derived products using tri(2,4-xylenyl)phosphate,[143] Pluronic L64,[146] Superox 20M,[148] and SE-52;[114] tobacco smoke condensate using OV-101;[149] and priority pollutants using SE-54.[150] The latter study used fused-silica open tubular columns and reported the observation that even for the acidic dinitrophenols and the highly chlorinated phenols, narrow response factor variation was observed.

Derivatization, such as alkylation, silylation, and acylation,[151] has been used to overcome tailing during chromatography of phenols. Recent studies include the chromatography of acetylated monoalkylphenols on open tubular columns coated with DC 550 (a phenylmethylsilicone oil) and Carbowax 20M,[152] acetylated urinary monohydric and dihydric phenols on an OV-101 open tubular column,[153] and acetylated chlorophenols in environmental samples on an SE-30 open tubular column.[154] The acetyl derivatives are generally more resistant to hydrolysis and therefore more useful for samples containing traces of water. The latter study used an electron capture detector (ECD) for detection. The less polar derivatives and extremely stable open tubular column facilitated the sensitivity of the ECD method.

Although open tubular column technology has advanced to the point that underivatized phenols can be chromatographed with minimal peak tailing, and derivatization for this purpose is no longer necessary, it may be necessary to produce specific phenol derivatives in order to exploit the ex-

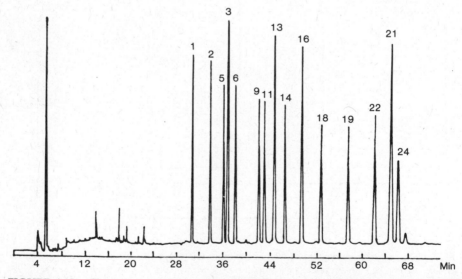

FIGURE 6.19. Chromatogram of a mixture of 2,4-dinitrophenol ethers. Note: 50 m SP-2100 fused-silica column, 220°C isothermal temperature after a 4 min initial 60°C temperature, He carrier gas. Selected peak assignments: (1) phenol; (2) 2-methylphenol; (3) 2,6-dimethylphenol; (5) 3-methylphenol; (6) 4-methylphenol; (9) 2,4-dimethylphenol; (22) 4-*tert*-butylphenol. (Reproduced with permission from ref. 159. Copyright Elsevier Scientific Publishing Company.)

tremely sensitive specific detectors such as the ECD for trace analysis. Esters of halogenated acids such as trifluoroacetates[155] and chloroacetates[156] have been used, although they suffer hydrolysis in the presence of water. On the other hand, α-bromo-2,3,4,5,6-pentafluorotoluene,[157] 4-chloro-α,α,α-trifluoro-3,5-dinitrotoluene,[157] heptafluorobutyrylimidazole,[158] and 1-fluoro-2,4-dinitrobenzene[157] form derivatives that are stable in aqueous solution and give good ECD response.

Lehtonen[159] has described the derivatization of phenols with 1-fluoro-2,4-dinitrobenzene to give the 2,4-dinitrophenol ethers for chromatography using the ECD. Open tubular fused-silica columns coated with OV-210 and SP-2100 were used. Figure 6.19 shows a chromatogram obtained from a mixture of 2,4-dinitrophenol ethers. The detection limit was 0.01–0.09 ng injected. Retention among positional isomers increased in the order *ortho*, *meta*, and *para*. Retention also increased with carbon number, was slightly lower for branched compared to straight chain substituents, and unsaturated substituents resulted in greater retention than the corresponding saturated groups. Both OV-210 and SP-2100 columns resolved all methylphenols and dimethylphenols from each other.

Fogelqvist et al.[160] described the rapid method of extractive alkylation of phenols from water using tetrabutylammonium ion as counterion and pentafluorobenzylbromide as alkylating agent. Chromatographic analysis was accomplished using an SE-52 glass open tubular column and ECD. The overall sensitivity of the method was in the range of 1–10 μg L^{-1}.

6.5 AMINES

The amines are among the most difficult chemicals to analyze by GC because of their strongly basic nature. Even the low concentration of hydroxyl groups on the surface of fused silica is enough to lead to irreversible adsorption of these compounds. The surfaces of leached glass and fused silica are acidic because of the presence of these hydroxyl groups. Figure 6.20 shows a chromatogram of six amines on a fused-silica capillary column coated with Carbowax 20M without any prior deactivation of the surface. Even though no peak tailing is observed, the primary aliphatic amines have been to some degree irreversibly adsorbed. Chromatograms of amines on nonpolar phases without prior surface deactivation show either complete compound adsorption or peaks exhibiting extreme tailing.

In order to properly analyze amines by open tubular column GC, either column surfaces must be completely deactivated, or the amines must be derivatized to form nonpolar products. There have been two approaches to the deactivation of open tubular column surfaces for the analysis of amines. First, the column surface can be modified by the addition of alkali metal salts. For example, the reaction of potassium hydroxide with surface hy-

droxyl groups leads to the formation of —Si OK functionality which imparts

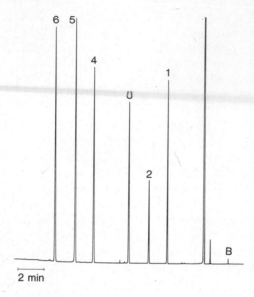

FIGURE 6.20. Chromatogram of amines on a polar phase fused-silica column without prior surface deactivation. Note: 20 m Carbowax 20M fused-silica column, temperature program from 100 to 200°C at 8°C min^{-1}, H$_2$ carrier gas. Peak assignments: (1) 1-aminooctane; (2) 1-aminononane; (3) 1-aminodecane; (4) 2,6-dimethylaniline; (5) aniline; (6) dicyclohexylamine. (Reproduced with permission from ref. 132. Copyright Friedr. Vieweg and Sohn.)

a basic character to the surface. Care must be exercised in controlling the amount of potassium hydroxide (or other salts) added to form only the "bound" potassium and not "free" potassium hydroxide. If excess base remains on the surface, catalytic decomposition of the stationary phase or analyte compounds occurs. In previous years, soda lime (soft) glass was used for the analysis of basic compounds such as the amines because of the basic nature of the glass surface. Figure 6.21 shows a chromatogram of the same amines as were analyzed in Figure 6.20, only on an alkali glass column that had been additionally treated with potassium hydroxide and coated with Carbowax 20M. The irreversible adsorption was greatly reduced in comparison.

Mistryukov et al.[161] described the *n*-hexane plug method for the application of basic salts onto the column surface prior to coating with the stationary phase PEG 40M. Surfaces containing potassium hydroxide, sodium orthophosphate, and potassium carbonate were compared. It was observed that the phosphate and carbonate columns showed better efficiency and deactivation for amines than did the potassium hydroxide column. Selectivity differences among the three columns were also observed.

The second method for surface deactivation for amines involves reactions such as the thermal polyethyleneglycol or alkylpolysiloxane degradation, cyclic siloxane reaction, or high-temperature silylation (see Section 3.3). Figure 6.22 shows chromatograms of amines on a nonpolar OV-101 stationary phase which was coated on fused-silica columns with no prior deactivation, a prior thermal alkylpolysiloxane deactivation, and a thermal Carbowax 20M deactivation, respectively. Both deactivated columns produced chromatograms with no tailing or adsorption, although it was found that the

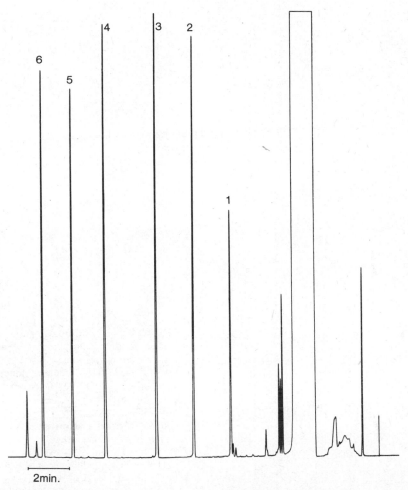

FIGURE 6.21. Chromatogram of amines on a polar phase alkali glass column additionally deactivated with KOH. Note: 20 m Carbowax 20M alkali glass column, temperature program from 60 to 200°C at 8°C min^{-1}, H$_2$ carrier gas. Peak assignments: same as in Figure 6.20. (Reproduced with permission from ref. 132. Copyright Friedr. Vieweg and Sohn.)

alkylpolysiloxane deactivated column was superior in terms of thermal stability and reproducible defined polarity.[132]

As just discussed, the chromatography of free amines requires stringent column surface deactivation in order to avoid adsorption problems. An alternative that is often employed is using suitable derivatives that reduce the highly polar character of the amine. Methods for forming the acetyl,[162-165] dimethyl,[165,166] trimethylsilyl,[165,167] trifluoroacetyl,[165,168-170] pentafluoropropionyl,[171] heptafluorobutyryl,[172] pentafluorobenzoyl,[173-175] 2,4-dinitrophenyl,[176,177] Schiff base,[178-182] isothiocyanate,[183-186] and N-dimethylthiophosphinyl[187] derivatives have been described. Many of these reactions have

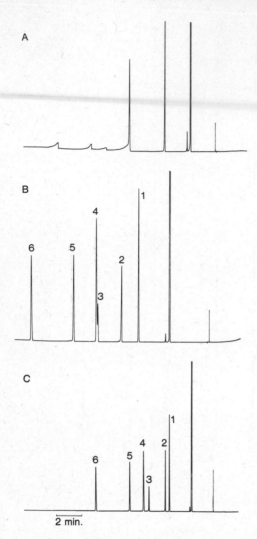

FIGURE 6.22. Chromatograms of amines on a nonpolar phase with various prior surface deactivations. Note: (A) 20 m OV-101 fused-silica column, nondeactivated; (B) 20 m OV-101 fused-silica column, thermal alkylpolysiloxane deactivation; (C) 12.5 m OV-101 fused-silica column, thermal Carbowax 20M deactivation; temperature program from 80 to 200°C at 8°C min^{-1}, N_2 carrier gas. Peak assignments: (1) aniline; (2) 1-octylamine; (3) 1-nonylamine; (4) 2,6-dimethylaniline; (5) 1-decylamine; (6) dicyclohexylamine. (Reproduced with permission from ref. 132. Copyright Friedr. Vieweg and Sohn.)

been used for determinations of the biologically important primary amines such as epinephrine, dopamine, catecholamines, phenethylamine, and amphetamine. The halogenated derivatives have been used for the electron capture detection of picogram amounts of primary amines. The N-dimethylthiophosphinyl derivative gave a detection limit of 500 fg using an alkali flame ionization detector, and a sulfur-containing Schiff base derivative[182] was analyzed using a flame photometric detector.

A number of problems exist with the different derivatization reactions. Silylation of amines is difficult and acylation of primary amines leaves a polar/NH functional group that can still lead to adsorption on the column wall. Successful derivatization of the lower molecular mass aliphatic pri-

mary amines can be difficult. For example, the reactions of pentafluorobenzoyl chloride[173] with methyl-, ethyl-, diethyl-, n-propyl-, and di-n-pro-
pylamines gave white, curdy precipitates and, therefore, were not quantitative. The procedure for forming the 2,4-dinitrophenylamines is also
complex. The formation of isothiocyanates leads to by-products that com-
plicate the chromatography. Schiff base formation, on the other hand, takes
place easily and rapidly at room temperature.[182] Figure 6.23 shows a chromatogram of sulfur-containing Schiff bases obtained on a PEG-20M glass
open tubular column.

Most of the applications in the literature describing the analysis of amines
with open tubular columns concern the aromatic amines. These compounds
are used as intermediates in the synthesis of a large number of organic compounds including dyes, drugs, pesticides, and plastics. Furthermore, they
are constituents of various coal-derived fuels and products. The known mutagenicity and carcinogenicity of the aromatic amines have stimulated much
effort to devise efficient methods for their analysis. The free amines have
generally been analyzed on polar phases such as Carbowax 20M,[188,189] PEG
1500,[190] PEG 400,[191] Amine 220 (1-ethanol-2-heptadecenyl-2-isoimida-
zole),[191,192] Reoplex 400 (polypropylene glycol adipate),[191] Versamid 900,[193]
and UCON 50 HB 2000.[194–196] All of these polar phases tend to neutralize
the adsorptive effects of the column wall and lead to more symmetrical
peaks. Generally speaking, compounds that contain functional groups in the
vicinity of the nitrogen atom are eluted preferentially on phases that enable
a hydrogen bond to be formed with the heteroatom of the molecule. The

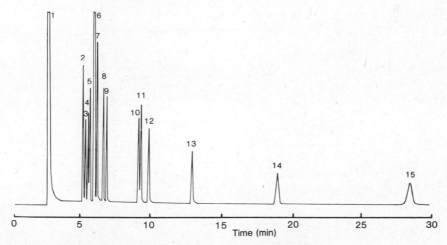

FIGURE 6.23. Chromatogram of sulfur-containing Schiff bases (derivatized amines). Note:
30 m PEG-20M glass column, 125°C isothermal column temperature, N₂ carrier gas, FID. Selected peak assignments: (2) isopropylamine; (3) methylamine; (8) n-propylamine; (10) n-bu-
tylamine; (13) n-amylamine; (14) n-heptylamine. (Reproduced with permission from ref. 182.
Copyright Elsevier Scientific Publishing Company.)

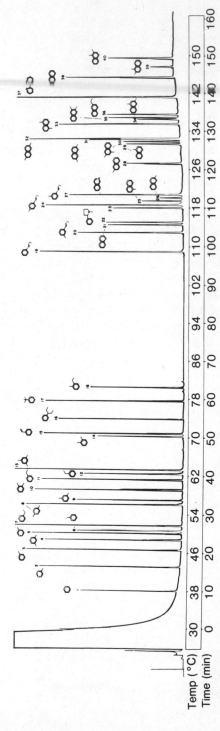

FIGURE 6.24. Chromatogram of aromatic nitrogen bases. Note: 50 m UCON 50 HB 2000 glass column. (Reproduced with permission from ref. 194. Copyright American Chemical Society.)

268

amines are more soluble in the basic phases, Amine 220 and Versamid 900, but these phases provide lower efficiencies. The major drawback of all of these phases is that they cannot be used at temperatures above about 200°C. This is not a problem for the analysis of the aromatic amines containing only one or two rings (see Figure 6.24), but higher-temperature phases must be used for the amino polycyclic aromatic compounds. Unfortunately, the higher-temperature phases have been restricted to the nonpolar silicone gum phases, offering little selectivity.

Open tubular columns have been used to analyze aromatic amines in air particulate matter,[189,193] coal-derived products,[191,192,194,196–202] and tobacco and marijuana smoke condensates.[195] In several cases,[198,200,201] derivatives (acetyl, trifluoroacetyl, and pentafluoropropionyl) were made in order to differentiate between primary and secondary amines, and tertiary amines by mass spectrometry or electron capture detection. The tertiary amines contain no protons on the nitrogen heteroatom and, therefore, do not react with the derivatizing reagent. Figure 6.25 shows a chromatogram of pentafluoropropionyl derivatives of several amino polycyclic aromatic compounds using an electron capture detector.[200] The selectivity of the ECD for the derivatized amines over the underivatized amines was greater than 1000:1.

Selective detection using a nitrogen-selective detector (NPD) is often used to differentiate the amines from other polar compounds that do not contain

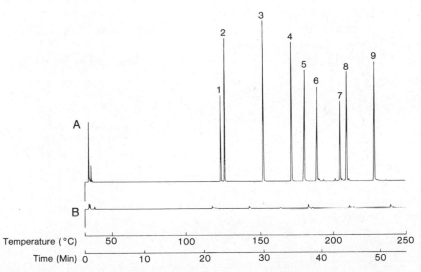

FIGURE 6.25. Chromatograms of (A) pentafluoropropionyl derivatives and (B) underivatized amino polycyclic aromatic standard compounds. Note: 30 m SE-52 fused-silica column, ECD. Selected peak assignments: (1) 1-aminonaphthalene; (2) 2-aminonaphthalene; (3) 4-aminobiphenyl; (6) 2-aminoanthracene; (8) 1-aminopyrene; (9) 6-aminochrysene. (Reproduced with permission from ref. 200. Copyright American Chemical Society.)

a nitrogen heteroatom.[198,201] A dual trace FID/NPD chromatogram of an aromatic nitrogen fraction of a coal liquid using a fused-silica effluent splitter is shown in the next chapter (Figure 7.27).[203] The splitter caused only a 5–10% efficiency loss due to dead volume.

Buchanan et al.[202] used ammonia chemical ionization mass spectrometry for differentiating between primary, secondary, and tertiary amines from the eluent of an open tubular column. When ammonia-d₃ was used as the ionization reagent, the amine was ionized by the transfer of a deuterium ion, and the hydrogen atoms on the nitrogen exchanged with the deuterium atoms of the ammonia. Thus, the resulting deuterated ion exhibited a change in mass of 1, 2, or 3 mass units depending on whether the compound was tertiary, secondary, or primary amine, respectively.

The separation of the enantiomers of chiral amines on optically active stationary phases has been reported by a number of researchers.[135,137,138,204–208] Oi et al.[135] reported the direct separation of racemic amines on optically active copper(II) complexes, although the peak shapes were broad and tailing. Generally, derivatives such as the trifluoroacetyl,[204–208] heptafluorobutyryl,[205] and pentafluoropropionyl[205,206] derivatives were prepared before chromatography on optically active stationary phases. Most of these phases have NH groups linked to the asymmetric carbon atoms, which form diastereoisomeric hydrogen bonds with solutes. These same phases are used for the separation of amino acid enantiomers and will be described in more detail in the next section.

Figure 6.26 shows a chromatogram of N-trifluoroacetylamines on a Pyrex open tubular column coated with a silicone polymer made by coupling L-valine-*tert*-butylamide to OV-225. The main advantage of this phase is its higher thermal stability compared to other reported phases.

The conversion of enantiomers into diastereoisomers using L-α-chloro-

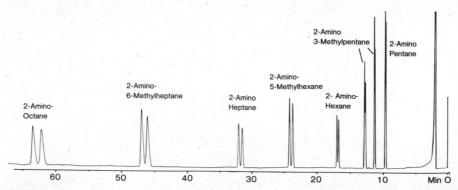

FIGURE 6.26. Chromatogram of N-trifluoroacetylamines on an optically active stationary phase. Note: 35 m Z-L-Val-OV-225 glass column, 80°C isothermal column temperature, H₂ carrier gas. (Reproduced with permission from ref. 207. Copyright Elsevier Scientific Publishing Company.)

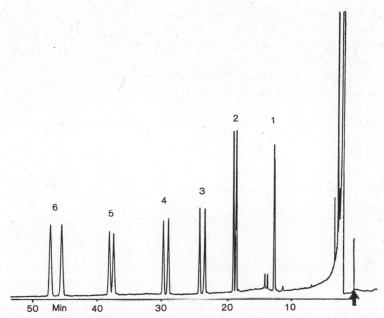

FIGURE 6.27. Chromatogram of N-(L-α-chloroisovaleryl)-($\pm$)amines on an optically active stationary phase. Note: 32 m N-TFA-L-phe-L-phe-cyclohexyl ester glass column, 150°C isothermal column temperature, H_2 carrier gas. Peak assignments: (1) ($\pm$)2-aminopentane; (2) ($\pm$)2-aminohexane; (3) ($\pm$)2-amino-5-methylhexane; (4) 2-aminoheptane; (5) 2-amino-6-methylheptane; (6) 2-aminooctane. (Reproduced with permission from ref. 137. Copyright Friedr. Vieweg and Sohn.)

isovaleryl chloride was reported by Konig et al. and Kruse et al.[137,138] These derivatives could be separated on either conventional or preferably with optically active stationary phases. Figure 6.27 shows a chromatogram of N-(L-α-chloroisovaleryl)($\pm$)amines on an open tubular column coated with N-TFA-L-phe-L-phe-cyclohexyl ester. Note the improved resolution as compared to the N-TFA derivatives (Figure 6.26).

6.6 AMINO ACIDS

Open tubular column GC of amino acids has been extensively studied because it is generally more sensitive, more rapid, and less expensive than conventional ion-exchange chromatography. The disadvantage is that the amino acids must be converted to stable, volatile derivatives. The most popular derivatives have been the trifluoroacetyl-n-butyl ester,[209] heptafluorobutyryl-n-propyl ester,[210,211] heptafluorobutyryl-isopentyl ester,[212,213] heptafluorobutyrylisobutyl ester,[214-220] and heptafluorobutyryl-n-butyl ester[221] derivatives. The latter two are preferable because they can be prepared by direct esterification and solvent evaporation at room temperature.

Once the amino acids are derivatized, the nonpolar methylpolysiloxane stationary phases are best suited for open tubular column GC of these derivatives. Using a glass column coated with OV-101, up to 32 amino acids can be analyzed in human plasma samples on a routine basis as their heptafluorobutyryl-isobutyl esters.[219] Open tubular GC was found to be about 30 times more sensitive than packed-column GC, allowing the analysis of plasma samples as small as 20 μL in the case of premature or newborn infants.

Poole and Verzele[221] found that the best separation of amino acids was obtained using the heptafluorobutyryl-*n*-butyl ester derivatives on an SE-30 column. Complementary information for identification purposes could be obtained by chromatographing the same derivatives on an OV-210 column.

In general, the elution order of the derivatized amino acids on methylpolysiloxane stationary phases depends on the boiling points of the derivatives. For isomers, structural features affect retention as follows:[219]

1. For amino group positional isomers, the amino group hydrogen exhibits an increase of polarity (increasing retention) with increasing distance from the carboxylic group.
2. The structure of the carbon backbone, such as the level of branching or cyclization, modifies the hydrophobic interactions.
3. Hydroxyl substitution by an amino group increases retention.

The method of injection is important in the quantitative analysis of amino acids. Excellent sensitivity and reproducibility were obtained using an all-glass solid injector.[219] A statistical comparison[220] between split and on-column injection, using a comparison of variances, showed that at the 5% level of significance, the precision obtained using on-column injection was better than that derived from a splitter.

Selective detection of amino acid derivatives has been demonstrated using an ECD[222,223] and an NPD.[224] Although ECD detection has been demonstrated to give significant improvement in sensitivity for certain derivatives, the use of the ECD requires special care to avoid contamination problems and to determine response factors. The combination of glass open tubular column GC with nitrogen-selective detection has permitted ready quantitation of picomole amounts of amino acids. Figure 6.28 shows a chromatogram of amino acids from a normal urine sample equivalent to 2 μL of urine.

The separation of amino acid enantiomers by open tubular column GC has received much attention in recent years. The determination of the configuration of amino acids is important in the investigation of meteorites, in the study of peptide antibiotics and other natural peptides, in the dating of amino acid-containing fossils, and in the determination of optical purity during peptide synthesis. The two approaches used are (a) the direct injection of derivatized enantiomeric mixtures onto optically active stationary phases

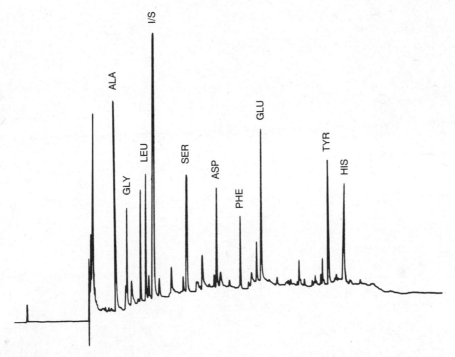

FIGURE 6.28. Chromatogram of amino acids in normal urine. Note: 50 m 1:1 mixture of Carbowax 20M and Silar 5CP glass column, temperature program from 110 to 190°C at 8°C min^{-1}, He carrier gas at 0.75 mL min^{-1}. I/S refers to the internal standard (norleucine). (Reproduced with permission from ref. 224. Copyright Preston Publications, Inc.)

and (b) converting enantiomers into diastereoisomeric derivatives by reaction with chiral reagents followed by chromatography on conventional stationary phases.

The separation of enantiomers by open tubular columns coated with chiral stationary phases has been reviewed by König.[225] Gil-Av and co-workers[226,227] first reported enantioselectivity of volatile amino acid derivatives on columns coated with derivatives of amino acids and dipeptides such as N-trifluoroacetyl-L-valyl-L-valine cyclohexyl ester (I, Figure 6.29). Subsequently, a large number of these types of phases have been synthesized and tested.[204,228–247] The main problems encountered with these phases are their limited thermostability and useable temperature range. Compound I, for example, can be used only between 100 and 110°C. The highest maximum operating temperature for the phases referenced above is 170°C; most are below 120°C.

Significant progress toward temperature stability was achieved with the synthesis of a polysiloxane with L-valine-*tert*-butylamide groups bound in the side chains[248–250] (II, Figure 6.29). This polysiloxane phase, called Chirasil-Val, is thermally stable to about 220°C. Figure 6.30 shows a separation

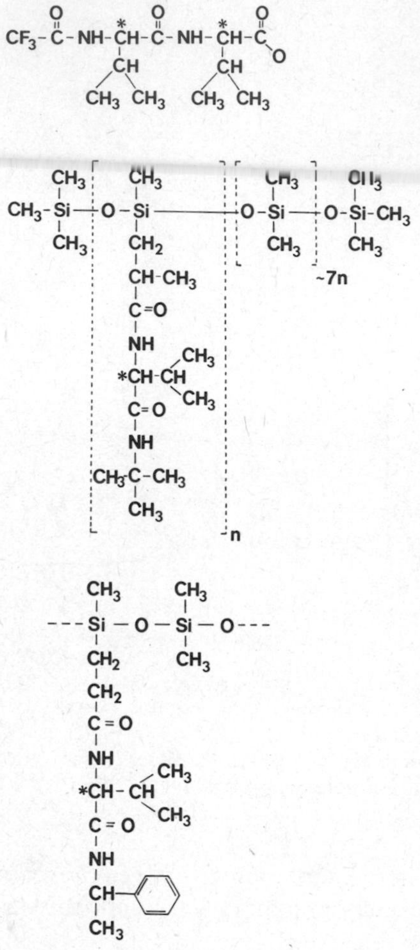

FIGURE 6.29. Chiral stationary phases.

of a derivatized racemic amino acid mixture on a glass open tubular column coated with this phase. It became possible for the first time to separate the enantiomers of all protein amino acids in a single chromatographic run.

Saeed et al.[251,252] synthesized phases very similar to Chirasil-Val by modifying OV-225 and Silar-10C by acidic hydrolysis of the cyanoalkyl side chains to yield carboxyalkyl groups and bonding L-valine-*tert*-butylamide via an amide linkage. A similar approach was reported by König et al.[246] in which XE-60 was modified by alkaline hydrolysis of the nitrile groups and coupling with L-valine-(S)-α-phenylethylamide (III, Figure 6.29). This polymer was found to be useful for the separation of a number of enantiomer types in addition to amino acids. In addition, benzyloxycarbonyl-L-valine and -L-leucine were coupled to the amino groups formed by reduction of the cyano groups in OV-225 with lithium aluminum hydride.[207]

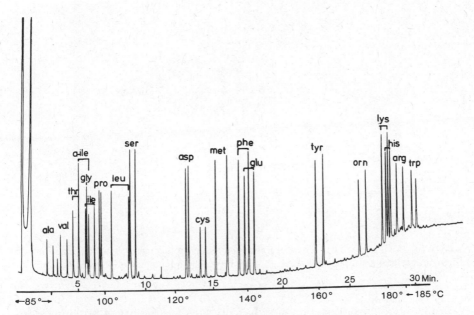

FIGURE 6.30. Chromatogram of derivatized racemic amino acid mixture. Note: 20 m Chirasil-Val glass column, temperature program from 85 to 185°C at 3.8°C min^{-1} after an initial 3 min isothermal period, H_2 carrier gas, N,O,S-pentafluoropropionyl isopropyl esters. (Reproduced with permission from ref. 250. Copyright Dr. Alfred Huethig Publishers.)

The first mechanism proposed for the separation of derivatized amino acids on dipeptide type phases was based on association between the enantiomeric solutes and the chiral solvent through a triply hydrogen-bonded complex[229,232] as shown in Figure 6.31. Corbin et al.[234] found that the amide portion (and not the ester portion) of the solvent molecule was the segment actually participating in complex formation, and other reports[241,253] verified these findings.

More recently, association complexes with two hydrogen bonds between the solute and solvent (between either planar "C_5" rings or folded "C_7" rings, see Figure 6.32) have been proposed[254-256] and are supported by IR and X-ray measurements.[257-260] On the other hand, Weinstein et al.[204] found that it was sufficient for a chiral stationary phase to contain an amide group with an asymmetric carbon atom attached to the nitrogen atom to show selectivity for N-trifluoroacetylamino acid esters. Furthermore, Oi et al.[206] found that α-hydroxycarboxylic acid esters were effective as optically active stationary phases for amino acids, supporting the fact that OH groups can contribute to the formation of diastereoisomeric association complexes.

Other types of intermolecular forces, such as dipole–dipole interactions or van der Waals forces between solvent and solute also contribute to enantiomer separation.[225] It was shown[137,226] that N-trifluoroacetyl-L-prolyl-

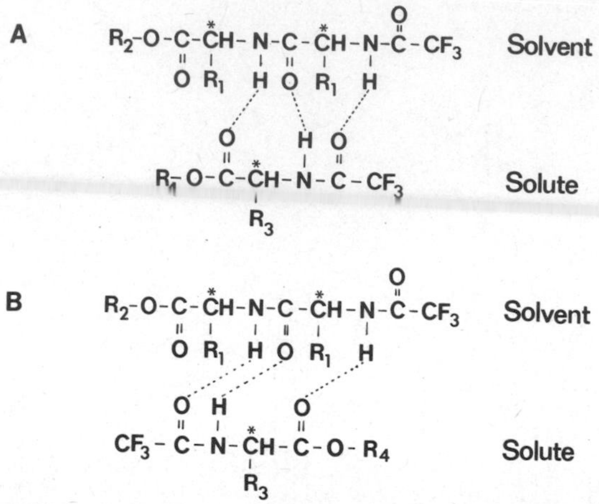

A R_2-O-C-$\overset{*}{C}$H- N-C- $\overset{*}{C}$H-N-C-CF$_3$ Solvent

B R_2-O-C-$\overset{*}{C}$H-N-C-$\overset{*}{C}$H-N-C-CF$_3$ Solvent

FIGURE 6.31. Proposed triply hydrogen-bonded association complex between enantiomeric amino acid solutes and dipeptide phases. A and B represent the complexes for both enantiomers.

L-proline cyclohexyl ester, which has no sites for hydrogen-bonding association (see Figure 6.33), effects complete separation of N-trifluoroacetyl-proline isopropyl ester enantiomers.

The mechanisms of resolution of enantiomers can be easily understood[204,261] by considering that the solvent is arranged in a somewhat orderly manner through intermolecular hydrogen bonding (or other forces). When a solute, the configuration of which is the same as the chiral solvent, is introduced into the solvent, it can readily assume a conformation analogous to the solvent, and thus lead to a good fit and close association. Insertion of the corresponding antipodic solute into the stack of solvent molecules leads to inversion of the positions of the hydrogen atom and the R_3-group attached to the asymmetric carbon atom. This distorts the solvent stack which leads to lower stability and reduced chromatographic retention.

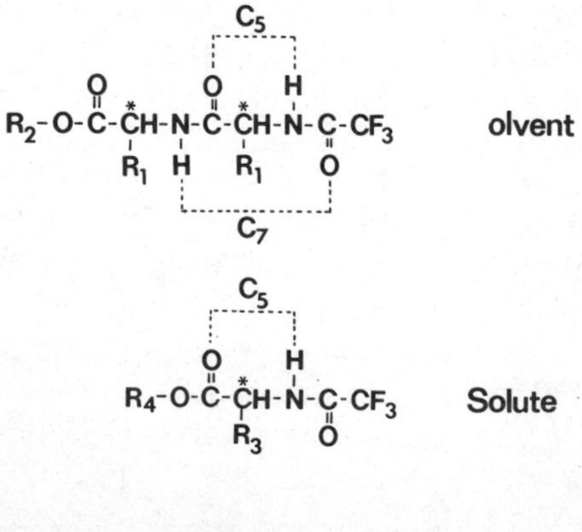

C_5

R_2-O-C-$\overset{*}{C}$H-N-C-$\overset{*}{C}$H-N-C-CF$_3$ olvent

C_7

C_5

R_4-O-C-$\overset{*}{C}$H-N-C-CF$_3$ Solute

FIGURE 6.32. Proposed doubly hydrogen-bonded association complexation sites between enantiomeric amino acid solutes and dipeptide phases.

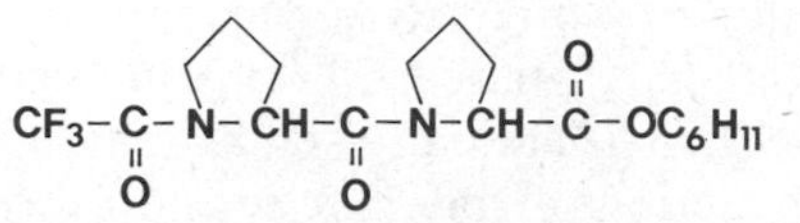

FIGURE 6.33. Chemical structure of N-trifluoroacetyl-L-prolyl-L-proline cyclohexyl ester.

For the separation of the N-trifluoroacetyl isopropyl ester derivatives of D,L amino acids on N-trifluoroacetyl-L-L-dipeptide cyclohexyl ester stationary phases, the hydrogen-bonded diastereoisomeric association complex with the L-isomer has enough stability to allow its retardation and subsequent longer elution time than the analogous D complex. The difference in free energy between the L,L- and the L,D-complexes results in the retention of one enantiomer relative to its antipode.

In studies of the effect of increasing the chain length of the R_1 group (see Figure 6.31) on the chiral carbon in the dipeptide stationary phase, it was found that when R_1 was lengthened, an increase in ease of solute–solvent complexation occurred, with the three and four carbon lengths competing for the highest degree of interaction.[238]

Selectivity between enantiomers on a chiral stationary phase is also influenced by the nature of the functional groups on the derivatized amino acids being analyzed. In the past, the main objective of the derivatization procedure was to provide volatile derivatives that could be analyzed on the relatively low-temperature phases. The N-trifluoroacetyl isopropyl esters have been most extensively used, although the pentafluoropropionyl derivatives have the highest volatility and, therefore, shortest analysis time.[233,235] More in-depth studies of the effect on resolution of the various N perfluoroacyl alkyl esters on Chirasil-Val have been published.[262,263] It was found that variations in the perfluoroacyl group have little effect on retention, while modifications of the alkyl group induce considerable changes in retention and elution order. In general, enantiomer resolution decreases with increasing size of the substituting groups.

The effect on the separation of amino acid enantiomers by converting the carboxyl group of the solute into an amide was studied by Oi et al.[264] It was found that the N-trifluoroacetyl-DL-proline isopropylamide was resolved with a high separation factor even though the same chromatographic conditions gave no detectable separation for N-trifluoroacetyl-DL-proline isopropyl ester.

With the presently available, more thermally stable polymeric phases, it is possible to use more polar and less volatile derivatives. For example, the N-methylamino acids are particularly susceptible to racemization.[265] In trying to form the acylated derivatives of N-methylamino acid esters, complete racemization occurs.[266] Reaction with isopropyl isocyanate yields N-methyl-N-ureidoamino acid amides (without racemization) which can be separated on the polymeric chiral phases.[225]

An alternative method for separating amino acid enantiomers is accomplished by introducing a second asymmetric center into the amino acid molecule by derivatization followed by separation of the resulting diastereoiso-mers on conventional, optically inactive stationary phases.[267–272] Although the use of open tubular columns with conventional thermally stable station-ary phases is an advantage to this approach, serious errors may result if the chiral reagents are not absolutely optically pure.[272] Furthermore, a decrease in accuracy may be caused by different reaction kinetics in the formation of diastereoisomers, since the reaction may proceed through transition states of different energy.[272]

6.7   CARBOXYLIC ACIDS

The importance of the open tubular column GC of carboxylic acids is em-phasized by the estimation that of all the papers that have been published on GC, nearly 25% are devoted to the analysis of fatty acids.[273] This is a result of the biological and economic importance of this class of compounds. Fatty acids containing carbon atoms from C_{12} to C_{24} are of particular interest because they are the main components of oils of animal and vegetable origin

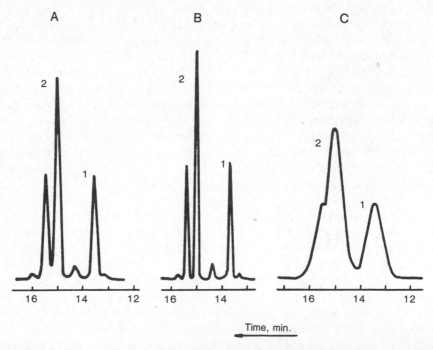

FIGURE 6.34.   Chromatograms of C_{18} fatty acid methyl esters on (A) support-coated open tubular, (B) wall-coated open tubular, and (C) packed column. Note: (A) 15.24 m × 0.50 mm i.d. DEGS SCOT column, (B) 45.72 m × 0.25 mm i.d. WCOT column, and (C) 2.44 m × 2.16 mm i.d. DEGS packed column. Peak assignments: (1) methylstearate; (2) methyl oleate. (Re-produced with permission from ref. 274. Copyright Preston Publications, Inc.)

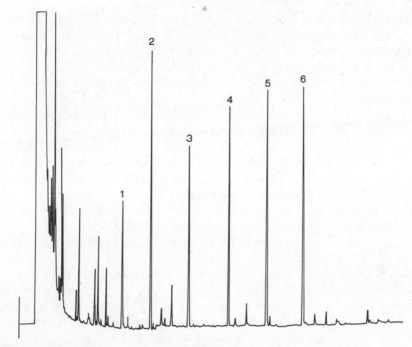

FIGURE 6.35. Chromatogram of normal C_3–C_8 free fatty acids. Note: 22 m Carbowax 1000 glass column, temperature program from 60 to 160°C at 4°C min^{-1}, He carrier gas. Peak assignments: (1) propionic acid; (2) butyric acid; (3) valeric acid; (4) caproic acid; (5) enanthic acid; (6) caprylic acid. (Reproduced with permission from ref. 282. Copyright Dr. Alfred Hucthig Publishers.)

which are used as such or after chemical modification for industrial purposes. The shorter chain, more volatile acids are difficult to analyze because of their high polarity. The longer chain, nonvolatile acids additionally stress the upper temperature limits of the chromatographic column.

The complexities of fatty acid mixtures are often magnified by the presence of unsaturation in the hydrocarbon chains. Since there exists a large number of possible positional and stereoisomeric ethylenic bonds, open tubular column GC is oftentimes necessary to achieve adequate resolution. Figure 6.34 compares chromatograms of methyl stearate and methyl oleate obtained on support-coated open tubular, wall-coated open tubular, and packed columns (see also Table 1.2).[274] It is obvious that given the same analysis time, open tubular columns give more information about the sample composition.

The separation of free fatty acids by open tubular column GC is difficult because of the high polarity of the carboxylic acid group and the tendency to form hydrogen bonds. Most of the separations in the past on open tubular columns have been achieved on acidic stationary phases or stationary phases that have been doped with a strong acid. These include trimer acid[275] (a mixture of C_{54} tricarboxylic and C_{36} dicarboxylic acids), a stationary phase containing orthophosphoric acid,[276] Ucon LB-550-X (polypropylene glycol) with phosphoric acid additive,[277,278] free fatty acid phase (FFAP),[279,280] and 1,15-pentadecane dicarboxylic acid.[281] The relatively poor thermal stabilities of these stationary phases and the necessity of acidic additives in many cases are unsatisfactory.

Hrivnac et al.[282] deposited a layer of silica on a leached Pyrex column and deactivated the surface with benzyltriphenylphosphonium chloride (BTPPC) before coating with various stationary phases (Carbowax 1000, XE-60, OV-73, UCON 50 HB 5100, and FFAP). Figure 6.35 shows a chromatogram of normal C_3–C_8 free fatty acids on a Carbowax 1000 column prepared in this manner. Stringent deactivation steps, such as these, must be undertaken in order to reduce peak tailing of free carboxylic acids to an acceptable level. Similarly, high-temperature silylation was used to deactivate leached Pyrex columns before coating with a mixture of OV-1 and FFAP.[283]

The introduction of fused-silica open tubular columns provided a surface that needs no further deactivation.[284] Free fatty acids exhibit only slight tailing on fused-silica columns coated with both nonpolar and polar stationary phases.[132] Figure 6.36 shows a chromatogram of the free fatty acids on a fused-silica column coated with Carbowax 20M.

Because of the difficulties encountered in open tubular column GC of free carboxylic acids, they are most often derivatized to form more volatile and less polar solutes. By far the most popular derivatives are the methyl esters. Much work was done in this area in the 1960s and has been reviewed by Krupcik et al.[273] More recent studies include the determination of the methyl ester derivatives of fatty acids in vegetable oils,[285–287] marine fish oils,[286,287]

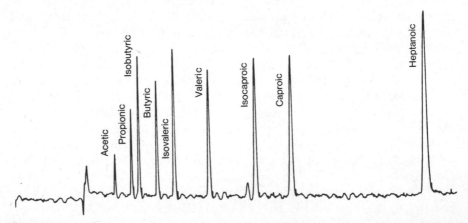

FIGURE 6.36. Chromatogram of free fatty acids. Note: 20–22 m Carbowax 20M fused-silica column, 100 pg level per component. (Reproduced with permission from ref. 284. Copyright Dr. Alfred Huethig Publishers.)

margarine and butter fats,[288,289] foods,[290] algae,[291] yeast,[292] marijuana smoke condensate,[293] fermentation products,[294] and animal and human biological material.[286,295-300]

Methylation with methanol and sulfuric acid[301] is one of the more desirable techniques because it is simple, rapid, reliable, and economical. Most other esterification techniques suffer from disadvantages: boron halides in methanol have a very limited shelf life; boron halides in butanol are not readily available; diazomethane is carcinogenic, toxic, and inflammable; and thionyl chloride is exceedingly corrosive and cannot form with formic acid an acyl chloride for subsequent methylation by methanol.[294]

Various stationary phases have been used in open tubular columns for the analysis of fatty acid methyl esters. Nonpolar phases included squalane,[302] Apiezon L,[273,285,288,291,292,303] SE-30,[273,303] OV-101,[298,304,305] OV-17,[298,306] SE-52,[273] SE-54,[298] SP-2100,[305] SP-1000,[294,307] and Pentasil.[295] In using these phases, extreme care is necessary to ensure proper surface deactivation.

In the past, the more polar phases that were used include butanediolsuccinate (BDS),[273,285,288,291,292,303,308] diethyleneglycolsuccinate (DEGS),[273,288,292] ethyleneglycolsuccinate (EGS),[273] polyethyleneglycol succinate (PEGS),[273] polyethyleneglycoladipate (PEGA),[273] FFAP,[289,293,309] Carbowax 20M,[273,286,310] Carbowax 1540,[302] polyphenylether,[311-313] and PZ-176 (a phenyl ether–phenyl sulfone polymer).[314,315] In more recent years, attention has focused on the use of cyano-containing polymers as stationary phases. These include Silar-5CP (50% cyanopropyl, 50% phenyl polysiloxane),[285,287,291,298,303,316] SP-2340 (75% cyanopropyl polysiloxane),[290,299,317] SS-4 (50% cyanoethyl, 50% methyl polysiloxane),[300] Silar 10C (100% cyanopropylpolysiloxane),[317] and OV 275 (cyanoallyl polysiloxane).[317] The cyanopolysiloxanes are of such high polarity that it is possible to separate many geometric and positional isomers of fatty acid methyl esters. In comparing the resolution of the cis and trans monounsaturated C_{18} fatty acid methyl esters on SP-2340 and FFAP, Grob et al.[289] found better resolution on the FFAP phase, thereby criticizing the high interest in the use of the cyanopolysiloxanes. One major disadvantage of all the polar phases used for this application is their relatively low thermal stabilities (upper temperature limits are approximately 200°C or lower).

In addition to the methyl esters, several other carboxylic acid derivatives have been studied. These include ethyl esters,[318] benzyl esters,[319,320] pentafluorobenzyl esters,[160] and trimethylsilyl derivatives.[321,322] The use of benzyl esters is convenient for determination of the lower acids because the lower volatility of these esters reduces evaporative losses. Furthermore, the presence of the benzyl group facilitates recognition of the mass fragmentation pattern. The benzyl esters can be prepared by direct action of phenyldiazomethane or via formation of the tetrabutylammonium salts. The former method[319] can be performed quantitatively even in the presence of water, making direct analysis in aqueous samples possible. The latter method[320] offers the advantages of rapidity and efficiency, while the former requires fresh reagents and many precautions. The pentafluorobenzyl esters can be detected at levels as low as 0.01–0.1 pg using an electron capture detector.[160] Flash-heater derivatization in the injector port using BSTFA to form the trimethylsilyl derivatives can be successfully used for carboxylic acids which do not decompose during the injection and which react rapidly with the derivatization reagent to form a single derivative.[221]

Both Kovats indices and equivalent chain length (ECL) values[273,323,324] have been used most successfully for tabulating retention data and for preliminary or tentative identification of branched and unsaturated acids. ECL values are calculated similarly to Kovats indices according to the equation

$$ECL = z + \frac{\log t'_R(x) - \log t'_R(z)}{\log t'_R(z + 1) - \log t'_R(z)} \qquad (6.5)$$

where x corresponds to an unknown fatty acid that is eluted between the normal saturated fatty acid with z and $z + 1$ carbon atoms. Almost identical standard deviation for ECL values of methyl esters of fatty acids with one double bond was obtained ($s = \pm 0.002$ ECL unit) as compared to Kovats indices of hydrocarbons ($s = \pm 0.2$ index unit).

The possibility of correlating retention values (ECL) of carboxylic acids with structure was advocated by Ackman and Hooper.[285,316,325,326] Structural increments or fractional chain length (FCL) values for branching and for double bonds were determined from ECL measurements, similar to work done on retention increments (ΔI) for hydrocarbons using the Kovats system (see Section 6.2). Generally, retention decreases with branching and increases with unsaturation, especially on polar phases.

In some cases, two acids that coelute at a given temperature can be resolved at a different operating temperature.[273,288,325] Within a sufficiently small temperature interval, the effect of temperature on adjusted retention time can be expressed as[273,327]

$$\log t'_R = \frac{H_e^s}{2.303 \; RT} + C \qquad (6.6)$$

From plots of $\log t'_R$ vs. $1/T$ for fatty acid esters, the enthalpies of solution (H_e^s) and optimum temperatures for separations can be obtained. The temperature dependence of ECL values can be used for preliminary estimation of the number of double bonds[288] or differentiation between some cis- and trans-isomers.[292]

The number and positions of double bonds can also be determined in some cases by reduction of the double bonds in the fatty acids between chromatographic analyses. Hydrazine reduction is a simple technique for the elimination of one or more polyenoic bonds without disturbing others.[291] Careful comparisons of retention data before and after reduction and correlations with FCL can be very useful in deducing structures. Complete hydrogenation followed by GC has been used in fatty acid methyl ester analysis to establish the carbon skeleton.[328] Placement of the platinum catalyst in the injection port just before the capillary column was reported to give excellent results.[329] Complex mixtures containing both unsaturated and branched compounds were simplified to only saturated compounds before chromatography.

Optical isomers of carboxylic acids have been studied using open tubular columns. Although the majority of these studies involve the analysis of chiral α-hydroxy acids,[208,330–333] Konig and Benecke[331] were able to resolve a number of enantiomers of 2-alkyl-substituted carboxylic acids as their (+)-3-methyl-2-butyl esters on an SE-30 open tubular column. In earlier studies,[334,335] enantiomers of acyclic isoprenoid acids were resolved as their (−)-menthyl esters on long open tubular columns coated with butanediol succinate polyester.

The triglycerides are carboxylic acids of particular interest to gas chromatographers. They are among the highest boiling compounds presently analyzed routinely by GC. Even so, relatively little has been published on their analysis by open tubular column GC. Grob et al.[336] have reviewed the most important considerations for their separation using open tubular columns. To be suitable for triglyceride analysis, columns should be stable to high temperatures (at least 330°C), extremely well-deactivated, relatively short (4–15 m), and coated with thin stationary phase films (~0.08–0.12 μm). Polar phases would be useful for the resolution of unsaturated compounds, but the low thermal stability of columns coated with presently available polar phases prevent their use. Schomburg et al.[337] reported the use of the polar Poly S 179 phase for the separation of triglycerides, but the temperature range was limited. Apolar phases, such as OV-1, SE-30, SE-52, and SE-54, are the most thermally stable and separate the triglycerides according to the total number of carbon atoms. They have good selectivity for structural and stereoisomers, but the separation of unsaturated species is less satisfactory: resolution may be high, but peaks are not grouped according to the number of double bonds as on polar phases.[336]

It was shown by Grob[338] that losses of triglycerides in open tubular columns occur mainly as a result of degradation, rather than polymerization, adsorption, and so on. Therefore, the chromatographic process must be optimized to provide minimum thermal stress to the compounds. The elution temperature is the most important parameter to control. The elution temperature can be minimized without loss in column efficiency by decreasing the stationary phase film thickness (0.08–0.12 μm is a good compromise between sufficient capacity and low retention[336]) and by choosing hydrogen as the carrier gas. On the other hand, increasing the carrier gas flow rate above the optimum and reducing the column length leads to reduced column efficiency in addition to lower elution temperature. Using these suggestions, complete separation of triglycerides differing by one double bond was obtained.[339]

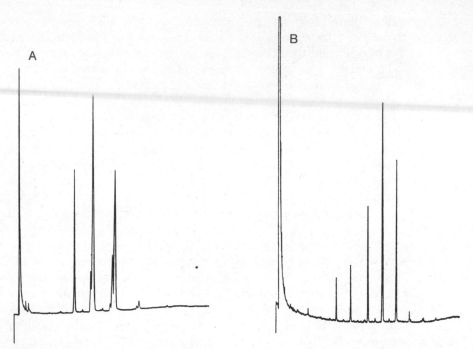

FIGURE 6.37. Chromatograms of Brazilian cocoa butter triglycerides (A) before and (B) after ozonolysis. Note: 15 m OV-1 (0.12 μm film thickness) glass column, temperature program (A) from 308 to 328°C at 1°C min^{-1} and (B) from 244 to 332°C at 4°C min^{-1} after a 1 min isothermal period, H_2 carrier gas. (Reproduced with permission from ref. 341. Copyright Dr. Alfred Huethig Publishers.)

The type of sample introduction affects the precision and accuracy for triglycerides. It was found[340] that split injection produced results with strong discrimination and high standard deviation, splitless injection gave results reflecting a discrimination resulting from insufficient elution out of the syringe needle, and cold on-column injection had little discrimination and standard deviations of 1–3%.

In order to obtain information concerning the position of double bonds in unsaturated triglycerides, Geeraert and De Schepper[341] cleaved the double bonds by ozonolysis, and measured the resultant carbon numbers or ECL values. Figure 6.37 shows chromatograms of Brazilian cocoa butter triglycerides before and after ozonolysis. Although there are limitations to this method (all double bonds are cleaved), much structural information can be obtained.

6.8   CARBOHYDRATES

Most carbohydrates are polymeric and must be hydrolyzed to mono- or oligosaccharides before analysis. Even then, derivatives must be formed in order to increase their volatility and decrease their polarity for analysis by GC. Trimethylsilyl (TMS),[342-344] acetyl,[345] and trifluoroacetyl (TFA)[346] derivatives of mono- and disaccharides have been analyzed using open tubular columns. The TFA derivatives proved to be useful for subpicogram detection of monosaccharides in sea water using an electron capture detector.[346] Even with highly efficient nonpolar (OV-101[342,345,346], SE-54,[346] and SE-30[342,343]) and polar (XE-60[342]) stationary phases, all of the anomers and ring structure isomers of the monosaccharides present in representative samples cannot be completely resolved. Figure 6.38 shows a chromatogram of nine monosaccharides as their TFA derivatives on a 50 m SE-54 open tubular column.[346]

Each monosaccharide formed by hydrolysis of a polysaccharide will be in at least two forms in solution due to the anomeric center at the glycosidic position. Hemiacetal formation creates two stereoisomers, called anomers, differing only in configuration at the hemiacetal carbon (see Figure 6.39).

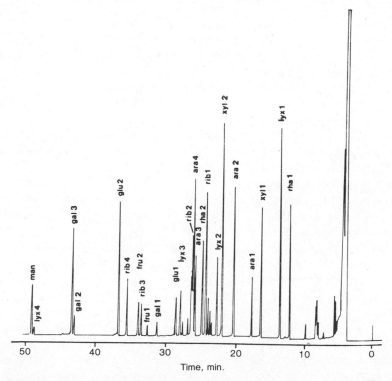

FIGURE 6.38. Chromatogram of nine monosaccharides as their TFA derivatives. Note: 50 m SE-54 glass column, temperature program from 74 to 130°C at 1°C min⁻¹ after an initial 4 min isothermal period. Peak identifications: rha = rhamnose, lyx = lyxose, xyl = xylose, ara = arabinose, rib = ribose, glu = glucose, fru = fructose, gal = galactose, man = mannose. (Reproduced with permission from ref. 346. Copyright Elsevier Scientific Publishing Company.)

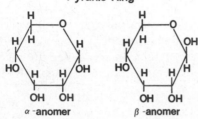

Furanic Ring

α-anomer   β-anomer

Pyranic Ring

α-anomer   β-anomer

FIGURE 6.39. Anomeric and ring structural isomers of D-ribose.

Furthermore, some monosaccharides, for example, D-ribose, exist as four isomers in solution, that is, as the α- and β-anomers of the furanose and pyranose forms (see Figure 6.39). Note the separation of all four D-ribose isomers in Figure 6.38, as well as many other monosaccharide isomers.

The large number of possible isomers present in complex monosaccharide mixtures places high demands for the maximum efficiency obtained with open tubular columns. Quantitation can be difficult if all peaks are not resolved and identified; for example, the proper quantitation of D-ribose (Figure 6.38) would be the sum of all four peaks.

An alternative method is to reduce the sugars to the corresponding polyols, followed by derivatization before GC. Reduction eliminates the possibility of hemiacetal formation and only one chromatographic peak for each monosaccharide results. The advantages of this approach are that (a) the chromatograms are not as complex; (b) quantitation of one chromatographic peak is easier than summing several; and (c) the signal-to-noise ratio for a monosaccharide producing a single peak is greater than for a sugar producing multiple peaks. The latter is important in applications involving trace analysis. The disadvantages are (a) the procedure cannot distinguish between certain pairs of sugars, for example, pairs such as lyxose-arabinose and gulose-glucose will give the same polyols upon reduction; (b) certain ketoses and aldoses, for example, sorbose and glucose, will also give the same polyol; and (c) ketoses always give two different products upon reduction, for example, fructose produces both mannitol and glucitol.[346]

Alditol acetates formed by reduction of the monosaccharides with sodium borohydride and acetylation with acetic anhydride have been analyzed on open tubular columns coated with Chirasil-Val (N-propionyl-L-valine-*tert*-

butyl-amide polysiloxane),[347] SP-2100,[348] FFAP,[348] and OV-275.[349] The best separations were obtained on the polar FFAP and OV-275 open tubular columns. Figure 6.40 shows a chromatogram of 17 standard alditol acetates on a 25 m FFAP fused-silica column.[348] Similarly, trimethylsilyl derivatives of the reduced monosaccharides have been chromatographed on nonpolar SF-96[350] and OV-101[351] open tubular columns. Figure 6.41 shows the separation obtained on a 50 m OV-101 column[351] for comparison with Figure 6.40.

Since the previously described reduction and derivatization methods were laborious and time-consuming, alternative methods have been published. Oximation prior to silylation[352–355] or acetylation[356] is often used, but each sugar gives rise to two chromatographic peaks which represent the *syn* and *anti* forms of the oxime. Honda et al.[357,358] recommended the use of diethyl dithioacetal trimethylsilyl derivatives of the aldoses for open tubular column

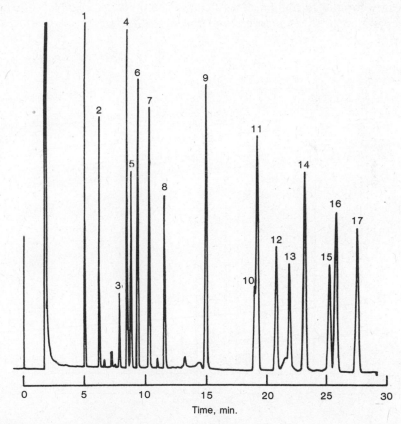

FIGURE 6.40. Chromatogram of standard alditol acetates. Note: 25 m FFAP fused-silica column, 205°C isothermal temperature. Selected peak assignments: (4) D-fucitol; (6) D-ribitol; (7) L-arabinitol; (14) D-mannitol; (15) L-glucitol; (16) D-galactitol. (Reproduced with permission from ref. 348. Copyright Elsevier Scientific Publishing Company.)

GC analysis. Although the total analysis time, including derivatization, was within 2 h, careful selection of the acid catalyst, its concentration, and the reaction time was necessary to avoid formation of isomeric monothioacetals and thioglycosides as by-products.

The enantiomers of monosaccharides have been separated on chiral polymers prepared by modification of the polysiloxanes XE-60 and OV-225. Side groups of S-valine-R-α-phenylethyl amide show high enantioselectivity for pentose and 6-deoxyhexose derivatives,[359] while corresponding stationary phases with S-α-phenylethyl amide groups have been used for the resolution of trifluoroacetylated hexose derivatives.[208] The acyl derivatives were essential for enantiomer separation; the corresponding trimethylsilyl derivatives were not separated.[208] It is most likely that the polar trifluoroacetyl groups support the formation of diastereoisomeric association complexes with the stationary phase, whereas the large trimethylsilyl groups prevent the interacting molecules from coming into close contact, which is necessary for the formation of association complexes. Figure 6.42 shows the separation of the enantiomers of the four different forms of ribose.[359]

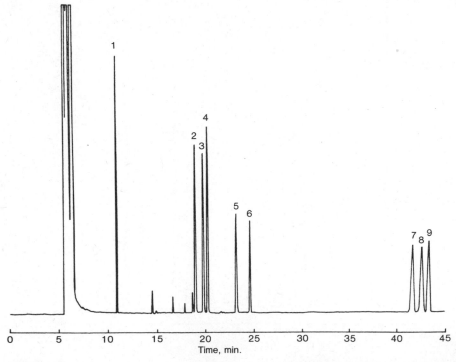

FIGURE 6.41. Chromatogram of standard alditol TMS ethers. Note: 50 m OV-101 fused-silica column, 190°C isothermal temperature. Selected peak assignments: (3) arabinitol; (4) ribitol; (5) fucitol; (7) mannitol; (8) glucitol; (9) galactitol. (Reproduced with permission from ref. 351. Copyright Elsevier Scientific Publishing Company.)

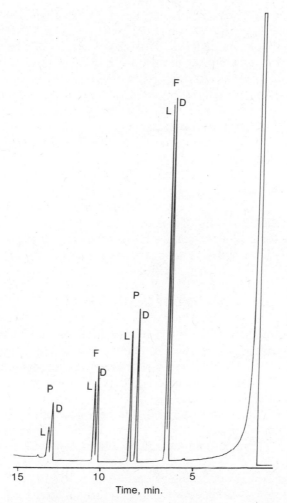

FIGURE 6.42. Chromatogram of the enantiomers of ribose trifluoroacetyl derivatives. Note: 40 m OV-225-S-valine-R-α-phenylethyl amide glass column, temperature program from 90 to 140°C at 3°C min⁻¹, labeled peaks refer to the L- and D-enantiomers and the pyranose and furanose ring structures. (Reproduced with permission from ref. 359. Copyright Dr. Alfred Huethig Publishers.)

Although the L-enantiomers are retarded in the case of ribose, the order of elution is not consistent for the different stereoisomers. In most cases, the separation factors for the pyranose derivatives were markedly greater than those for the furanose derivatives. This suggests that the enantioselectivity of the chiral phase is effected by the ring size.

The determination of the disaccharide, sucrose, in sugar cane juice and in sugar factory products is of great importance to the sugar industry. Nurok and Reardon[344,360] reported the successful analysis of sucrose as the TMS derivative on an OV-17 open tubular column. The relative standard deviation was under 0.1%, and the analysis time was over three times faster than reported packed-column analyses. The disadvantages of using the TMS derivatives are the following: (a) the proportions of each anomer of fructose and glucose, which are also of interest, will depend on solvent composition, temperature, and the length of time the sugar has been dissolved; (b) the overlap between fructose and glucose leads to inaccurate results; (c) the overlap of the two major monosaccharides with other minor constituents in cane molasses will also give inaccurate results; (d) the signal-to-noise ratio for a monosaccharide producing multiple peaks is lower than for a sugar producing a single peak, which limits the sensitivity of the determination. To overcome these problems, the use of a buffered oximation reagent to form the oxime before silylation was recommended[355] before analysis on a SP-2250 open tubular column. Using this procedure, aqueous sugar solutions containing fructose, glucose, and sucrose can be derivatized rapidly and reproducibly without hydrolyzing sucrose or affecting its silylation.

6.9 POLYCHLORINATED BIPHENYLS

Polychlorinated biphenyls (PCBs) are ubiquitous pollutants in the environment, and, due to their lipophilic and persistent characteristics, they exhibit bioaccumulating and biomagnifying properties. Gas chromatography using an electron capture detector (ECD) has been the most widely used method for the analysis of PCBs. Because of the large number of compounds present in commercial PCB mixtures (the total number of possible products resulting from chlorination of biphenyl is 209), open tubular column GC is required in order to provide maximum structural and quantitative information (see Figure 6.43).

The gas chromatographic determination of PCBs in various samples is commonly accomplished by comparing the chromatogram of the sample with the chromatograms of technical Aroclor or Clophen mixtures. In some cases, the total peak area of the sample is compared with the total peak area of the closest matching technical PCB mixture: alternatively, only a few similar peaks are used for quantitation. Erroneous results may be obtained using these approaches because of several reasons. Often, the PCB distribution in the sample is different from those of the commercial mixtures as a result of weathering or bioaccumulation. Furthermore, the response of the ECD has been shown to vary as much as two to four orders of magnitude between monochloro- and polychloro-species,[361] and its sensitivity varies with positional isomerism, for example, different sensitivity for PCB isomers having the same degree of chlorination.[362]

In order to simplify the determination of complex PCB mixtures, methods involving either perchlorination[363-365] or hydrodechlorination[366-368] of PCBs to give a single compound (decachlorobiphenyl or biphenyl) for subsequent

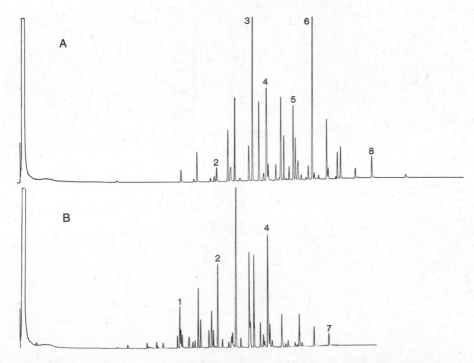

FIGURE 6.43.   Chromatograms of technical PCB mixtures: (A) Aroclor 1260 and (B) Aroclor 1254. Note: 40 m SE-30 glass column, temperature program from 40 to 140°C at 50°C min⁻¹, 3 min at 140°C, then 140 to 190°C at 1.6°C min⁻¹ after an initial 2 min isothermal period, H_2 carrier gas. Selected peak assignments: (1) 2,3',4',5-tetrachlorobiphenyl; (2) 3,3',4,4'-tetrachlorobiphenyl; (3) 2,2',4,4',5,5'-hexachlorobiphenyl; (4) 2,2',3,4,4',5'-hexachlorobiphenyl; (5) 2,2',3,3',4,5,6-heptachlorobiphenyl; (6) 2,2',3,4,4',5,5'-heptachlorobiphenyl; (7) 2,2',3, 3',4,4',5-heptachlorobiphenyl; (8) 2,2',3,3',4,4',5,5'-octachlorobiphenyl. (Reproduced with permission from ref. 376. Copyright Springer-Verlag.)

analysis have been demonstrated. Perchlorination, however, leads to a species of lower volatility for analysis, and extensive fractionation and cleanup steps are required to prevent the chlorination of contaminant compounds. Hydrodechlorination is easily accomplished by injecting the sample into the injection port of a gas chromatograph which has been modified to accept a reaction tube containing a palladium or platinum catalyst. Figure 6.44 shows chromatograms of a mixture of Aroclor 1248, D88 (a polychlorinated naphthalene mixture), aldrin, dieldrin, heptachlor, p,p'-DDT, p,p'-DDE, and p,p'-TDE both before (Figure 6.44A) and after (Figure 6.44B) dechlorination.[367] The chromatogram was simplified to three major peaks: naphthalene from the polychlorinated naphthalenes, biphenyl from the PCBs, and diphenylethane from DDT, DDE, and TDE. Although dechlorination of PCBs to biphenyl simplifies the determination to only one species, quantification may be more difficult because of the nonselective nature of the FID and the

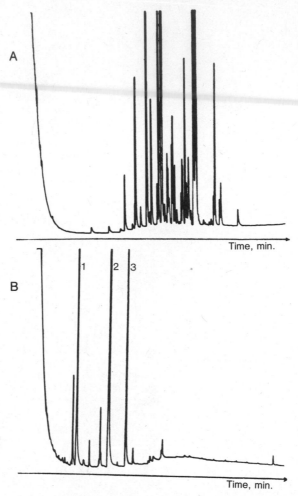

FIGURE 6.44. Chromatograms of a mixture of Aroclor 1248, D88, aldrin, dieldrin, heptachlor, p,p'-DDT, p,p'-DDE, and p,p'-TDE obtained (A) before and (B) after catalytic hydrodechlorination. Note: 50 m OV-1 column, temperature program from 60 to 250°C at 8°C min^{-1}. Peak assignments: (1) naphthalene; (2) biphenyl; (3) diphenylethane. (Reproduced with permission from ref. 367. Copyright Elsevier Scientific Publishing Company.)

complexity of the sample matrix. The advantages of selectivity and sensitivity of the ECD can no longer be used.

It is generally agreed that the most valid PCB quantitation would be obtained if each component could be individually measured and summed (taking into account the different detector response factors) to obtain a "total" PCB content. This approach has a significant advantage over "chemical summation," for example, perchlorination or hydrodechlorination, in that information can be obtained concerning levels of the more highly toxic or biologically active isomers that may be present.

The separation of PCBs by open tubular column GC has most often been done using low polarity stationary phases such as SF-96, Dexsil 300, Apiezon L, OV-101, OV-1, and SE-30. The reason for this is that the column technology is well developed for these phases. Columns that are well-deactivated, highly efficient, thermally and mechanically stable, and reproducible can be made. Unfortunately, many PCBs cannot be resolved on these columns. In order to distinguish between 2,2'- and 2,6-dichlorobiphenyls, an OV-101 column with an efficiency of 600,000 theoretical plates would be required.[369]

PCBs do not chromatograph well on highly polar stationary phases (i.e., OV-275, FFAP) because of their low solubility in these phases. The most polar liquid phase that has been found effective for PCB analysis is poly-MPE,[370] but the high efficiency obtained with the methylpolysiloxane phases was lacking. Albro et al.[371] preferred the use of an efficient moderately polar Dexsil 410 open tubular column as the primary column for resolving Aroclor mixtures. Selected pairs of isomers which were not resolved on Dexsil 410 could be easily distinguished on short, less efficient, moderately polar columns coated with Silar 5C, Apiezon L, or OV-25. Krupcik et al.[372] achieved the best separation of the Aroclor 1242 mixture on an Apiezon L open tubular column. Although, no one capillary column has, thus far, provided sufficient resolution to distinguish all of the possible PCB isomers, the retention behavior of PCBs is sufficiently different on different liquid phases that one can generally find a phase capable of resolving any given PCB pair.[370]

The resolution of PCBs from organochlorine pesticides is another complication in the analysis of environmental samples. Overlap of these compounds can affect both the quantitative and qualitative analysis of either class of compounds. Using a mixed phase (4:3 OV-17/QF-1) open tubular column,[373] Aroclor 1242 was found to elute completely before p,p'-DDE, and ca. 95% of the components of Aroclor 1260 eluted later than p,p'-DDE. All of the other organochlorine pesticides (γ-BHC, aldrin, dieldrin, o,p'-DDT, p,p'-DDT, and Mirex) were well resolved from the PCB compounds.

The identification of individual PCB compounds in technical mixtures and in environmental samples must depend heavily on chromatographic retention measurements. GC-MS[369,372,374,375] can help to provide molecular mass data and number of chlorines, but isomer differentiation is almost always impossible. Retention indices (RIs) have been measured for PCBs using the n-alkanes as reference homologs for FID detection.[369,371,372,376,377] Similarly, retention indices can be measured for electron capture detection using the n-alkyl-trichloroacetates as reference homologs.[378,379] The measured RI values depend on a number of factors including the column deactivation, stationary phase type and film thickness, temperature program rate, and so on. The reproducibility of RI measurements of PCBs was found to be insufficient for interlaboratory data exchange,[369] and no standard indices could be indexed at present.[376] Nevertheless, standardizing of equipment and procedures within each laboratory could ensure proper determinations.

Approximately only one-third of the possible 209 PCBs are available to use as reference compounds for RI comparisons. Therefore, the identification of PCBs that are not available must be performed by calculation of their RIs. Sissons and Welti[380] reasoned that the RI of a compound is directly proportional to its free energy of solution in a stationary phase, which in turn is an approximately additive function of the groups constituting the molecule. Consequently, any PCB molecule can be thought of as consisting of two chloro-substituted phenyl groups each with its own $\frac{1}{2}$ RI value. All PCBs are combinations of 20 such groups, and the RI of any PCB can be estimated by adding together the $\frac{1}{2}$ RI values of the two-component phenyl groups. Any electronic and steric effects must be allowed for when predicting the RIs of PCBs by using appropriate $\frac{1}{2}$ RI values derived from a similar type of PCB molecule. Calculated RI values have been tabulated and used to identify specific PCBs.[376,378,380,381] The agreement of the calculated values with measured values is in the range of ± 2 retention index units for the same chromatographic system.

The identification of the source of PCB contamination in an environmental sample can be accomplished to some degree of success by comparing the chromatographic profiles of the standard Aroclor mixtures with that of the unknown sample. Environmental samples, however, seldom give a single mixture pattern, although a certain type of pattern might predominate. Mixed patterns can be determined by using the ratios of diagnostic PCB components that are only or predominantly present in technical Aroclor mixtures.[376] These compounds should be resistant to weathering in order to minimize problems of rationalizing degraded PCB mixtures in environmental samples. Additionally, a computer program has been developed which looks for the closest fit to a given PCB pattern in a sample on the basis of the internal ratios of all PCB components in the various technical PCB mixtures.[376]

Quantitation using a best-fit mixture can be done by comparing a few diagnostic peaks in the sample with the best-fit mixture. There are inherent errors in this method resulting from the extent of sample degradation. The best quantitation is obtained by the summation of all PCB peaks corrected by their individual detector response factors.

Quantitation of individual components of PCB mixtures has been done using an FID.[371,372,382] The FID response factors increased with increasing number of chlorine atoms in the molecule. Mean values of the response factors for dichlorobiphenyls ($f_i = 1.29$), trichlorobiphenyls ($f_i = 1.48$), and tetrachlorobiphenyls ($f_i = 1.62$) were used for quantitation of the compounds which were not available as standards.[372]

Quantitative analysis using the ECD is most desirable since PCB mixtures isolated in trace amounts from environmental or biological samples require the halogen-selectivity and sensitivity of the ECD. Relative molar responses of the ECD to chlorobiphenyls have been published,[376] but the ECD cannot be used to estimate amounts of any individual component in the absence of a calibration standard of that component. Albro et al.[371] calculated relative

ECD/FID responses of the separated components of three different Aroclor mixtures for use in quantitation of environmental samples. Boe and Egaas[377] developed a mathematical model for relating the ECD/FID responses to the structure of PCBs, which permits the calculation of detector responses for all PCBs.

6.10   INORGANIC AND ORGANOMETALLIC COMPOUNDS

It has been known for some time that metal ions could be separated by GC if they were converted to suitable volatile chelates.[383,384] Ligands such as the acetylacetonates, thioacetylacetonates, bidentate and tetradentate β-ketoamines and their thio analogues, dialkyldithiocarbamates, and dialkyldithiophosphates have been used.[383–387] Limitations in the past have stemmed from insufficient deactivation of the chromatographic column. The recent development of low surface area, well-deactivated, fused-silica open tubular columns, has re-opened this area of GC to further research and development.

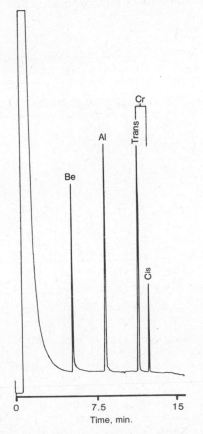

FIGURE 6.45.   Chromatogram of the trifluoroacetylacetonate complexes of Be, Al, and Cr. Note: 11 m Dexsil-300 glass column, temperature program from 75 to 140°C at 5°C min^{-1} after a 2 min initial isothermal period. (Reproduced with permission from ref. 390. Copyright Dr. Alfred Huethig Publishers.)

Open tubular column chromatographic studies of metal chelates have primarily featured the dithiocarbamates,[388-391] but a number of other derivatives have also been studied.[387,390,391] Short columns containing nonpolar methylpolysiloxane stationary phases (OV-1, OV-101, and SE-30) have been most successful. More polar Carbowax 20M columns failed to elute some complexes, and columns longer than 10 m resulted in longer analysis times without a gain in resolution.[390] Short, highly efficient columns and less-dense carrier gases (H₂ or He) yielded the best results. Figure 6.45 shows a chromatogram of the trifluoroacetylacetonate complexes of Be, Al, and Cr on an 11 m Dexsil-300 glass open tubular column.[390] The geometric isomers of the Cr(III) complex were well resolved on this column. However, the Cu(II) complex could only be eluted from an SE-30 fused-silica column.

The injection temperature and conditions must be carefully controlled when analyzing metal chelates. Many are not stable at the usual injector operating temperatures, and cold on-column injection is often the best method of choice.

Although the flame ionization detector usually suffices for most analyses, the use of various selective detectors for metal chelates and organometallic compounds has been reported. Estes et al.[392-394] have published the cali-

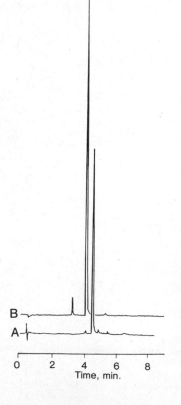

FIGURE 6.46. Chromatograms of an organometallic compound mixture using a microwave-excited helium plasma detector monitoring (A) chromium at 267.7 nm and (B) manganese at 257.6 nm. Note: 12.5 m SP-2100 fused-silica column, temperature program from 80 to 116°C at 4°C min⁻¹. (Reproduced with permission from ref. 392. Copyright Dr. Alfred Huethig Publishers.)

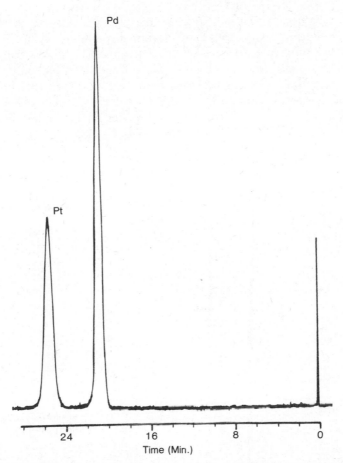

FIGURE 6.47. Chromatogram of the Pd and Pt diisopropyldithiophosphates using an FPD in the sulfur mode. Note: 10 m PS-300 glass column, 170°C isothermal temperature. (Reproduced with permission from ref. 387. Copyright Elsevier Scientific Publishing Company.)

bration curves, selectivity ratios, and detection limits for over 29 elements using fused-silica open tubular columns and microwave-excited helium plasma detection. Figure 6.46 shows chromatograms of an organometallic compound mixture monitoring chromium at 267.7 nm (Figure 6.46A) and manganese at 257.6 nm (Figure 6.46B).[392] Detection limits better than 10 pg of triphenyltin methyl were obtained using a tin-selective FPD.[395]

Selective detection based on the properties of the ligand has also been reported.[387,390] Figure 6.47 shows a chromatogram of the Pd and Pt diiso-propyldithiophosphates using an FPD in the sulfur mode.[387] Sucre and Jennings[390] have published chromatographic results and detection limits of a number of metal chelates using the ECD, FPD, and NPD.

The recent analysis of the bis(trimethylsiloxy)silicon(IV) derivatives of alkyl porphyrins is an excellent example of the solution to a problem that could only be done using inert fused-silica columns.[396] Volatile derivatives of the porphyrins were formed by introducing dihydroxy Si(IV) into the center of the molecule, followed by derivatization to the bis(trimethylsilyl) ether (Figure 6.48). Packed columns were found to yield poor sensitivity due to adsorption effects, and inadequate resolution. Furthermore, flash-vaporization injection onto a Pyrex open tubular column of OV-1 coated on BaCO₃ was not successful. Incomplete transfer of the sample from the sy-

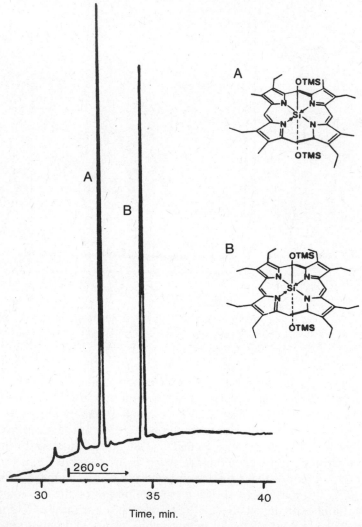

FIGURE 6.48. Chromatogram of the bis(trimethylsiloxy)-Si(IV) derivatives of (A) aetioporphyrin-1 and (B) octaethylporphyrin. Note: 25 m OV-73 fused-silica column, temperature program from 70 to 260°C at 6°C min⁻¹, H₂ carrier gas. (Reproduced with permission from ref. 396. Copyright Dr. Alfred Huethig Publishers.)

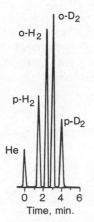

FIGURE 6.49. Chromatogram of the nuclear spin isomers of hydrogen and deuterium. Note: 82 m soft-glass column roughened by purging with 10% solution of NaOH at 100°C, 47.0 K isothermal temperature, Ne carrier gas. (Reproduced with permission from ref. 400. Copyright Preston Publications, Inc.)

ringe needle, decomposition in the injector at the required temperature, and adsorption on the column wall all contributed to the problem. These difficulties were overcome by resorting to on-column injection into an inert fused-silica column as shown in Figure 6.48.

The gas chromatographic analysis of silicate anions has been reported.[397,398] Trimethylsilylated derivatives of the silicic acid products released from the reaction of acids on silicate materials have been chromatographed on nonpolar open tubular columns. It was found that retention characteristics depended mainly on the number of trimethylsilyl groups located on the surface of the molecule, while the size and structure of the molecular skeleton had only minor effects. Compounds of considerably different molecular weights and structures showed similar GC retention. The considerable shielding effect of the voluminous trimethylsilyl groups may be considered as a possible explanation of this unexpected phenomenon.[397]

Permanent gases are generally not suitable for analysis by open tubular column GC. The capacity ratios are extremely low for these compounds, and many cannot be detected with conventional open tubular column GC detectors. There are, however, cases where open tubular columns can be advantageous. For example, the analysis of air samples for gaseous sulfur compounds at the low and sub-ppb concentrations are complicated by the sensitivity requirements, the large number of chemical compounds present at these ultra-trace levels, and the adsorption affinities of many sulfur compounds. Packed-column analysis using a sulfur-selective FPD has been the main approach in the past, but it suffers for several reasons: the large surface areas of chromatographic supports cause adsorption losses; certain compounds (e.g., H_2S and COS) are difficult to resolve; high pressure drops are obtained across long packed columns; and quenching of the FPD response can result from unresolved components. Farwell et al.[399] have described the separation of a number of sulfur-containing gases (H_2S, COS, CH_3SH, CH_3SCH_3, CS_2, CH_3SSCH_3, and other organosulfur species) on well-deactivated glass open tubular columns starting at cryogenic temperatures ($-70°C$). The low surface activity and high efficiency of the column and the selectivity of the FPD proved to be essential to the success of this approach. In a novel application of open tubular column GC, the nuclear spin isomers of hydrogen and deuterium were separated on a surface-roughened 82 m glass column at temperatures below 60 K (Figure 6.49).

REFERENCES

1. A. Matukuma, in *Gas Chromatography 1968*, C. L. A. Harbourn and R. Stock, editors. Institute of Petroleum, London, 1969, p. 55.
2. N. T. Sultanov and L. G. Arustamova, *J. Chromatogr.* **115**, 553 (1975).
3. C. A. Cramers, J. A. Rijks, V. Pacakova, and I. Ribeiro de Andrade, *J. Chromatogr.* **51**, 13 (1970).
4. J. A. Rijks, Ph.D. Thesis, University of Technology, Eindhoven, 1973.
5. J. A. Rijks and C. A. Cramers, *Chromatographia* **7**, 99 (1974).
6. G. Schomburg and G. Dielmann, *J. Chromatogr. Sci.* **11**, 151 (1973).
7. E. Kovats, *Helv. Chim. Acta* **41**, 1915 (1958).
8. G. D. Mitra, G. Mohan, and A. Sinha, *J. Chromatogr.* **91**, 633 (1974).
9. N. C. Saha and G. D. Mitra, *J. Chromatogr. Sci.* **11**, 419 (1973).
10. G. D. Mitra and N. C. Saha, *Chromatographia* **6**, 93 (1973).
11. N. G. Johansen and L. S. Ettre, *Chromatographia* **15**, 625 (1982).
12. J. Simekova, N. Pronayova, R. Pies, and M. Ciha, *J. Chromatogr.* **51**, 91 (1970).
13. K. Petrovic and D. Vitorovic, *J. Chromatogr.* **65**, 155 (1972).
14. E. J. Gallegos, I. M. Whittemore, and R. F. Klaver, *Anal. Chem.* **46**, 157 (1974).
15. H. Wehner and M. Teschner, *J. Chromatogr.* **204**, 481 (1981).
16. F. Berthou, Y. Gourmelun, Y. Dreano, and M. P. Friocourt, *J. Chromatogr.* **203**, 279 (1981).
17. R. Jeltes, E. Burghardt, Th. R. Thijsse, and W. A. M. den Tonkelaar, *Chromatographia* **10**, 430 (1977).
18. K. D. Bartle, M. L. Lee, and M. Novotny, *Int. J. Environ. Anal. Chem.* **3**, 349 (1974).
19. S. A. Wise, S. N. Chesler, F. R. Guenther, H. S. Hertz, L. R. Hilpert, W. E. May, and R. M. Parris, *Anal. Chem.* **52**, 1828 (1980).
20. F. P. DiSanzo, P. C. Uden, and S. Siggia, *Anal. Chem.* **51**, 1529 (1979).
21. F. P. DiSanzo, P. C. Uden, and S. Siggia, *Anal. Chem.* **52**, 906 (1980).
22. R. J. Crowley, S. Siggia, and P. C. Uden, *Anal. Chem.* **52**, 1224 (1980).
23. D. W. Later, M. L. Lee, K. D. Bartle, R. C. Kong, and D. L. Vassilaros, *Anal. Chem.* **53**, 1612 (1981).
24. R. F. Severson, R. F. Arrendale, and O. T. Chortyk, *J. High Resoln. Chromatogr./ Chromatogr. Commun.* **3**, 11 (1980).
25. E. Gelpi and J. Oro, *J. Chromatogr. Sci.* **8**, 210 (1970).
26. J. L. Glajch and R. A. Schunn, *J. High Resoln. Chromatogr./Chromatogr. Commun.* **4**, 333 (1981).
27. O. L. Hollis, *J. Chromatogr. Sci.* **11**, 335 (1973).
28. G. Schomburg, R. Dielmann, H. Borwitzky, and H. Husmann, *J. Chromatogr.* **167**, 337 (1978).

29. K. Grob and K. Grob, Jr., *J. High Resoln. Chromatogr./Chromatogr. Commun.* **1,** 57 (1978).

30. K. Grob, *J. High Resoln. Chromatogr./Chromatogr. Commun.* **1,** 263 (1978).

31. B. W. Wright and M. L. Lee, *J. High Resoln. Chromatogr./Chromatogr. Commun.* **3,** 352 (1980).

32. T. H. Gouw, I. M. Whittemore, and R. E. Jentoft, *Anal. Chem.* **42,** 1394 (1970).

33. R. R. Freeman, *High Resolution Gas Chromatography*, 2nd edition. Hewlett-Packard, Avondale, Pennsylvania, 1981, p. 168.

34. L. Sojak and A. Bucinska, *J. Chromatogr.* **51,** 75 (1970).

35. M. Ryba, *Chromatographia* **5,** 23 (1972).

36. L. Sojak and J. Hrivnak, *J. Chromatogr. Sci.* **10,** 701 (1972).

37. L. Sojak, P. Majer, P. Skalak, and J. Janak, *J. Chromatogr.* **65,** 137 (1972).

38. O. Eisen, A. Orav, and S. Rang, *Chromatographia* **5,** 229 (1972).

39. L. Sojak, J. Hrivnak, P. Majer, and J. Janak, *Anal. Chem.* **45,** 293 (1973).

40. S. Rang, K. Kuningas, A. Orav, and O. Eisen, *Chromatographia,* **10,** 55 (1977).

41. L. Sojak, J. Hrivnak, I. Ostrovsky, and J. Janak, *J. Chromatogr.* **91,** 613 (1974).

42. L. Sojak, J. Krupcik, and J. Janak, *J. Chromatogr.* **191,** 199 (1980).

43. L. Sojak, J. Krupcik, and J. Janak, *J. Chromatogr.* **195,** 43 (1980).

44. S. Rang, K. Kuningas, A. Orav, and O. Eisen, *J. Chromatogr.* **119,** 451 (1976).

45. Th. Welsch, W. Engewald, and P. Berger, *Chromatographia* **11,** 5 (1978).

46. L. Sojak, V. G. Berezkin, and J. Janak, *J. Chromatogr.* **209,** 15 (1981).

47. W. Meltzow, S. Warwel, and B. Fell, *Chromatographia* **6,** 183 (1973).

48. A. Zlatkis and I. M. R. de Andrade, *Chromatographia* **2,** 298 (1969).

49. V. Schurig, R. C. Chang, A. Zlatkis, E. Gil-Av, and F. Mikes, *Chromatographia* **6,** 223 (1973).

50. K. Muller, *Functional Group Determination of Olefinic and Acetylenic Unsaturation.* Academic Press, New York, 1975.

51. T. S. Ma and A. S. Ladas, *Organic Functional Group Analysis by Gas Chromatography.* Academic Press, New York, 1976, p. 119.

52. G. W. Francis and T. Taude, *J. Chromatogr.* **150,** 139 (1978).

53. S. Rang, K. Kuningas, A. Orav, and O. Eisen, *J. Chromatogr.* **119,** 451 (1976).

54. Th. Welsch, W. Engewald, and P. Berger, *Chromatographia* **11,** 5 (1978).

55. G. Schomburg and G. Dielmann, *Anal. Chem.* **45,** 1647 (1973).

56. G. Dielmann, D. Schwengers, and G. Schomburg, *Chromatographia* **7,** 215 (1974).

57. S. Rang, A. Orav, K. Kuningas, and O. Eisen, *Chromatographia* **10,** 115 (1977).

58. G. Firpo, M. Gassiot, M. Martin, R. Carbo, X. Guardino, and J. Albaiges, *J. Chromatogr.* **117,** 105 (1976).

59. J. Albaiges and X. Guardino, *Chromatographia* **13,** 755 (1980).

60. W. F. Tully, *J. Chromatogr. Sci.* **9,** 635 (1971).

61. I. Stopp, W. Engewald, H. Kuhn, and Th. Welsch, *J. Chromatogr.* **147,** 21 (1978).

62. S. Hala, J. Eyem, J. Burkhard, and S. Landa, *J. Chromatogr. Sci.* **8,** 203 (1970).

63. H. Maarse and R. E. Kepner, *J. Agric. Food Chem.* **18,** 1095 (1970).

64. E. Gelpi, P. C. Wszolek, E. Yang, and A. L. Burlingame, *J. Chromatogr. Sci.* **9,** 147 (1971).

65. E. J. Gallegos, *Anal. Chem.* **43,** 1151 (1971).

66. E. J. Gallegos, *Anal. Chem.* **45,** 1399 (1973).

67. J. G. Pym, J. E. Ray, G. W. Smith, and E. V. Whitehead, *Anal. Chem.* **47**, 1617 (1975).

68. R. M. Seifert and R. G. Buttery, *J. Agric. Food Chem.* **26**, 181 (1978).

69. T. Saeed, G. Redant, and P. Sandra, *J. High Resoln. Chromatogr./Chromatogr. Commun.* **2**, 75 (1979).

70. L. Sojak and J. A. Rijks, *J. Chromatogr.* **119**, 505 (1976).

71. L. Sojak, J. Janak, and J. A. Rijks, *J. Chromatogr.* **135**, 71 (1977).

72. L. Sojak, J. Janak, and J. A. Rijks, *J. Chromatogr.* **138**, 119 (1977).

73. L. Sojak, J. Janak, and J. A. Rijks, *J. Chromatogr.* **142**, 177 (1977).

74. W. Engewald and L. Wennrich, *Chromatographia* **9**, 540 (1976).

75. J. Lesko, S. Holotik, J. Krupcik, and V. Vesely, *J. Chromatogr.* **119**, 293 (1976).

76. C. L. Stuckey, *J. Chromatogr. Sci.* **9**, 575 (1971).

77. R. K. Kuchhal, B. Kumar, P. Kumar, and P. L. Gupta, *J. High Resoln. Chromatogr./Chromatogr. Commun.* **3**, 497 (1980).

78. W. Bertsch, A. Zlatkis, J. L. Laseter, and G. W. Griffin, *Chromatographia* **5**, 324 (1972).

79. E. Grushka and J. F. Solsky, *Anal. Chem.* **45**, 1836 (1973).

80. L. Sojak, G. Kraus, I. Ostrovsky, E. Kralovicova, and J. Krupcik, *J. Chromatogr.* **206**, 475 (1981).

81. G. Kraus and M. Schierhorn, *J. High Resoln. Chromatogr./Chromatogr. Commun.* **4**, 123 (1981).

82. M. R. Tirgan and N. Sharifi-Sandjani, *J. Chromatogr.* **193**, 397 (1980).

83. M. L. Lee, M. V. Novotny, and K. D. Bartle, *Analytical Chemistry of Polycyclic Aromatic Compounds*. Academic Press, New York, 1981.

84. M. L. Lee and B. W. Wright, *J. Chromatogr. Sci.* **18**, 345 (1980).

85. H. Borwitzky and G. Schomburg, *J. Chromatogr.* **170**, 99 (1979).

86. L. S. Lysyuk and A. N. Korol, *Chromatographia* **10**, 712 (1977).

87. H. Beernaert, *J. Chromatogr.* **173**, 109 (1979).

88. M. L. Lee, D. L. Vassilaros, C. M. White, and M. Novotny, *Anal. Chem.* **51**, 768 (1979).

89. D. L. Vassilaros, R. C. Kong, D. W. Later, and M. L. Lee, *J. Chromatogr.* **252**, 1 (1982).

90. K. D. Bartle, M. L. Lee, and S. A. Wise, *Chromatographia* **14**, 69 (1981).

91. A. Radecki, H. Lamparczyk, and R. Kaliszan, *Chromatographia* **12**, 545 (1979).

92. J. Mostecky, M. Popl, and J. Kriz, *Anal. Chem.* **42**, 1132 (1970).

93. J. Kriz, M. Popl, and J. Mostecky, *J. Chromatogr.* **97**, 3 (1974).

94. K. Tesarik, J. Frycka, and S. Ghyczy, *J. Chromatogr.* **148**, 223 (1978).

95. J. Novrocik and M. Novrocikova, *Collection Czechoslov. Chem. Commun.* **45**, 2919 (1980).

96. W. L. Zielinski, Jr., R. A. Scanlan, and M. M. Miller, *J. Chromatogr.* **209**, 87 (1981).

97. R. J. Laub, W. L. Roberts, and C. A. Smith, *J. High Resoln. Chromatogr./Chromatogr. Commun.* **3**, 355 (1980).

98. R. J. Laub, W. L. Roberts, and C. A. Smith, in *Polynuclear Aromatic Hydrocarbons: Chemistry and Biological Fate*, M. Cooke and A. J. Dennis, editors. Battelle Press, Columbus, Ohio, 1981, p. 287.

99. R. C. Kong, M. L. Lee, Y. Tominaga, R. Pratap, M. Iwao, R. M. Castle, and S. A. Wise, *J. Chromatogr. Sci.* **20**, 502 (1982).

100. M. A. Apfel, H. Finkelmann, R. J. Laub, B.-H. Luhmann, A. Price, W. L. Roberts, and C. A. Smith, *Makromol. Chem., Rapid Commun.* submitted.

101. R. C. Kong, M. L. Lee, Y. Tominaga, R. Pratap, M. Iwao, and R. N. Castle, *Anal Chem.* **54**, 1802 (1982).

102. R. J. Laub, *Am. Lab.* **13**, 47 (1981).

103. K. Grob, *Chromatographia* **7**, 94 (1974).

104. T. Romanowski, W. Funcke, J. Konig, and E. Balfanz, *J. High Resoln. Chromatogr./ Chromatogr. Commun.* **4**, 209 (1981).

105. L. Blomberg, J. Buijten, J. Gawdzik, and T. Wannman, *Chromatographia* **11**, 521 (1978).

106. P. A. Peaden, B. W. Wright, and M. L. Lee, *Chromatographia* **15**, 335 (1982).

107. G. Grimmer, J. Jacob, and K.-W. Naujack, *Fres. Z. Anal. Chem.* **306**, 347 (1981).

108. E. Balfanz, J. Konig, W. Funcke, and T. Romanowski, *Fres. Z. Anal. Chem.* **306**, 340 (1981).

109. M. L. Lee, D. L. Vassilaros, and D. W. Later, *Int. J. Environ. Anal. Chem.* **11**, 251 (1982).

110. Y. L. Tan, *J. Chromatogr.* **176**, 319 (1979).

111. L. Szepesy, K. Lakszner, L. Ackermann, L. Podmaniczky, and P. Literathy, *J. Chromatogr.* **206**, 611 (1981).

112. L. S. Ramos and P. G. Prohaska, *J. Chromatogr.* **211**, 284 (1981).

113. D. L. Vassilaros, P. W. Stoker, G. M. Booth, and M. L. Lee, *Anal. Chem.* **54**, 106 (1982).

114. D. W. Later, M. L. Lee, K. D. Bartle, R. C. Kong, and D. L. Vassilaros, *Anal. Chem.* **53**, 1612 (1981).

115. V. Cantuti, G. P. Cartoni, A. Liberti, and A. G. Torri, *J. Chromatogr.* **17**, 60 (1965).

116. N. Carugno and S. Rossi, *J. Gas Chromatogr.* **5**, 103 (1967).

117. G. Alberini, V. Cantuti, and G. P. Cartoni, in *Gas Chromatography 1966*, A. B. Littlewood, editor. Institute of Petroleum, London, 1967, p. 258.

118. A. Bjørseth and G. Eklund, *J. High Resoln. Chromatogr./Chromatogr. Commun.* **2**, 22 (1979).

119. S. Kapila and C. R. Vogt, *J. High Resoln. Chromatogr./Chromatogr. Commun.* **4**, 233 (1981).

120. F. J. Schwende, M. Novotny, and J. E. Purcell, *Chromatogr. Newslett.* **8**, 1 (1980).

121. D. M. Hembree, A. A. Garrison, R. A. Crocombe, R. A. Yokley, E. L. Wehry, and G. Mamantov, *Anal. Chem.* **53**, 1783 (1981).

122. M. L. Lee, B. W. Wright, and K. D. Bartle, in *Proceedings of the Fourth International Symposium on Capillary Chromatography*, R. E. Kaiser, editor. Hüthig, Heidelberg, 1981, p. 505.

123. B. W. Wright, unpublished results.

124. J. G. Moncur, *J. High Resoln. Chromatogr./Chromatogr. Commun.* **5**, 53 (1982).

125. A. B. Littlewood, *Gas Chromatography*, second edition. Academic Press, New York, 1970.

126. J. Krupcik, K. Tesarik, and J. Hrivnak, *Chromatographia* **8**, 533 (1975).

127. T. Shibamoto, K. Harada, K. Yamaguchi, and A. Aitoku, *J. Chromatogr.* **194**, 277 (1980).

128. K. Grob, Jr., H. P. Neukom, and H. Kaderli, *J. High Resoln. Chromatogr./Chromatogr. Commun.* **1**, 98 (1978).

129. S. Holotik, J. Lesko, J. Krupcik, and K. Tesarik, *Chromatographia* **9**, 443 (1976).

130. S. Boneva and N. Dimov, *Chromatographia* **14**, 601 (1981).

131. B. W. Wright, Ph.D. Dissertation, Brigham Young University, Provo, Utah, 1982.

132. G. Schomburg, H. Husmann, and H. Behlau, *Chromatographia* **13**, 321 (1980).

133. E. L. Plummer, T. E. Stewart, K. Byrne, G. T. Pearce, and R. M. Silverstein, *J. Chem. Ecol.* **2**, 307 (1976).

134. B. Gerken, S. Grune, J. P. Vite, and K. Mori, *Naturwissenschaften* **65**, 110 (1978).

135. N. Oi, K. Shiba, T. Tani, H. Kitahara, and T. Doi, *J. Chromatogr.* **211**, 274 (1981).

136. N. Oi, T. Doi, H. Kitahara, and Y. Inda, *J. Chromatogr.* **208**, 404 (1981).

137. W. A. Konig, K. Stoelting, and K. Kruse, *Chromatographia* **10**, 444 (1977).

138. K. Kruse, W. Francke, and W. A. Konig, *J. Chromatogr.* **170**, 423 (1979).

139. P. Kalo, *J. Chromatogr.* **205**, 39 (1981).

140. J. Hrivnak, *J. Chromatogr. Sci.* **8**, 602 (1970).

141. J. Hrivnak and J. Macak, *Anal. Chem.* **43**, 1039 (1971).

142. J. Hrivnak and E. Beska, *J. Chromatogr.* **54**, 277 (1971).

143. P. Buryan, J. Macak, and V. M. Nabivach, *J. Chromatogr.* **148**, 203 (1978).

144. J. Macak, P. Buryan, and J. Hrivnak, *J. Chromatogr.* **89**, 309 (1974).

145. G. Alexander, G. Garzo, and G. Palyi, *J. Chromatogr.* **91**, 25 (1974).

146. F. R. Guenther, R. M. Parris, S. N. Chesler, and L. R. Hilpert, *J. Chromatogr.* **207**, 256 (1981).

147. C. M. White and N. C. Li, *Anal. Chem.* **54**, 1564 (1982).

148. C. M. White and N. C. Li, *Anal. Chem.* **54**, 1570 (1982).

149. M. Malaterre, J. Loheac, N. Sellier, and G. Guiochon, *Chromatographia* **8**, 624 (1975).

150. A. D. Sauter, L. D. Betowski, T. R. Smith, V. A. Strickler, R. G. Beimer, B. N. Colby, and J. E. Wilkinson, *J. High Resoln. Chromatogr./Chromatogr. Commun.* **4**, 366 (1981).

151. K. Blau and G. King, editors, *Handbook of Derivatives for Chromatography*. Heyden, London, 1977.

152. V. Raverdino and P. Sassetti, *J. Chromatogr.* **153**, 181 (1978).

153. V. Fell and C. R. Lee, *J. Chromatogr.* **121**, 41 (1976).

154. W. Krijgsman and C. G. Van de Kamp, *J. Chromatogr.* **131**, 412 (1977).

155. A. T. Shulgin, *Anal. Chem.* **36**, 920 (1964).

156. R. J. Argauer, *Anal. Chem.* **40**, 122 (1968).

157. J. N. Seiber, D. G. Crosby, H. Fouda, and C. J. Soderquist, *J. Chromatogr.* **73**, 89 (1972).

158. L. L. Lamparski and T. J. Nestrick, *J. Chromatogr.* **156,**, 143 (1978).

159. M. Lehtonen, *J. Chromatogr.* **202**, 413 (1980).

160. E. Fogelqvist, B. Josefsson, and C. Roos, *J. High Resoln. Chromatogr./Chromatogr. Commun.* **3**, 568 (1980).

161. E. A. Mistryukov, A. L. Samusenko, and R. V. Golovnya, *J. Chromatogr.* **169**, 391 (1979).

162. C. J. W. Brooks and E. C. Horning, *Anal. Chem.* **36**, 1540 (1964).

163. E. C. Horning, M. G. Horning, W. J. A. VandenHeuvel, K. L. Knox, B. Holmstedt, and C. J. W. Brooks, *Anal. Chem.* **36**, 1546 (1964).

164. W. J. A. VandenHeuvel, W. J. Gardiner, and E. C. Horning, *Anal. Chem.* **36**, 1550 (1964).

165. L. D. Metcalfe and R. J. Martin, *Anal. Chem.* **44**, 403 (1972).

166. M. W. Scoggins, L. Skurcenski, and D. S. Weinberg, *J. Chromatogr. Sci.* **10**, 678 (1972).

167. N. P. Sen and P. L. McGeer, *Biochem. Biophys. Res. Commun.* **13**, 390 (1963).

168. L. M. Bertani, S. W. Dziedzic, D. D. Clarke, and S. E. Gitlow, *Clin. Chim. Acta* **30**, 227 (1970).

169. K. Imai, M. Segiura, and Z. Tamura, *Chem. Pharm. Bull.* **19**, 409 (1971).

170. J. P. Chaytor, B. Crathorne, and M. J. Saxby, *J. Chromatogr.* **70**, 141 (1972).

171. E. Anggard and G. Sedvall, *Anal. Chem.* **41**, 1250 (1969).

172. S. Kawai and Z. Tamura, *Chem. Pharm. Bull.* **16**, 699 (1968).

173. A. R. Mosier, C. E. Andre, and F. G. Viets, Jr., *Environ. Sci. Technol.* **7,** 642 (1973).

174. S. B. Matin and M. Rowland, *J. Pharm. Sci.* **61,** 1235 (1972).

175. A. C. Moffat, E. C. Horning, S. B. Matin, and M. Rowland, *J. Chromatogr.* **66,** 255 (1972).

176. R. E. Weston and B. B. Wheals, *Analyst* **95,** 680 (1970).

177. E. W. Day, Jr., T. Golab, and J. R. Koons, *Anal. Chem.* **38,** 1053 (1966).

178. A. C. Moffat and E. C. Horning, *Biochim. Biophys. Acta* **222,** 248 (1970).

179. A. C. Moffat and E. C. Horning, *Anal. Lett.* **3,** 205 (1970).

180. T. Uno, T. Nakagawa, and R. Toyoda, *Bunseki Kagaku* **21,** 993 (1972).

181. R. Toyoda, T. Nakagawa, and T. Uno, *Bunseki Kagaku* **22,** 914 (1973).

182. Y. Hoshika, *J. Chromatogr.* **136,** 253 (1977).

183. H. Brandenberger and E. Helbach, *Helv. Chim. Acta* **50,** 958 (1967).

184. H. Gross and F. Franzen, *Biochem. Z.* **340,** 403 (1964).

185. G. Blotny, J. Szafranek, and J. Kusmierz, *Anal. Lett.* **11,** 1063 (1978).

186. G. Blotny, J. Kusmierz, E. Malinski, and J. Szafranek, *J. Chromatogr.* **193,** 61 (1980).

187. K. Jacob, C. Falkner, and W. Vogt, *J. Chromatogr.* **167,** 67 (1978).

188. B. Olufsen, *J. Chromatogr.* **179,** 97, (1979).

189. G. Becher, *J. Chromatogr.* **211,** 103 (1981).

190. G. Goretti, M. Ciardi, and C. Di Palo, *J. High Resoln. Chromatogr./Chromatogr. Commun.* **3,** 523 (1980).

191. K. Tesarik and S. Ghyczy, *J. Chromatogr.* **91,** 723 (1974).

192. J. Macak, V. M. Nabivach, P. Buryan, and J. S. Berlizov, *J. Chromatogr.* **209,** 472 (1981).

193. D. Brocco, A. Cimmino, and M. Possanzini, *J. Chromatogr.* **84,** 371 (1973).

194. M. Novotny, R. Kump, F. Merli, and L. J. Todd, *Anal. Chem.* **52,** 401 (1980).

195. F. Merli, D. Wiesler, M. P. Maskarinec, M. Novotny, D. L. Vassilaros, and M. L. Lee, *Anal. Chem.* **53,** 1929 (1981).

196. M. Novotny, J. W. Strand, S. L. Smith, D. Wiesler, and F. J. Schwende, *Fuel* **60,** 213 (1981).

197. W. W. Paudler and M. Cheplen, *Fuel* **58,** 775 (1979).

198. M. R. Guerin, C.-h. Ho, T. K. Rao, B. R. Clark, and J. L. Epler, *Environ. Res.* **23,** 42 (1980).

199. D. W. Later, M. L. Lee, K. D. Bartle, R. C. Kong, and D. L. Vassilaros, *Anal. Chem.* **53,** 1612 (1981).

200. D. W. Later, M. L. Lee, and B. W. Wilson, *Anal. Chem.* **54,** 117 (1982).

201. L. J. Felice, *Anal. Chem.* **54,** 869 (1982).

202. M. V. Buchanan, C.-h. Ho, M. R. Guerin, and B. R. Clark, in *Polynuclear Aromatic Hydrocarbons: Chemistry and Biological Fate*, M. Cooke and A. J. Dennis, editors. Battelle Press, Columbus, Ohio, 1981, p. 133.

203. D. W. Later, B. W. Wright, and M. L. Lee, *J. High Resoln. Chromatogr./Chromatogr. Commun.* **4,** 406 (1981).

204. S. Weinstein, B. Feibush, and E. Gil-Av, *J. Chromatogr.* **126,** 97 (1976).

205. C. H. Lochmüller and J. V. Hinshaw, Jr., *J. Chromatogr.* **202,** 363 (1980).

206. N. Oi, H. Kitahara, and T. Doi, *J. Chromatogr.* **207,** 252 (1981).

207. W. A. Konig and I. Benecke, *J. Chromatogr.* **209,** 91 (1981).

208. W. A. Konig, I. Benecke, and S. Sievers, *J. Chromatogr.* **217,** 71 (1981).

209. R. W. Zumwalt, D. Roach, and C. W. Gehrke, *J. Chromatogr.* **53**, 171 (1970).

210. C. W. Moss, M. A. Lambert, and F. J. Diaz, *J. Chromatogr.* **60**, 134 (1971).

211. C. W. Moss and M. A. Lambert, *Anal. Biochem.* **59**, 259 (1974).

212. J. P. Zanetta and G. Vincendon, *J. Chromatogr.* **76**, 91 (1973).

213. P. Felker and R. Bandurski, Anal. Biochem. **67**, 245 (1975).

214. S. L. MacKenzie and D. Tenaschuk, *J. Chromatogr.* **97**, 19 (1974).

215. M. A. Kirkman, *J. Chromatogr.* **97**, 175 (1974).

216. S. L. MacKenzie and D. Tenaschuk, *J. Chromatogr.* **111**, 413 (1975).

217. J. F. March, *Anal. Biochem.* **69**, 420 (1975).

218. R. J. Siezen and T. H. Mague, *J. Chromatogr.* **130**, 151 (1977).

219. J. Desgres, D. Boisson, and P. Padieu, *J. Chromatogr.* **162**, 133 (1979).

220. I. M. Moodie and J. Burger, *J. High Resoln. Chromatogr./Chromatogr. Commun.* **4**, 218 (1981).

221. C. F. Poole and M. Verzele, *J. Chromatogr.* **150**, 439 (1978).

222. R. W. Zumwalt, K. Kuo, and C. W. Gehrke, *J. Chromatogr.* **57**, 193 (1971).

223. B. Brechbühler, L. Gay, and H. Jaeger, *Chromatographia* **10**, 478 (1977).

224. R. F. Adams, F. L. Vandemark, and G. J. Schmidt, *J. Chromatogr. Sci.* **15**, 63 (1977).

225. W. A. König, *J. High Resoln. Chromatogr./Chromatogr. Commun.* **5**, 588 (1982).

226. E. Gil-Av, B. Feibush, and R. Charles-Sigler, in *Gas Chromatography 1966*, A. B. Littlewood, editor. Institute of Petroleum, London, 1966, p. 227.

227. E. Gil-Av and B. Feibush, *Tetrahedron Lett.* **1967**, 3345 (1967).

228. W. A. König, W. Parr, H. A. Lichtenstein, E. Bayer, and J. Oro, *J. Chromatogr. Sci.* **8**, 183 (1970).

229. B. Feibush and E. Gil-Av, *Tetrahedron* **26**, 1361 (1970).

230. S. Nakaparksin, P. Birrell, E. Gil-Av, and J. Oro, *J. Chromatogr. Sci.* **8**, 183 (1970).

231. J. A. Corbin and L. B. Rogers, *Anal. Chem.* **42**, 1786 (1970).

232. W. Parr, C. Yang, E. Bayer, and E. Gil-Av, *J. Chromatogr. Sci.* **8**, 591 (1970).

233. W. Parr, C. Yang, J. Pleterski, and E. Bayer, *J. Chromatogr.* **50**, 510 (1970).

234. J. A. Corbin, J. E. Rhoad, and L. B. Rogers, *Anal. Chem.* **43**, 327 (1971).

235. W. Parr, J. Pleterski, C. Yang, and E. Bayer, *J. Chromatogr. Sci.* **9**, 141 (1971).

236. W. Parr and P. Y. Howard, *Chromatographia* **4**, 162 (1971).

237. W. Parr and P. Y. Howard, *Angew. Chem. Int. Ed. Engl.* **11**, 314 (1972).

238. W. Parr and P. Y. Howard, *Anal. Chem.* **45**, 711 (1973).

239. P. Y. Howard and W. Parr, *Chromatographia* **7**, 283 (1974).

240. W. A. König and G. J. Nicholson, *Anal. Chem.* **47**, 951 (1975).

241. R. Brazell, W. Parr, F. Andrawes, and A. Zlatkis, *Chromatographia* **9**, 57 (1976).

242. W. A. König, *Chromatographia* **9**, 72 (1976).

243. I. Abe, T. Kohno, and S. Musha, *Chromatographia* **11**, 393 (1978).

244. I. Abe and S. Musha, *J. Chromatogr.* **200**, 195 (1980).

245. S. Weinstein, B. Feibush, and E. Gil-Av, *J. Chromatogr.* **126**, 97 (1981).

246. W. A. König, S. Sievers, and I. Benecke, in *Proceedings of the Fourth International Symposium on Capillary Chromatography*, R. E. Kaiser, editor. Huethig, Heidelberg, 1981, p. 703.

247. N. Oi, H. Kitahara, Y. Inda, and T. Doi, *J. Chromatogr.* **213**, 137 (1981).

248. H. Frank, G. J. Nicholson, and E. Bayer, *J. Chromatogr. Sci.* **15**, 174 (1977).

249. H. Frank, G. J. Nicholson, and E. Bayer, *J. Chromatogr.* **167**, 187 (1978).

250. G. J. Nicholson, H. Frank, and E. Bayer, *J. High Resoln. Chromatogr./Chromatogr. Commun.* **2**, 411 (1979).

251. T. Saeed, P. Sandra, and M. Verzele, *J. Chromatogr.* **186**, 611 (1979).

252. T. Saeed, P. Sandra, and M. Verzele, *J. High Resoln. Chromatogr./Chromatogr. Commun.* **3**, 35 (1980).

253. K. Grohmann and W. Parr, *Chromatographia* **5**, 18 (1972).

254. U. Beitler and B. Feibush, *J. Chromatogr.* **123**, 149 (1976).

255. R. Charles and E. Gil-Av, *J. Chromatogr.* **195**, 317 (1980).

256. B. Feibush, A. Balan, B. Altman, and E. Gil-Av, *J. Chem. Soc., Perkin Trans. II*, 1230 (1979).

257. S. Mizushima, T. Shimanouchi, M. Tsuboi, and T. Arakawa, *J. Am. Chem. Soc.* **79**, 5357 (1957).

258. M. Tsuboi, T. Shimanouchi, and S. Mizushima, *J. Am. Chem. Soc.* **81**, 1406 (1959).

259. T. Ichikawa and Y. Iitaka, *Acta Cryst.* **B25**, 1824 (1969).

260. Y. Harada and Y. Iitaka, *Acta Cryst.* **B30**, 1452 (1974).

261. S.-C. Chang, R. Charles, and E. Gil-Av, *J. Chromatogr.* **202**, 247 (1980).

262. R. Liardon and S. Ledermann, *J. High Resoln. Chromatogr./Chromatogr. Commun.* **3**, 475 (1980).

263. I. Abe, K. Izumi, S. Kuramoto, and S. Musha, *J. High Resoln. Chromatogr./Chromatogr. Commun.* **4**, 549 (1981).

264. N. Oi, M. Horiba, and H. Kitahara, *J. Chromatogr.* **202**, 299 (1980).

265. J. R. McDermott and N. L. Benoiton, *Can. J. Chem.* **51**, 2562 (1973).

266. W. A. König and U. Hess, *Liebigs Ann. Chem.* 1087 (1977).

267. E. Gil-Av, R. Charles-Sigler, G. Fischer, and D. Nurok, *J. Gas Chromatogr.* **4**, 51 (1966).

268. G. E. Pollock, *Anal. Chem.* **44**, 2368 (1972).

269. W. A. Bonner, *J. Chromatogr. Sci.* **10**, 159 (1972).

270. F. Raulin and B. N. Khare, *J. Chromatogr.* **75**, 13 (1973).

271. M. A. VanDort and W. A. Bonner, *J. Chromatogr.* **133**, 210 (1977).

272. W. A. König, W. Rahn, and J. Eyem, *J. Chromatogr.* **133**, 141 (1977).

273. J. Krupcik, J. Hrivnak, and J. Janak, *J. Chromatogr. Sci.* **14**, 4 (1976).

274. L. S. Ettre and E. W. March, *J. Chromatogr.* **91**, 5 (1974).

275. L. Zoccolillo, A. Liberti, and G. Goretti, *J. Chromatogr.* **43**, 497 (1969).

276. J. Hrivnak, *J. Chromatogr. Sci.* **8**, 602 (1970).

277. J. Hrivnak, L. Sojak, E. Beska, and J. Janak, *J. Chromatogr.* **68**, 55 (1972).

278. P. Sandra, M. Verstappe, and M. Verzele, *J. High Resoln. Chromatogr./Chromatogr. Commun.* **1**, 28 (1978).

279. G. Goretti and A. Liberti, *J. Chromatogr.* **61**, 334 (1971).

280. M. Verzele and P. Sandra, *J. High Resoln. Chromatogr./Chromatogr. Commun.* **2**, 303 (1979).

281. J. L. Marshall and D. A. Parker, *J. Chromatogr.* **122**, 425 (1976).

282. M. Hrivnac, L. Sykora-Cechova, and M. Muller-Aerne, *J. High Resoln. Chromatogr./Chromatogr. Commun.* **4**, 323 (1981).

283. Gy. Vigh, J. Hlavay, Z. Varga-Puchony, and T. Welsch, *J. High Resoln. Chromatogr./Chromatogr. Commun.* **5**, 124 (1982).

284. R. Dandeneau and E. Zerenner, *J. High Resoln. Chromatogr./Chromatogr. Commun.* **2**, 351 (1979).

285. R. G. Ackman and S. N. Hooper, *J. Chromatogr. Sci.* **12**, 131 (1974).

286. J. Flanzy, M. Boudon, C. Leger, and J. Pihet, *J. Chromatogr. Sci.* **14**, 18 (1976).

287. R. G. Ackman, S. M. Barlow, and I. F. Duthie, *J. Chromatogr. Sci.* **15**, 290 (1977).

288. J. Krupcik, J. Hrivnak, L. Barnoky, and J. Janak, *J. Chromatogr.* **65**, 323 (1972).

289. K. Grob, Jr., H. P. Neukom, D. Frohlich, and R. Battaglia, *J. High Resoln. Chromatogr./ Chromatogr. Commun.* **1**, 94 (1978).

290. H. I. 3lover and F. Lanza, *J. Am. Oil Chem. Soc.* **56**, 933 (1979).

291. R. G. Ackman, A. Manzer, and J. Joseph, *Chromatographia* **7**, 107 (1974).

292. J. Hrivnak, L. Sojak, J. Krupcik, and Y. P. Duchesne, *J. Am. Oil Chem. Soc.* **50**, 68 (1973).

293. M. P. Maskarinec, G. Alexander, and M. Novotny, *J. Chromatogr.* **126**, 559 (1976).

294. D. B. Drucker, *J. Chromatogr.* **208**, 279 (1981).

295. J. M. B. Apon and N. Nicolaides, *J. Chromatogr. Sci.* **13**, 467 (1975).

296. S.-N. Lin and E. C. Horning, *J. Chromatogr.* **112**, 465 (1975).

297. S.-N. Lin, C. D. Pfaffenberger, and E. C. Horning, *J. Chromatogr.* **104**, 319 (1975).

298. H. Jaeger, W. Wagner, J. Homoki, H. U. Klör, and H. Ditschuneit, *Chromatographia* **10**, 492 (1977).

299. H. Heckers, F. W. Melcher, and U. Schloeder, *J. Chromatogr.* **136**, 311 (1977).

300. T. Kobayashi, *J. Chromatogr.* **194**, 404 (1980).

301. L. V. Holdeman, E. P. Cato, and W. E. C. Moore, *Anaerobe Laboratory Manual*, 4th ed. Virginia Polytechnic Institute and State University, Blacksburg, Virginia, 1977.

302. F. S. Calixto and A. G. Raso, *Chromatographia* **14**, 143 (1981).

303. R. G. Ackman and C. A. Eaton, *J. Chromatogr. Sci.* **13**, 509 (1975).

304. R. G. Riley, K. Shiosaki, R. M. Bean, and D. M. Schoengold, *Anal. Chem.* **51**, 1995 (1979).

305. R. H. Fish, A. S. Newton, and P. C. Babbitt, *Fuel* **61**, 227 (1982).

306. H. M. Liebich, A. Pickert, U. Stierle, and J. Wöll, *J. Chromatogr.* **199**, 181 (1980).

307. J. Shen, *Anal. Chem.* **53**, 475 (1981).

308. A. Smith and A. K. Lough, *J. Chromatogr. Sci.* **13**, 486 (1975).

309. H. Jaeger, H.-U. Klör, G. Blos, and H. Ditschuneit, *Chromatographia* **8**, 507 (1975).

310. M. J. Waechter, M. C. Pfaff-Dessalles, and G. Mahuzier, *J. Chromatogr.* **204**, 245 (1981).

311. C. R. Scholfield and H. J. Dutton, *J. Am. Oil Chem. Soc.* **48**, 228 (1971).

312. C. R. Scholfield, *J. Am. Oil Chem. Soc.* **49**, 583 (1972).

313. N. S. Nikitina, N. I. Vikhrestyuk, and A. E. Mysak, *J. Chromatogr.* **91**, 775 (1974).

314. S.-N. Lin, C. D. Pfaffenberger, and E. C. Horning, *J. Chromatogr.* **104**, 319 (1975).

315. S.-N. Lin and E. C. Horning, *J. Chromatogr.* **112**, 465 (1975).

316. R. G. Ackman and S. N. Hooper, *J. Chromatogr.* **86**, 83 (1973).

317. E. S. Van Vleet and J. G. Quinn, *J. Chromatogr.* **151**, 396 (1978).

318. R. Gloor and H. Leidner, *Chromatographia* **9**, 618 (1976).

319. R. Liardon and U. Kühn, *J. High Resoln. Chromatogr./Chromatogr. Commun.* **1**, 47 (1978).

320. X. Monseur, J. Walravens, P. Dourte, and M. Termonia, *J. High Resoln. Chromatogr./ Chromatogr. Commun.* **4**, 49 (1981).

321. A. S. Christophersen, K. E. Rasmussen, and F. Tonnesen, *J. Chromatogr.* **179**, 87 (1979).

322. S. Lewis, C. N. Kenyon, J. Meili, and A. L. Burlingame, *Anal. Chem.* **51**, 1275 (1979).

323. F. P. Woodford and C. M. van Gent, *J. Lipid Res.* **1**, 188 (1960).

324. T. K. Miwa, K. L. Mikolajczak, F. R. Earle, and I. A. Wolf, *Anal. Chem.* **32**, 1739 (1960).

325. R. G. Ackman, *J. Chromatogr.* **28**, 225 (1967).

326. R. G. Ackman and S. N. Hooper, *J. Chromatogr.* **86**, 73 (1973).

327. G. Blu, L. Jacob, and G. Guiochon, *J. Chromatogr.* **50**, 1 (1970).

328. M. Beroza, *Accounts Chem. Res.* **3**, 33 (1970).

329. T. E. Kuzmenko, A. L. Samusenko, V. P. Uralets, and R. V. Golovnya, *J. High Resoln. Chromatogr./Chromatogr. Commun.* **2**, 43 (1979).

330. W. A. Konig and S. Sievers, *J. Chromatogr.* **200**, 189 (1980).

331. W. A. Konig and I. Benecke, *J. Chromatogr.* **195**, 292 (1980).

332. N. Oi, M. Horiba, H. Kitahara, T. Doi, T. Tani, and T. Sakakibara, *J. Chromatogr.* **202**, 305 (1980).

333. N. Oi, H. Kitahara, M. Horiba, and T. Doi, *J. Chromatogr.* **206**, 143 (1981).

334. R. G. Ackman, R. E. Cox, G. Eglinton, S. N. Hooper, and J. R. Maxwell, *J. Chromatogr. Sci.* **10**, 392 (1972).

335. M. Kates, A. J. Hancock, and R. G. Ackman, *J. Chromatogr. Sci.* **15**, 177 (1977).

336. K. Grob, Jr., H. P. Neukom, and R. Battaglia, *J. Am. Oil Chem. Soc.* **57**, 282 (1980).

337. G. Schomburg, R. Dielmann, H. Husmann, and F. Weeke, *J. Chromatogr.* **122**, 55 (1976).

338. K. Grob, Jr., *J. Chromatogr.* **205**, 289 (1981).

339. H. Traitler and A. Prevot, *J. High Resoln. Chromatogr./Chromatogr. Commun.* **4**, 109 (1981).

340. K. Grob, Jr., *J. Chromatogr.* **178**, 387 (1979).

341. E. Geeraert and D. De Schepper, *J. High Resoln. Chromatogr./Chromatogr. Commun.* **5**, 80 (1982).

342. K. Tesarik, *J. Chromatogr.* **65**, 295 (1972).

343. J. W. Mourits, H. G. Merkus, and L. deGalan, *Anal. Chem.* **48**, 1557 (1976).

344. D. Nurok and T. J. Reardon, *Anal. Chem.* **50**, 855 (1978).

345. J. Szafranek and A. Wisniewski, *J. Chromatogr.* **161**, 213 (1978).

346. G. Eklund, B. Josefsson, and C. Roos, *J. Chromatogr.* **142**, 575 (1977).

347. G. Holzer, J. Oro, S. J. Smith, and V. M. Doctor, *J. Chromatogr.* **194**, 410 (1980).

348. R. Oshima, A. Yoshikawa, and J. Kumanotani, *J. Chromatogr.* **213**, 142 (1981).

349. J. Klok, E. H. Nieberg-Van Velzen, J. W. De Leeuw, and P. A. Schenck, *J. Chromatogr.* **207**, 273 (1981).

350. G. Schomburg, H. Husmann, and F. Weeke, *J. Chromatogr.* **99**, 63 (1974).

351. A. G. W. Bradbury, D. J. Halliday, and D. G. Medcalf, *J. Chromatogr.* **213**, 146 (1981).

352. S. Adam and W. G. Jennings, *J. Chromatogr.* **115**, 218 (1975).

353. D. Anderle, J. Königstein, and V. Kovacik, *Anal. Chem.* **49**, 137, (1977).

354. H. Zegota, *J. Chromatogr.* **192**, 446 (1980).

355. K. J. Schaffler and P. G. Morel Du Boil, *J. Chromatogr.* **207**, 221 (1981).

356. C. D. Pfaffenberger, J. Szafranek, and E. C. Horning, *J. Chromatogr.* **126**, 535 (1976).

357. S. Honda, N. Yamauchi, and K. Kakehi, *J. Chromatogr.* **169**, 287 (1979).

358. S. Honda, M. Nagata, and K. Kakehi, *J. Chromatogr.* **209**, 299 (1981).

359. I. Benecke, E. Schmidt, and W. A. König, *J. High Resoln. Chromatogr./Chromatogr. Commun.* **4**, 553 (1981).

360. D. Nurok and T. J. Reardon, *Carbohydr. Res.* **56**, 165 (1977).

361. J. W. Rote and P. G. Murphy, *Bull. Environ. Contam. Toxicol.* **6**, 377 (1971).

362. T. Cairns and E. G. Siegmund, *Anal. Chem.* **53**, 1183A (1981).

363. O. W. Berg, P. L. Diosady, and G. A. V. Rees, *Bull. Environ. Contam. Toxicol.* **13,** 338 (1972).

364. L. R. Kamps, W. J. Trotter, S. J. Young, L. J. Carson, J. A. G. Roach, J. A. Sphon, J. T. Tanner, and B. McMahon, *Bull. Environ. Contam. Toxicol.* **20,** 589 (1978).

365. A. L. Robbins and C. R. Whillhite, *Bull. Environ. Contam. Toxicol.* **21,** 428 (1979).

366. B. Zimmerli, *J. Chromatogr.* **88,** 65 (1974).

367. M. Cooke, G. Nickless, and D. J. Roberts, *J. Chromatogr.* **187,** 47 (1980).

368. M. Cooke and D. J. Roberts, *J. Chromatogr.* **193,** 437 (1980).

369. J. Krupcik, P. A. Leclercq, A. Simova, P. Suchanek, M. Collak, and J. Hrivnak, *J. Chromatogr.* **119,** 271 (1976).

370. P. W. Albro, J. K. Haseman, T. A. Clemmer, and B. J. Corbett, *J. Chromatogr.* **136,** 147 (1977).

371. P. W. Albro, J. T. Corbett, and J. L. Schroeder, *J. Chromatogr.* **205,** 103 (1981).

372. J. Krupcik, P. A. Leclercq, J. Garaj, and A. Simova, *J. Chromatogr.* **191,** 207 (1980).

373. M. Cooke and A. G. Ober, *J. Chromatogr.* **195,** 265 (1980).

374. H. Tausch, G. Stehlik, and H. Wihlidal, *Chromatographia* **14,** 403 (1981).

375. F. W. Crow, A. Bjorseth, K. T. Knapp, and R. Bennett, *Anal. Chem.* **53,** 619 (1981).

376. K. Ballschmiter and M. Zell, *Fres. Z. Anal. Chem.* **302,** 20 (1980).

377. B. Boe and E. Egaas, *J. Chromatogr.* **180,** 127 (1979).

378. M. Zell, H. J. Neu, and K. Ballschmiter, *Fres. Z. Anal. Chem.* **292,** 97 (1978).

379. H. Neu, M. Zell, and K. Ballschmiter, *Fres. Z. Anal. Chem.* **293,** 193 (1978).

380. D. Sissons and D. Welti, *J. Chromatogr.* **60,** 15 (1971).

381. P. W. Albro, J. K. Haseman, T. A. Clemmer, and B. J. Corbett, *J. Chromatogr.* **136,** 147 (1977).

382. M. Zell, J. H. Neu, and K. Ballschmiter, *Chemosphere* **2/3,** 69 (1977).

383. R. W. Moshier and R. E. Sievers, *Gas Chromatography of Metal Chelates.* Pergamon, Oxford, 1965.

384. G. Guiochon and C. Pommier, *Gas Chromatography in Inorganics and Organometallics.* Ann Arbor Scientific Publishers, Ann Arbor, Michigan, 1973.

385. P. C. Uden and D. E. Henderson, *Analyst (London)* **102,** 889 (1977).

386. D. N. Sokolov, *Russ. Chem. Rev.* **46,** 388 (1977).

387. P. J. Marriott and T. J. Cardwell, *J. Chromatogr.* **234,** 157 (1982).

388. A. Tavlaridis and R. Neeb, *Fres. Z. Anal. Chem.* **292,** 199 (1978).

389. J. Krupcik, P. A. Leclercq, J. Garaj, and J. Masaryk, *J. Chromatogr.* **171,** 285 (1979).

390. L. Sucre and W. Jennings, *J. High Resoln. Chromatogr./Chromatogr. Commun.* **3,** 452 (1980).

391. P. C. Uden, D. E. Henderson, F. P. DiSanzo, R. J. Lloyd, and T. Tetu, *J. Chromatogr.* **196,** 403 (1980).

392. S. A. Estes, P. C. Uden, M. D. Rausch, and R. M. Barnes, *J. High Resoln. Chromatogr./Chromatogr. Commun.* **3,** 471 (1980).

393. S. A. Estes, P. C. Uden, and R. M. Barnes, *Anal. Chem.* **53,** 1829 (1981).

394. S. A. Estes, P. C. Uden, and R. M. Barnes, *Anal. Chem.* **53,** 1336 (1981).

395. B. W. Wright, M. L. Lee, and G. M. Booth, *J. High Resoln. Chromatogr./Chromatogr. Commun.* **2,** 189 (1979).

396. R. Alexander, G. Eglinton, J. P. Gill, and J. K. Volkman, *J. High Resoln. Chromatogr./Chromatogr. Commun.* **3,** 521 (1980).

397. G. Garzo and D. Hoebbel, *J. Chromatogr.* **119**, 173 (1976).
398. G. Garzo, D. Hoebbel, Z. J. Ecsery, and K. Ujszaszi, *J. Chromatogr.* **167**, 321 (1978).
399. S. O. Farwell, S. J. Gluck, W. L. Bamesberger, T. M. Schutte, and D. F. Adams, *Anal. Chem.* **51**, 609 (1979).
400. A. Purer and R. L. Kaplan, *J. Chromatogr. Sci.* **9**, 59 (1971).

SEVEN

APPLICATIONS: SAMPLE TYPE

7.1 INTRODUCTION

The major advantage of open tubular column GC lies in the extensive information obtained by the resolution of individual components in complex mixtures. Valuable qualitative and quantitative information about mixture components can be provided. The explosive growth of open tubular column GC during the past few years is due in large part to the recent improvements in column technology and the demand for more complete and precise analyses. This technique is currently being applied to the analysis of complex matrices such as environmental samples, fossil fuels, foods, cosmetics, and biological material. Open tubular column technology and methodology developed for the GC analysis of these types of samples and others will be discussed in the following sections. This chapter is not intended, however, to be a comprehensive review of all available literature on each subject covered, but rather to provide information that illustrates the application of open tubular column GC to the solution of a wide range of different analytical problems.

7.2 ENVIRONMENTAL

Air Pollution

Increasing realization of the dangers associated with environmental pollution has led to the focusing of modern analytical chemistry in the determination of chemical species in the air of both cities and workplaces. Gas chromatography with open tubular columns and MS detection is particularly applicable to the range of volatile substances containing up to 20 carbon atoms which may be present in air as vapor and aerosols, as well as to compounds such as polycyclic aromatic compounds (PAC) which are adsorbed on particulate matter suspended in air.

The high dilution of most organic substances present as vapors in air (10^{-4}–10^{-8}% by volume) means that a concentration step is necessary, usually on a suitable adsorbent by a purge-and-trap procedure (Section 7.4). The adsorbed compounds are then either extracted with a suitable solvent or, better, thermally desorbed into the cooled open tubular column. Of course, concentration of the extracting solvent would lead to losses of volatiles, and splitless injection of the fairly dilute solution is necessary. Problems from impurities in solvents are also inevitable.[1] In thermal desorption, holding the early part of the column at liquid nitrogen temperature and then programming from subambient temperature results in sharp peaks and useful separations even for C_3–C_4 hydrocarbons and chloromethane.

Table 7.1 summarizes some of the methods used and the results obtained in analyzing air for organic pollutant vapors by open tubular column GC. Sampling through Tenax GC is particularly popular. Experiments[3,7] with tubes packed with this adsorbent and arranged in series showed that among common air pollutants only benzaldehyde and acetophenone could not be retained with trapping efficiency greater than 90%. Tenax GC has high thermal stability (so that high-temperature desorption is possible), does not react with adsorbed compounds, is relatively free of artifactual contaminants, and is hydrophobic enough to allow millionfold excesses of water to pass through without interfering with the sorption of organic vapors.[8,9] However, rep-

TABLE 7.1. Analysis of air for organic pollutant vapors.

Air type	Location	Sample volume (m^3)	Adsorbent	Number of compounds identified	Reference
Urban	Zurich	25	Activated carbon	108	2
Urban	Houston	0.12–0.2	Tenax GC	98	3
Urban	Paris	0.2–2	Graphitized carbon black	72	4
Urban	Leningrad	0.01	Polysorbimide and carbochrom	136	5
Rural	Talladega National Forest, Ala. U.S.A.	0.025	Tenax GC	40	6

resentation of a given compound in a chromatogram may not reflect its concentration in air because of irreversible adsorption and "breakthrough effects";[8] caution is necessary in making quantitative interpretations. Other carbonaceous adsorbents in addition to those in Table 7.1 have been investigated, and among these Ambersorb XE-340 and Carbopack BHT have been found promising.[9] Amberlite resins may be contaminated with alkyl derivatives of aromatic hydrocarbons and require clean-up before use.[10]

The majority of pollutants detected in air by the above methods and identified from mass spectra are aliphatic and aromatic hydrocarbons, especially the alkylated benzenes; chlorinated compounds are also prominent. Nonpolar stationary phases are therefore preferred. A comparison of urban and rural air shows[6] common pollutants, but with a reduction of approximately two orders of magnitude in the latter. Other compounds derived from natural sources, particularly terpenes, are also found in rural air.[6]

Open tubular column GC has an important role to play in the measurement and control of chemical hazards in the working environment. A good example is the analysis of air in film-processing laboratories for aromatic amines which present considerable toxicological dangers.[11] Sampling was carried out on silica-gel adsorption tubes which were extracted with 2-butanone. The amines were separated on glass open tubular columns coated with Carbowax 20M containing alkali. Nitrogen-selective detection allowed quantitation of amines at an airborne concentration of 3–13 μg m^{-3} (Figure 7.1), after identification by GC-MS.

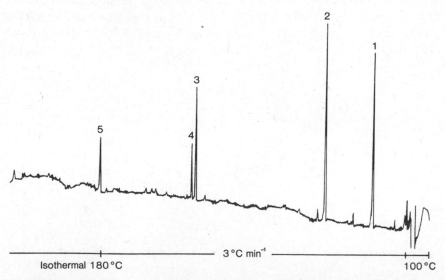

FIGURE 7.1. Chromatogram of airborne aromatic amines from a film-processing laboratory. Note: Carbowax 20M glass column, nitrogen-selective detection. Selected peak assignments: (1) *N,N*-diethylaniline; (2) 2,6-dimethylaniline; (3) *N,N*-diethyl-1,4-diaminobenzene; (4) *N,N*-diisopropyl-1,4-diaminobenzene. (Reproduced with permission from ref. 11. Copyright Elsevier Scientific Publishing Company)

The widespread interest in the analysis of the polycyclic aromatic compounds (PAC) of urban air and workplace atmospheres stems from the
marked carcinogenicity of these compounds. PAC comprise the largest class
of chemical carcinogens known and are widely distributed in the environment.[12,13] They generally originate from combustion sources (see later sections) and are present in air, being adsorbed on suspended particulate matter
with average diameter less than 10 μm. Determination of PAC in aerosols
is intrinsically important for reasons associated with the long-term health
risks of airborne PAC, and also because these compounds can provide "fingerprints" of transported air pollutants.[14] After extraction from air particulate matter and clean-up,[12,13] PAC are best analyzed on short well-deactivated glass or fused-silica open tubular columns coated with SE-52 (see
Figure 7.2 and Chapter 6).[13,15] The retention index system of Lee et al.[16]
allows identifications, although MS is best used as detection; over 100 PAC
have been identified in airborne particulates in this way.[17] The main problem
here is that the mass spectra of PAC isomers are often indistinguishable. A
novel open tubular column GC-MS method uses a mixed-charge exchange
chemical ionization reagent gas which allows differentiation of isomers.[18]
For example, Figure 7.3 shows plots of m/z 192 and 193 ions for open tubular
column GC-MS (methane + argon reagent gas) of an air particulate PAC
sample. These peaks could correspond to the M^+ and $(M + 1)^+$ ions of
either methylanthracenes or methylphenanthrenes; in fact, the measured (M
+ 1)/M abundance ratios (Table 7.2) of all four peaks correspond to those
of methylphenanthrenes. The FPD can allow[19] selective detection of sulfur-
containing PAC, while the nitrogen-selective thermionic detector has been

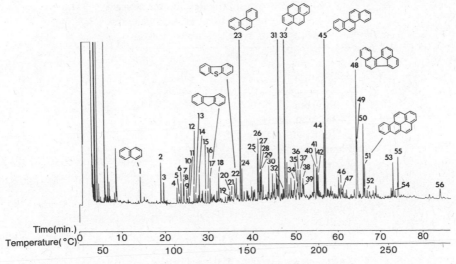

FIGURE 7.2. Chromatogram of the polycyclic aromatic compound fraction from St. Louis
air particulate matter. Note: 20 m SE-52 fused-silica column, H_2 carrier gas.

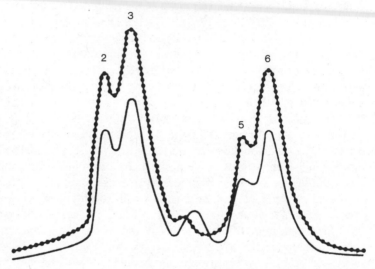

FIGURE 7.3. Selected ion plots of *m/z* 192 and 193 of the polycyclic aromatic compound fraction of Utah County air particulate matter. Note: Peak numbers refer to peaks in Table 7.2. (Reproduced with permission from ref. 18. Copyright U.S. Government Printing Office.)

TABLE 7.2. Abundance ratios for several anthracenes, phenanthrenes, and selected peaks from air particulate matter.

Compound[a]	$\dfrac{M + 1}{M}$ Ratio[b]
Anthracene	0.82
2-Methylanthracene	0.77
9-Methylanthracene	0.79
Phenanthrene	1.57
1-Methylphenanthrene	1.20
9-Methylphenanthrene	1.20
Peak No. 2	1.26
Peak No. 3	1.20
Peak No. 5	1.24
Peak No. 6	1.24

[a] Peak numbers refer to peaks in Figure 7.3.
[b] The ratios have been corrected for the natural abundance of ^{13}C.

used for the selective detection of nitrogen heterocycles in air particulate matter.[20] Attention has been also focused on the detection of PAC with molecular masses of up to 400 in airborne particulate matter by the use of highly thermostable fused-silica columns.[21]

Thermal desorption of the total organic fraction of air particulate matter from the filter onto the open tubular column allows simultaneous analysis of some of the other constituents, such as alkanes, phthalates, and so on, in addition to the PAC.[22] Analysis on an SE-52 coated glass open tubular column of the alkane fraction separated by column chromatography from air particulates has also been reported.[23] A novel application of open tubular column GC is the identification by GC-MS of dimethyl sulfate, a potent mutagen and carcinogen, in a polar extract of fly ash.[24]

Water Pollution

The high sensitivity and resolution of open tubular column GC make this method as applicable to the pollutants in water as to those in polluted air. Again, MS detection is generally required, and there are the same requirements for the concentration of organics, before introduction onto the column.

At least three procedures are available and all have implications for the mode of injection. Grob and co-workers have shown[25,26] how liquid–liquid extraction is successful as long as the water sample is not turbid. Extraction of 1 dm^3 water with 0.5–1.0 cm^3 solvent allows determination of pollutants at low levels. The implication for GC is, of course, that splitless injection is necessary, and the solvent must be extremely pure: pentane is most readily obtained sufficiently free from impurities. Concentration by adsorption[27] on active carbon or organic resins such as the Amberlite XAD macroreticular resins or Tenax GC, followed by washing off with pure solvent, is commonly applied. The third method is to purge and trap in a head-space procedure adapted from that applied to body fluids.[28,29] Volatile organics are stripped from the heated water sample with ultrapure helium or in a closed-circuit system and adsorbed on Tenax GC, from which they are thermally desorbed and cryogenically focused in the open tubular column (Figure 7.4).

Analyses of water for pollutants at the ppb–ppt level or even lower are possible by these methods. The highest sensitivity and selectivity for halogen compounds is obtained with ECD or microwave emission detection. Chlorine compounds in water can arise from pollution or in water-treatment processes. Pesticides present at the microgram per cubic decimeter or even picogram per cubic decimeter level in water were extracted with pentane and the concentrated extract was analyzed on a 60 m glass SE-30 coated open tubular column with low split.[30] Chlorobenzenes in waste water were preconcentrated on a small column of Chromosorb 102, eluted with pentane, and analyzed on Carbowax 20M and SP-2100 glass columns with splitless injection, and as above, ECD detection.[31] Lower limits varied between 0.1

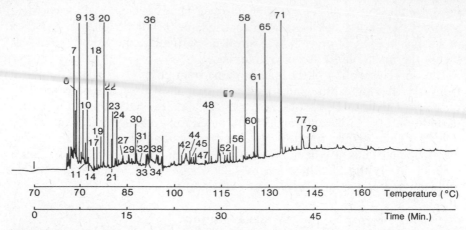

FIGURE 7.4. Chromatogram of volatiles from the Black Warrior River, Alabama. Note: Headspace volatiles were sparged from a 1.8 dm³ sample onto Tenax GC, 59 m Emulphor ON 870 column. Selected peak assignments: (9) acetone; (20) ethylbenzene; (36) 1-chlorooctane; (58,65,71) anethole isomers. (Reproduced with permission from ref. 28. Copyright Elsevier Scientific Publishing Company.)

ng dm^{-3} (penta- and hexachlorobenzenes) and 500 ng dm^{-3} for monochlorobenzenes. Spent bleach liquors from a sulfite wood-pulp mill contain chlorinated and brominated lipophilic compounds extractable by cyclohexane and identified by open tubular column GC-MS.[32] Volatile halogenated hydrocarbons in water were extracted by pentane before analysis on a Carbowax 20M deactivated SE-52 coated glass column with ECD;[33] the maximum sensitivity was 1 fg for CCl$_4$. The halogenated compounds of polluted water resulting from chlorination include[33,34] the haloforms (bromine and iodine compounds via hypohalite generated from Br$^-$ and I$^-$). These pollutants may be separated on a short (10 m) thick-film (1.5 μm) SE-52 column in only 30 s.[33] The products of chlorination of the aromatic fraction of diesel fuel,[35] and of humic and fulvic acids,[34] which include the haloforms, have also been separated on open tubular columns. Microwave emission detection was used in the latter analysis.[34]

Hydrocarbons comprise one of the largest groups of pollutants emitted into the environment, and their analysis presents a considerable challenge because of the complexity of the mixtures. The source of oil spills on the sea can be established from open tubular column chromatograms in which the pattern of both *n*-alkanes and branched/cyclic alkanes along with the ratios of the acyclic isoprenoids, pristane and phytane, can all be distinctive.[36,37] Figure 7.5 compares the chromatograms of a pollutant oil sample from the coast of Holland, with an oil from a suspected ship's cargo.[36] Even though the light ends have evaporated from the polluting oil, the distribution of *n*-alkanes shows it to correspond to residues of a heavy Middle-Eastern crude, and not to the lighter Sahara crude of the cargo. Even if *n*-alkanes have been lost by biodegradation, the ratio of pristane to phytane is constant, for example, in fresh and weathered oil pollution from a harbor. Mathematical treatments have also allowed identification of oil spills from the branched cyclic alkane pattern in the open tubular column chromatogram.[37] A further, sulfur-compound fingerprint of crude oil is also generated if FPD detection is used.[38]

PAC extracted from potable and waste waters by a solvent, or by adsorption on a polymer,[12] are also regularly analyzed by open tubular column GC by methods similar to those applied in air-pollutant work.[39] A variety of other organic pollutants of water have been identified,[40] including volatile sulfur compounds such as dimethyl di-, tri-, and tetrasulfides, by open tubular column GC-MS. These compounds are precursors of the sulfuric acid responsible for the corrosion of municipal sewage systems. Carboxylic acids and phenols may be determined on an SE-52 open tubular column with ECD as pentafluorobenzyl esters after extractive alkylation.[41]

All of the priority pollutant fractions of water have been analyzed[42] in a single chromatographic run. The extractable organics are first injected, and the volatile components are desorbed onto the fused-silica SE-54 column with cryogenic focusing; detection is by GC-MS (Figure 7.6).

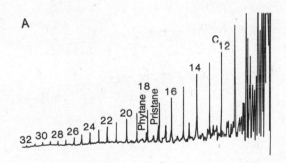

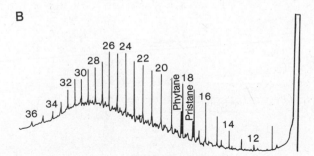

FIGURE 7.5. Chromatograms of (A) ship cargo oil and (B) sea pollutant. Note: 60 m SE-30 column, FID detection, temperature program from 70 to 290°C at 4°C min^{-1}. Peak numbers denote carbon numbers of *n*-alkanes. (Reproduced with permission from ref. 36. Copyright Friedr. Vieweg and Sohn.)

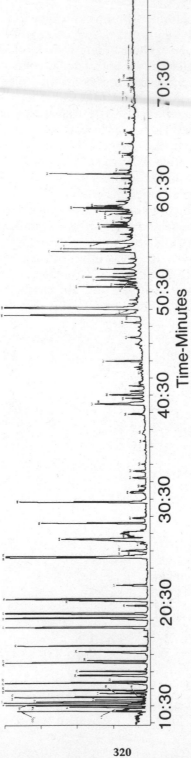

FIGURE 7.6. Chromatogram of simultaneous analysis of all five organic priority pollutant fractions from water. Note: 1 μg dm^{-3}, 30 m SE-54 fused-silica column. (Reproduced with permission from ref. 42. Copyright Dr. Alfred Huethig Publishers.)

Pollution of Soils, Sediments, and Solid Waste

Most analyses under this heading refer to nonvolatile organics, but solid waste materials can be analyzed for volatiles by head-space techniques such as concentration on Tenax GC followed by thermal desorption and analysis on glass or fused-silica open tubular columns.[43] Nonvolatile pollutants are solvent extracted and then analyzed on open tubular columns coated with nonpolar or weakly polar phases. The problem of the presence of different pollutant types can be at least partially solved[44] by separation on open tubular columns followed by simultaneous detection with a (nonspecific) FID and with an element-specific detector (e.g., FPD for sulfur compounds or phosphorus-containing pesticides, ECD for chlorine-containing insecticides, and thermionic NP detection).

As in the case of water pollution, petroleum hydrocarbons are often incorporated into marine sediments, and there is concern about such pollution during offshore oil production. Again, glass open tubular column GC allows the identification of petroleum origin through the characteristic patterns of n-alkanes, branched and cyclic alkanes, and isoprenoids in sediment extracts,[45] although there are often only small differences between contaminated and pristine sediments; an SE-52 column with splitless injection proved effective here.[45] The petroleum aromatics, analyzed by GC-MS with open tubular columns of the unsaturated fraction provide a further means of identification,[46] but interlaboratory comparisons have shown[47] large differences in results for both alkanes and aromatics and better clean-up methods have been proposed.[47]

The problem of distinguishing between pollutants and indigenous, naturally occurring compounds is magnified in the case of PAC in soils and sediments. For example, open tubular column GC is the preferred method in the analysis of these materials which shows the presence of many groups of PAC derived from both diagenetic transformations of biogenic precursors and combustion of fossil fuels and forest fires (see below). Sediment from the Charles River (New England) contains[48] PAC with up to 11 different aromatic ring structures with abundant alkyl-substituted derivatives. Among these, perylene, various phenanthrene homologs derived from diterpenes, and tetra- and pentacyclic PAC (hydrogenated chrysenes and picenes) from triterpenes, all arise biogenically and are distributed evenly with depth and at fairly low concentration in sediments.[49,50] Much more abundant[51] in surface sediments are PAC of anthropogenic (mainly combustion) origin. The standard procedures for the analysis of sediments for PAC after clean-up is open tubular column chromatography with (usually low resolution) MS detection on columns coated with SE-52 and either splitless or on-column injection.[12,15] In this way, cores from lake sediments have been studied, and the contribution of biogenesis,[49,50] water run-off,[48,51] and long-range atmospheric transport[50] as sources of PAC have been assessed.

After petroleum and PAC pollution, the presence of chlorinated com-

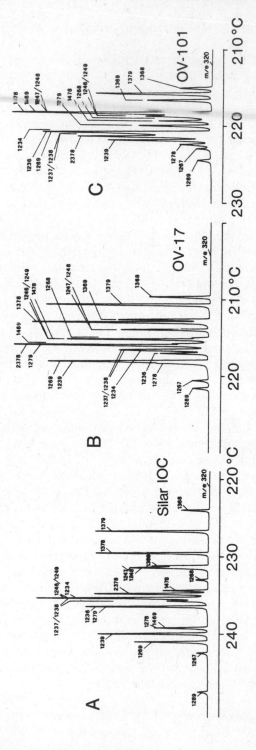

FIGURE 7.7. Mass chromatograms (*m/z* 320) of a composite sample showing elution of all 22 TCDD isomers. Note: (A) 55 m Silar 10C column, (B) 50 m OV-17 column, and (C) 50 m OV-101 column. (Reproduced with permission from ref. 53. Copyright American Chemical Society.)

pounds in soils and sediments generates the most interest. The highly toxic, mutagenic and teratogenic 2,3,7,8-tetrachloro-*p*-dioxin (TCDD) is formed as a by-product in the manufacture of trichlorophenol, and hence may be present in the herbicide 2,4,5-trichlorophenoxyacetic acid, which is made from trichlorophenol, and in its combustion products. Methods capable of detecting TCDD at the low ppt concentration level are required, and it seems that open tubular columns, for example, 30 m glass columns coated with SE-30 interfaced with medium- or high-resolution MS, provide the required resolution, sensitivity, and specificity.[52] ^{37}Cl-TCCD has been used as added internal standard with M$^+$ at *m/z* 327.8847. Interference from a polychlorobiphenyl (PCB) contaminant with M$^+$ at *m/z* 327.8775 is corrected for[52] by making use of the PCB peak at nominal *m/z* 326.

The 22 tetrachloro-*p*-dioxin isomers can be separated on 50 m long Silar 10C, OV-17, and OV-101 glass columns (Figure 7.7), although none of these provides complete resolution.[53] Single (*m/z* 320), or multiple-ion (*m/z* 320, 322, and 324) monitoring as detector with coinjection of standards proved that the chief TCDD isomer in contaminated soil from Seveso, Italy,[53,54] and from various horse arenas in the United States[53] is the toxic 2,3,7,8-TCDD. Indeed, this was the only isomer detected in a contaminated fish sample.[53] On the other hand, fly ash from a municipal incinerator showed[53] the presence of at least 17 TCDDs, but with 2,3,7,8-TCDD only comprising 1% of the total.

Pesticides, Herbicides, and Fungicides

The extensive use of agricultural chemicals of varying degrees of persistence has generated the need for analytical methods capable of defining the presence of residues and metabolites in a variety of matrices. While the low thermal stability and polar nature of many pesticides, herbicides, and fungicides has posed many problems for the use of open tubular column GC, there has also been considerable scope for the application of selective detectors. Coupling of the latter to deactivated glass or fused-silica columns now offers versatile analytical methods for many compounds in this area.

The determination of organochlorine pesticides is difficult because of the large number of compounds that may be present, along with metabolites, and the ubiquitous presence of polychlorobiphenyls (PCBs) adds a further complication. The clean-up procedure is therefore of vital importance, highly stable and inactive open tubular columns are necessary, and selective detection methods are often employed. After Franken and Vader[55] had pioneered the separation of chlorinated pesticides on a glass SE-30 column with ECD, and Steinwandter reported[56] the baseline separation of 17 such compounds on an OV-17 coated column after clean-up on silica gel. However, the officially recommended phase for the packed-column GC of pesticides is a 4:3 mixture of OV-17 and QF-1, and this has been shown[57] to give resolution in open tubular column work superior to either OV-17 or SE-

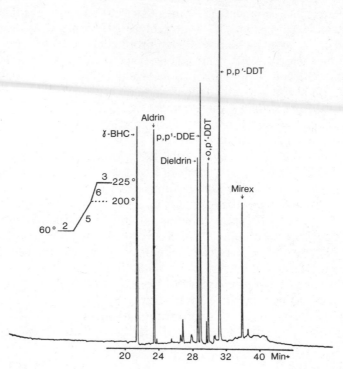

FIGURE 7.8. Chromatogram of a mixture of organochlorine pesticides. Note: 20 m OV-17-QF-1 glass column. (Reproduced with permission from ref. 57. Copyright Elsevier Scientific Publishing Company.)

30. Crucial separations not possible on packed columns, such as p,p-DDE and dieldrin, can be made with this mixed phase (Figure 7.8). Resolution of organochlorine pesticides from PCBs is also possible on such a column.[57]

Analysis of PCBs, however, is more usually carried out on nonpolar phases; only glass or fused-silica columns are suitable.[58] Zell et al.[59] identified 83 structurally defined PCBs with between three and eight chlorine atoms in fish-liver oil (Figure 7.9) on an SE-30 column with [63]Ni ECD. Theoretically, 190 such compounds are possible, among the 209 PCBs. Compound identification was made on the basis of retention indices based on a homologous series of n-alkyl trichloroacetic acid esters. The same group has reported[60] detailed analyses of seven technical PCB mixtures on glass SE-30 and Apiezon columns. A total of 201 peaks were either identified from retention indices of known compounds or established from rules relating index to structure. Tausch et al.[61] preferred an SE-52 column, with MS detection, which allows the number of chlorine substituents to be determined. Tuinstra and co-workers[62,63] have developed a procedure for determining the major individual PCBs in milkfat and other agricultural products using a precolumn in the injector and cold on-column injection. Separation of alkylchlorobiphenyls, a possible replacement for PCBs, has been recommended on Apiezon L-coated open tubular columns.[64]

The insecticide toxaphene, a mixture of chlorinated camphenes with empirical formula $C_{10}H_{10}Cl_8$, has been fractionated by column chromatography and TLC, and the fractions analyzed by open tubular column GC on a very

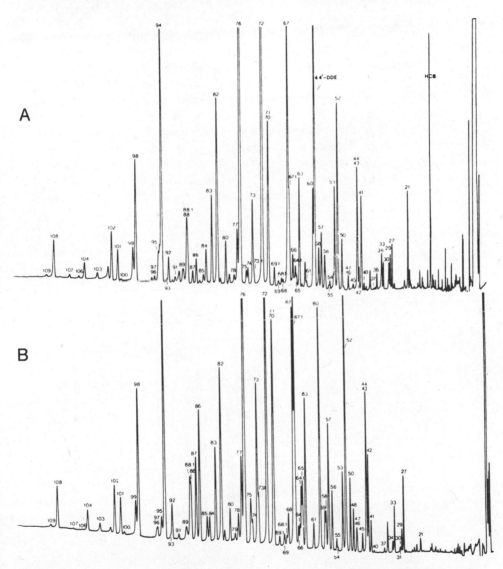

FIGURE 7.9. Chromatograms of (A) hexane extract of cod liver and (B) PCB standard simulating the environmental input of 54% Cl-PCB and 60% Cl-PCB mixtures (Clophen A50: Clophen A60 = 1:2.2). Note: 40 m SE-30 column, temperature program from 150 to 200°C at 1°C min^{-1}. Selected peak assignments:(44) 2,2′3,4′,6; (52) 2,2′,4,5,5′; (60) 3,3′,4,4′; (67) 2,2′,3,4′,5′,6; (72) 2,2′,4,4′,5,5′; (76) 2,2′,3,4,4′5′; (82) 2,2′,3,3′,4,4′; (94) 2,2′,3,4,4′,5,5′; (98) 2,2′,3,3′,4,4′,5. (Reproduced with permission from ref. 59. Copyright Springer-Verlag.)

long (305 m) OV-101 column.[65] The presence of at least 177 C_{10} polychlorinated compounds was shown, of which approximately two-thirds were $C_{10}H_{11}Cl_7$, $C_{10}H_{10}Cl_8$, and $C_{10}H_9Cl_9$ isomers. Twenty-six of these are present in 1–2.5% quantities, making up about 40% of the material. 2,5,6-*exo*,8,8,9,10-heptachlorodihydrocamphene and 2,2,5-*endo*,6-*exo*,8,9,10-heptachlorobornene have been identified[66] in toxaphene. The latter, one of the most toxic components, undergoes[67] reductive dechlorination under a variety of photochemical, chemical, and metabolic conditions to products identified by open tubular column GC-MS.

The persistence of organochlorine pesticides has led to progressive replacement by organophosphorus compounds. The preferred method for analysis of such pesticides and their metabolites involves thermionic[68] or flame photometric detection.[69] The FPD is generally more selective and sensitive than the former in this context, however, MS detection has also been applied.[70] Wolf *et al.*[68] showed the separation on a short glass OV-1 column of 23 organophosphorus insecticides (Figure 7.10), while Krijgsman and van de Kamp[69] reported retention data on SE-30 for over 60 such compounds. Methods for the analysis of organophosphorus pesticides and their metab-

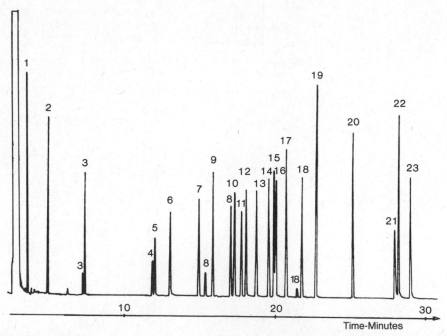

FIGURE 7.10. Chromatogram of organophosphorus insecticides. Note: 14.5 m OV-1 (0.4 μm film) glass column, ethyl acetate solution, splitless injection, temperature program from 115 to 190°C at 4°C min^{-1}, and 8°C min^{-1} to 260°C, thermionic detector in the phosphorus mode. Selected peak assignments: (2) Dichlorvos; (9) Diazinon; (17) Bromophos; (19) Tetrachlorvinphos; (22) Azinphos-methyl. (Reproduced with permission from ref. 68. Copyright Dr. Alfred Huethig Publishers.)

olites in plant material[71] and in the serum of poisoned human patients,[72] respectively, at the 50 ng g^{-1} and 2 ng cm^{-3} level, have been described, both with thermionic detection.

This detection method was also used in the analysis of N-methylcarbamate insecticides and related compounds.[73] A fused-silica SE-52 column was sufficiently inert to allow elution without prior derivatization, as is necessary in packed-column work with these compounds.[73]

The s-triazines, used extensively as herbicides, may be analyzed on open tubular columns coated with SE-52,[74] OV-101, or better, Carbowax 20M on etched soda glass.[74,75] Degradation products, particularly the hydroxy derivatives, can be analyzed after methylation under the same conditions as the parent compounds.[76] Substituted phenylureas are also widely used as herbicides, but direct analysis by GC is difficult because of low thermal stability which results in ready decomposition to amines and isocyanates. Claims to have chromatographed intact phenylureas on glass open tubular columns with conventional (high-temperature) injection[74] have been disputed.[77] Rapid catalytic hydrolysis to anilines, followed by derivatization with heptafluorobutyric anhydride and chromatography on open tubular columns, coated with a nonpolar phase, and with ECD has been described with sensitivity at the 1 pg level. Derivatization without prior hydrolysis is also effective for certain phenylureas.[78] Grob has compared different columns for the separation of intact phenylurea herbicides.[77] Even with cold on-column injection there is decomposition on barium carbonate-coated glass, pyrolyzed polymethylsiloxane treated glass, and most of all on Carbowax-deactivated fused-silica columns. A high-temperature silanized glass OV-73 column showed[77] least catalytic activity and allowed chromatography of most phenylurea herbicides (Figure 7.11) even though standard test mixtures suggested an adsorptive surface.

The fungicides bupirimate, fenarimol, vinclozolin, and triadimefon in grape juice and wine from treated vines can be determined[79,80] by adsorption onto XAD-2 resin which is then extracted with methylene chloride. Glass open tubular column chromatography with ECD allowed recoveries at the 80–100% level to be demonstrated[79] for concentrations in juice or wine of 0.1–5 ppb.

A number of organotin pesticides have been used extensively as fungicides and molluscicides. Triphenyltin hydroxide, used to control rice diseases, along with its degradation products may be converted to methyl derivatives and analyzed[81] on a short (12 m) OV-101 glass open tubular column with a flame photometric detector modified so as to be tin-selective (Chapter 4).

Combustion Products and Emissions

As discussed earlier, many of the pollutants for which analyses are made in a variety of matrices arise from combustion sources. There is therefore an extensive literature on the organic components of emissions from combus-

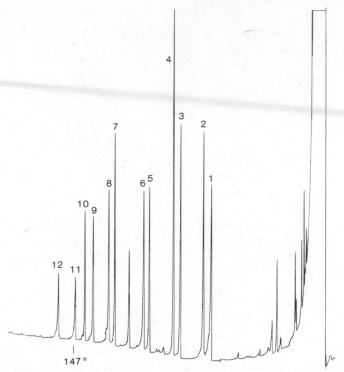

FIGURE 7.11. Chromatogram of phenylurea herbicides. Note: 15 m OV-73 (0.04 μm film) glass column, on-column injection, temperature program from 80°C at 5°C min^{-1}. Selected peak assignments: (1) Fenuron; (3) Cycluron; (6) Monuron; (8) Chlortoluron; (9) Buturon; (12) Metoxuron. (Reproduced with permission from ref. 77. Copyright Elsevier Scientific Publishing Company.

tion.[12,13,15,82] These compounds, although present in low concentration, may be hazardous to health. PAC represent the most important group of such compounds, and are not only important in determining the total burden of carcinogenic substances in the environment, but also in identifying the origin of pollutants. When adsorbed on aerosol particles, PAC may be transported over long distances[14,83] and hence can act as fingerprints for the identification of sources of other combustion-derived pollutants, such as acid rain.

The greatest known current and projected emissions of PAC arise from burning coal refuse, coke production, forest fires, and emissions from coal-fired residential and power plant furnaces.[12] Automobile and motorcycle exhausts and oil-fired furnaces may also be significant in urban environments.[12]

PAC fractions from the combustion of three common fuels chosen to represent some of the above sources were analyzed by Lee et al.[84] by open tubular column GC using SE-52 coated glass columns (e.g., Figure 7.12). Comparisons were made with the distributions of PAC in air particulate matter from high- and low-coal consumption areas of the United States.

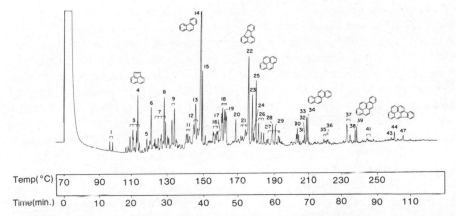

FIGURE 7.12. Chromatogram of the PAH fraction of coal combustion products. Note: 19 m SE-52 glass column. (Reproduced with permission from ref. 84. Copyright Heyden and Son, Ltd.)

Greater relative concentrations of alkylated PAC and of sulfur-containing compounds were found in coal soot as compared with either wood or kerosene combustion products. Alkyl homolog plots (e.g., Figure 7.13) show close similarity between curves for coal-soot PAC and Indianapolis air particulates, and between kerosene-soot PAC and Boston air particulates.

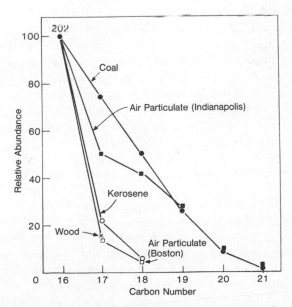

FIGURE 7.13. Alkyl homolog distribution plots for the pyrene-type series ($Z = -22$) in the combustion products of coal, wood, and kerosene, and in air particulate matter from Indianapolis and Boston. The abundance of the parent compound in each series was normalized to 100. (Reproduced with permission from ref. 84. Copyright Heyden and Son, Ltd.)

Other commonly used emission source indicators are the ratios of pyrene to benzo[a]pyrene[85] and of benzo[a]pyrene to benzo[ghi]perylene.[85,86] Values of the latter ratio are generally much higher for coal-combustion emissions than for those from automobiles, incinerators, and oil-burning power plants.[85,86] High resolution is required in gas chromatographic determinations of these compounds especially for the benzo[a]pyrene/benzo[e]pyrene separation. Cold on-column injection is also necessary to avoid discrimination against the higher molecular weight compounds, but this possible source of error is removed if isomer ratios are determined.[87] The ratio of chrysene to triphenylene is significantly higher for emissions from the combustion of wood and kerosene than for coal-combustion emissions and can be determined if the SE-52 phase is modified with N,N'-bis(p-n-hexyloxybenzylidene)-α,α'-bi-p-toluidine liquid crystal.[87] The polyphenyl ether sulfone phase Poly S-179 also permits this separation,[88] although a very long column is necessary. A further possibility is a specially synthesized mesogenic phase (see Figure 7.21).

The environmental impact of the increasing use of coal is a particularly important area of pollution analysis.[89] the PAC products of coal combustion have been compared[90] with those from gasification by chromatography on an OV-101 glass column with FID/FPD and FID/NPD and with splitless injection. The effect of combustion conditions on PAC production is well illustrated by comparison of the complexity of open tubular column chromatograms of products from uncontrolled coal combustion[84] and combustion in a high-temperature entrained flow gasifier:[91] the former produced a complex mixture with alkyl compounds prominent, whereas the latter produced mainly stable parent PAC by complete combustion.

The effect of fuel type (toluene, heptane) and combustion parameters on the content of individual PAC in the product soot has been investigated[92] with an SE-52 coated fused-silica open tubular column; acenaphthylene was found to be the most abundant product from all the flames. Benz[e]acenaphthylene has been positively identified in a number of combustion effluents;[93] chromatography on open tubular columns is required to resolve this compound from its isomers pyrene and fluoranthene.

Carbon blacks are industrial materials produced in large quantities by the incomplete combustion of carbonaceous gases. There is concern about the hazards associated with their use (e.g., in automobile tires) particularly in view of the potent carcinogenicity of a number of PAC which may be adsorbed on the particles. High-temperature chromatography on a well-deactivated glass SE-52 column[94] allowed elution of PAC with molecular weight up to 376 at temperatures up to 350°C. Bonded polymethylsiloxane phase glass columns also showed good baseline stability at very high temperatures in the separation of the PAC extracted from carbon black.[94,95] Seven sulfur-containing PAC were identified in carbon blacks from petroleum feedstocks by computerized GC-MS with a 19 m SE-52 glass column.[96]

Organic compounds other than PAC may also be important pollutants in combustion emissions. A homologous series of alkyl-9-fluorenones was identified[97] in a fraction of diesel exhaust particulate extracts by a range of techniques that included GC-MS (SE-30 open tubular column) and the novel method of GC-FTIR (Chapter 4) (SCOT SE-30 open tubular column); the latter allowed the isomeric 9-fluorenones and benzo[c]cinnolines to be distinguished. Aldehydes and ketones were also identified[98] in auto exhaust and tobacco smoke as 2,4-dinitrophenylhydrazones which were chromatographed on a 20 m glass column coated with SF-96. Two mutagenic fractions of an extract of diesel-engine soot were shown[99] by GC-MS with a 15 m glass SP 2100 column to contain alkyl fluorenones, aldehydes, and quinones as well as a range of polycyclic aromatic hydrocarbons; one nitro polycyclic aromatic compound was also identified.

Marine and Fresh-Water Biota

The procedures developed for the analysis of sediments for pollutants by open tubular column GC are also applicable to biota, although different clean-up methods may be necessary. Thus the standard method for the PAC of sediments, air particulates, and so on (short glass columns coated with SE-52 after acid leaching and high-temperature silylation) may also be applied to shellfish.[100] Clean-up on Sephadex LH-20 is necessary, however, to remove the interfering polyunsaturated aliphatic hydrocarbons.[101]

A comprehensive study of the levels of aliphatic and aromatic hydrocarbons in oysters from beds near the Amoco Cadiz oil spill off Northern France in 1978 was carried out by Berthou et al.[102] Extracts cleaned up on Florisil and silica gel were analyzed on 50 m Carbowax 20M-deactivated OV-1 and SE-52 columns with splitless injection. Quantification was carried out by adding deuterated hydrocarbons $C_{24}D_{50}$ and $C_{16}D_{10}$ before clean-up; these compounds were completely resolved on the columns employed.

Interlaboratory comparisons of trace-level hydrocarbons in tissue homogenates from mussels from different sites have been made.[103] Analyses included total aliphatic (C_{10}–C_{30}) and aromatic hydrocarbons, pristane/phytane ratios, and identities and amounts of most abundant PAC. Gas chromatography with open tubular columns was the final stage in all analyses, and, in spite of widely differing extraction and clean-up methods, a 1σ precision of $\pm 40\%$ was found.

Tobacco and Tobacco Smoke

The widespread use of tobacco and the complexity of its head space and smoke, taken with the associated health risks, have generated an extensive literature devoted to these topics, with open tubular column GC analysis in the forefront. It has been estimated that cigarette smoke contains at least 10^4 compounds of which approximately half have been identified.

Grob and Grob[104] analyzed the aroma of tobacco and compared its composition with that of the smoke by concentrating both, in turn, on active carbon before desorbing onto the column. The nonvolatile, extractable constituents of tobacco have also received attention; for example, nicotine alkaloids were analyzed[105] on a Carbowax 20M column with NPD detection. The sterol fractions of tobacco and marijuana plant material were resolved as TMS derivatives on glass columns coated with SE-30, SE-52, and Poly I-110 phases.[106]

By far the greatest number of applications in this area have been for tobacco (usually cigarette) mainstream smoke. Three sample types are generally analyzed: the "vapor" phase, the "semivolatiles," and the condensate. The vapor phase is defined as that portion of the smoke which passes through a Cambridge filter (i.e., particles less than 0.3 μm in diameter). This material comprises no more than 2% of the whole smoke and is highly diluted but reactive, so that there are special problems in introducing it into an open tubular column. A freezing method,[107] in which the early part of the column is held at about $-70°C$ or as low as liquid-nitrogen temperature, is generally used after the smoke has been transferred from the cigarette, either simply with a heated syringe[107] or more reproducibly via through a heated sample loop connected to a smoking machine.[108,109] Intermediate carbon or porous-polymer adsorbents may be interposed.[104] Methods have been considered sufficiently standard to allow differences in smoke composition between "normal," freeze-dried,[110] and heat-treated[111] tobacco to be detected. The composition of exhaled smoke has also been studied.[112]

The semivolatiles are the more volatile (boiling temperature 180–350°C) portion of the particulate smoke phase and result from simple solvent extraction of the Cambridge filter, or a solution of smoke. Concentration of the extract is difficult without loss of solutes, and splitless injection of the dilute solution is necessary.[113,114] The presence of both acidic (e.g., phenols) and basic components among a whole range of different functional-group types in the vapor phase and semivolatile fractions makes choice of stationary phase difficult. Nonpolar oil phases such as SF-96 on deactivated (Carbowax 20 M or silylation) glass surfaces allow separations in the "cryogenic" region,[107,115] that is, as the column is warmed after the trapping of the vapor phase (Figure 7.14). Polar phases, often in mixtures, have also been applied, for example, the Emulphor phases,[116] Ucon oils,[109] and dinonylphthalate/tetraethylene glycol dimethyl ether.[113] Prior separation of the smoke into fractions on a packed polar column (2,2'-oxydipropionitrile) before analysis on a nonpolar open tubular column,[117] or heart-cutting techniques with open tubular columns[116] are both interesting solutions to the problem of varying polarity of smoke components.

Mass spectrometry coupling is generally essential in coping with the enormous number of peaks separated in tobacco smoke analysis. A semiautomated microprocessor-controlled glass open tubular column-based procedure for the sampling, injection, separation, and digitization of mass spectrometric detector data has been described.[108] The data are transmitted

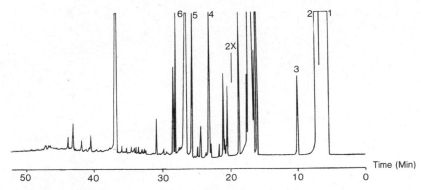

FIGURE 7.14. Chromatogram (FPD) of the gas phase of tobacco smoke. Note: 90 m SF-96 glass column, 1:7 split ratio, temperature program from -70 to $195°C$ at $5°C \ min^{-1}$. Selected peak assignments: (1) hydrogen sulfide; (2) carbonyl sulfide; (3) sulfur dioxide; (4) thiophene; (6) dimethyldisulfide. (Reproduced with permission from ref. 115. Copyright Elsevier Scientific Publishing Company.)

to a host computer for post-run processing (i.e., profile averaging and subtraction and plotting). This procedure allows pattern comparison for correlation with sensory evaluation by a panel. The sulfur compounds of cigarette smoke, important in "flavor" considerations, are amenable to selective detection by FPD[115,116,118] (Figure 7.14).

Separation of tobacco smoke condensate yields a variety of subfractions for study of the health risks associated with smoking. Thus, polynuclear aromatic hydrocarbons (PAH) have been credited with the major cancer-inducing activity of tobacco smoke.[12] As discussed in Chapter 6, the carcinogenic potency of PAH is strongly dependent on the presence and position of alkyl substitution and only 1–3% of the activity of tobacco smoke can be explained by the presence of unsubstituted PAH[119]. High-resolution chromatography is clearly necessary and complete methodologies involving glass open tubular columns with coupled MS as the chief identification step have been described.[12,120]

For example, after solvent partitioning of the condensate, the PAH fraction is separated on a lipophilic gel and the collected fractions are further chromatographed by HPLC on a polar chemically bonded phase.[119,120] The final step is GC-MS on efficient SE-52 coated columns. The procedure has been applied to PAH fractions of tobacco and marijuana smoke condensates.[119,121] Some 150 PAH in each material were identified and quantitated[119] (Figure 7.15).

The higher mutagenicity of marijuana smoke as compared with tobacco smoke has been attributed to the larger proportion of heavier PAH with molecular weight > 250 in the former.[121] Nine isomeric compounds of molecular mass 276 with a total concentration of 49 ng per marijuana cigarette were detected,[121] as opposed to three such compounds totaling 8 ng per cigarette for tobacco.

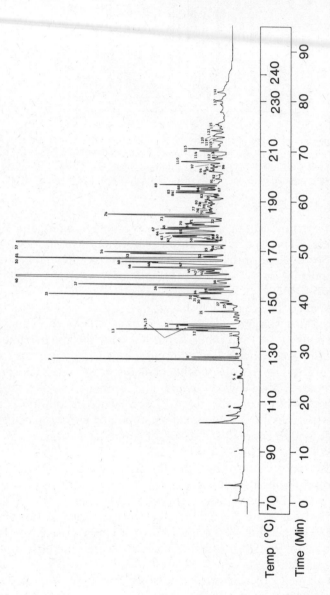

FIGURE 7.15. Chromatogram of the polycyclic aromatic hydrocarbon fraction of tobacco smoke condensate. Note: 11 m SE-52 glass column. Selected peak assignments: (7) phenanthrene; (8) anthracene; (33) fluoranthene; (37) pyrene; (73) benz[a]anthracene; (74) chrysene; (114) benzo[e]pyrene; (115) benzo[a]pyrene. (Reproduced with permission from ref. 119. Copyright American Chemical Society.)

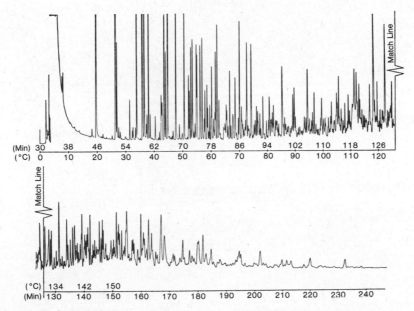

FIGURE 7.16. Chromatogram of a concentrate of the basic fraction of cannabis smoke condensate. Note: 50 m UCON 50-HB-2000 glass column. (Reproduced with permission from ref. 126. Copyright Elsevier Scientific Publishing Company.)

The acidic,[122] phenolic,[123] basic,[124] and alkane[125] fractions of tobacco smoke condensate have also been analyzed by open tubular column GC; Figure 7.16 shows the chromatogram[126] of the basic fraction of marijuana smoke condensate on a 50 m glass UCON 50-HB-2000 column. A detailed study of the water-soluble fraction of cigarette smoke condensate included analysis of GC-MS (30 m FFAP-coated glass open tubular column) of the acid fraction.[122] A number of new components including amino pyridines have been identified[124] in the basic fraction of tobacco tar on a 65 m Carbowax 20M-coated open tubular column.

7.3 FOSSIL FUELS

Petroleum

Due to the chemical complexity of crude oil and derived products such as gasoline, paraffin waxes, and lubricating oils, open tubular column GC has been extensively used as an analytical tool in the petroleum industry. While the saturated hydrocarbons are the major components in petroleum, olefins, alkenes, isoprenoids, naphthenes, and aromatic hydrocarbons as well as nitrogen-, oxygen-, and sulfur-containing classes of compounds also occur in crude oils or are produced during manufacturing processes. Quality and

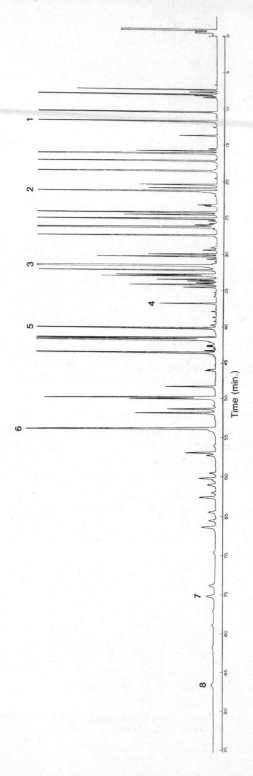

FIGURE 7.17. Chromatogram of a full-range gasoline. Note: 70 m squalane stainless steel column, temperature program from 0 to 95°C at 2°C min^{-1}, pressure program yielding column flow rates of 0.5–2.0 cm^3 min^{-1}, 1:60 split ratio. Selected peak assignments: (1) *n*-pentane; (2) benzene; (3) toluene; (4) *n*-octane; (5) ethylbenzene; (6) 1,2,4-trimethylbenzene; (8) naphthalene. (Reproduced with permission from ref. 129. Copyright Elsevier Scientific Publishing Company.)

process control of the chemical composition of articles of commerce and by-products of petroleum refining are essential for industrial and environmental reasons. Open tubular column GC has thus found increasing application in the petroleum industry and in the evaluation of environmental pollution attributed to petroleum sources.

Analysis of motor gasolines as reviewed by Sanders and Maynard[127] has traditionally been done on stainless steel open tubular columns 50–200 m in length and coated with squalane,[127–129] Apiezon,[130] or comparable apolar methylsilicone phases such as SE-30, SF-96, or SP-2100. Generally, retention index systems (e.g., Kovats) are used for the detailed identification of hydrocarbons (see Chapter 6) in petroleum samples. Since problems of compound adsorption are negligible with most hydrocarbons of major concentration found in petroleum materials, stainless steel was long preferred as column material over glass in spite of its inherently higher surface activity. Adlard et al[129] described the use of an automated system for the high-resolution gas chromatographic analysis of gasoline-range hydrocarbon mixtures on squalane-coated stainless steel capillary columns. Approximately 125 hydrocarbons ranging from C_3 to naphthalene were quantitatively and qualitatively determined, and an example is displayed in Figure 7.17.

High coating efficiencies are obtainable with glass columns.[131] Furthermore, the inherently more inert surfaces of glass, and especially fused silica, enable the fabrication of higher quality columns capable of more complete separations of not only the complex hydrocarbon mixtures but also reactive sulfur-, nitrogen-, and oxygen-containing compounds. Open tubular columns with thick films are required for the separation of the more volatile gasoline components (Figure 5.10).[132,133] Application of this principle has even enabled the separation of methane, ethane, propane, and other low molecular weight hydrocarbons in petroleum gas mixtures.[134] Most gasoline-range hydrocarbon analyses require subambient initial temperatures that can result in loss of efficiency due to operation below the lower temperature limit of the stationary phase. An added benefit of using thicker stationary films is that the analysis can be performed at initial temperatures near or above room temperature.[132–134] This eliminates the need for cryogenic cooling and improves column efficiency by working well within the temperature specifications of the stationary phase.

Analysis methods for naphthas (a cut from crude oil containing hydrocarbons in the C_4–C_{14} range) have been described by Kumar and co-workers[135] and are similar to those used for gasolines. Herrera et al.[136] showed that the use of a 25 m × 0.25 mm i.d. glass open tubular column coated with SE-30 sufficed for the qualitative and quantitative analysis of components in pyrolysis gasolines. Internal standards and retention information from standard blends of known components were used. Gallegos et al.[137] performed a similar gas chromatographic analysis using a squalane-coated open tubular column coupled to a mass spectrometer to provide mass spectral identification of about 150 naphthenes, olefines, and aromatic hydrocarbons in a pyrolysis naphtha. Other workers[138,139] have reported the separation of alkylnaphthalenes and alkylbiphenyls using open tubular columns coated with polar phases of BPB [m-bis(m-phenoxiphenoxi)benzene] and PEGA (polyethylene glycol adipate) which provide better selectivity for isomeric separation of these aromatics in pyronaphthas. Another process of concern in petroleum refining is reforming. Kennard demonstrated the use of a SP-2100 coated glass open tubular column for the analysis of reformate feed and product.[140]

The higher boiling compounds obtained from crude oil refining can also be analyzed by open tubular column GC. Paraffin waxes with carbon numbers greater than C_{60} have been successfully chromatographed on open tubular columns. Special attention to the following is necessary: (a) use of on-column injection; (b) use of large-diameter capillaries with thin films resulting in increased phase ratios and decreased capacity ratios; (c) use of short capillaries (7–15 m) coated with thermally stable phases which can be programmed to temperatures exceeding 300°C; and (d) use of carrier gas flow programming after reaching the final temperature of the column. The elution of hydrocarbons up to C_{65} in a paraffin wax was obtained on a fused-silica, SE-54 (5% phenyl, 1% vinyl methylpolysiloxane) open tubular column (Figure 6.4).[140] Dielmann et al.[141] used similar techniques in conjunction with a quadrupole mass spectrometer for the separation and identification of C_{30} to C_{55} hydrocarbons in industrial and natural waxes.

An important analytical aspect in the petroleum industry is the determination of physical and chemical properties of fuels, naphthas, crudes, and generally most end-products of refining. Anderson and co-workers[142] discussed the advantages of using gas–liquid chromatography to replace several ASTM inspection tests. For example, the relative concentrations of n-heptane and isoctane are traditionally used to derive the octane number of a fuel. Compositional data obtained from single open tubular gas chromatographic runs were used to calculate the research octane number for several motor gasolines. Typically, 31 groups of compounds from the entire boiling point range of the sample were used for their calculations. Excellent agreement with ASTM octane number methods were reported, and it was stated that the determination of as many as six bulk properties could be achieved by one open tubular GC analysis. Similarly, Jackson et al.[143] utilized a short OV-101 coated open tubular column for the determination of boiling point distributions of crude oils. The GC method compared well with the ASTM method.

Several open tubular column gas chromatographic analyses of petroleum compounds that contain various functional groups have been reported. Generally, preseparation methods such as liquid–liquid extraction or adsorption column chromatography followed by HPLC[144] are necessary for the enrichment of the components of interest. The carboxylic acids of petroleum crudes have been analyzed on WCOT open tubular columns coated with SP-1000 (chlorophenylsilicone)[145] and SCOT columns with SE-30, OV-101,

Dexsil 300, and PE (ethylene-propylene copolymer) stationary phases.[146] Nitrogen-containing compounds, such as anilines and pyridines, were isolated from crude oil and differentiated by selective acetylation of the anilines followed by separation on an SE-54 coated fused-silica column and mass spectrometric detection.[147] Polycyclic aromatic hydrocarbons and nitrogen-, oxygen-, and sulfur-containing polycyclic aromatic compounds have been isolated from used engine oils,[148,149] lubricating oils,[150] and waxy distillates and raffinates[151] and analyzed by open tubular column GC using methylsilicone stationary phases (SE-52, CP-Sil 5, and OV-101). The use of element selective detectors greatly aids in the GC determination of petroleum compounds containing a functional group. Sulfur-sensitive detectors such as the flame photometric detector or the Hall electrolytic conductivity detector (in the sulfur mode) have been employed for the analysis of crude oils.[152,153] The NPD was used by Lee et al.[148] in the fingerprint analysis of used engine oils for the detection of nitrogen-containing polycyclic aromatic compounds, while McCarthy et al.[153] used the Hall detector in the nitrogen-selective mode for the same class of compounds in polar isolates of crude oils. In the latter case, a novel effluent splitter was used for simultaneous dual trace gas chromatographic analysis. A GC-UV detection system was designed and implemented by Novotny and co-workers[154,155] for the analysis of UV-absorbing compounds in gasoline. The variable-wavelength detector could be adjusted to different settings for selective detection of various groups of polycyclic aromatic compounds with specific absorption maxima (see Figure 7.18). To increase the capacity of the open tubular columns so that enough sample for detection could be chromatographed, wide-bore columns (0.7 mm i.d.) with "whisker" modified surfaces were prepared with different stationary phases and in different lengths. A 50 μL detection cell was used as a compromise between detector sensitivity and the necessity to preserve resolution.

The modification of GC injectors for the use of adsorbent preconcentration devices for purge-and-trap introduction of compounds into an open tubular column, as described in depth in previous sections (4.3 and 7.2), can also be used for analysis of volatile petroleum hydrocarbons. Novotny et al.[156] applied this technique to the analysis of natural gases and used dual trace open tubular column GC with flame ionization–flame photometric detectors to show the presence of sulfur-containing species (presumably sulfides). Picker and Sievers[157] coupled a cartridge containing a nonvolatile coordination complex formed from reaction of a europium salt with a bis(β-diketonate) ligand behind a Tenax GC adsorption tube for the purpose of preseparating the oxidized portion of gasoline. After desorption of all compounds collected on the Tenax onto the open tubular column through the europium polymer, the compounds with oxygen functionality (aldehydes, ketones, alcohols, acids, etc.) yield coordination complexes with the lanthanide while the nonfunctional hydrocarbons are passed onto the column for analysis. Later, the retained compounds can be thermally desorbed and

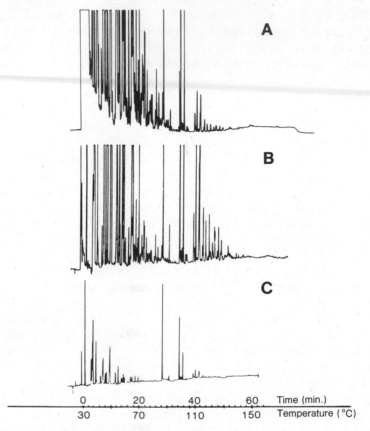

FIGURE 7.18. Chromatograms of a lead-free gasoline using (A) flame ionization detector, (B) UV detector at 200 nm, and (C) UV detector at 260 nm. Note: 32 m UCON 50-HB-2000 glass column. (Reproduced with permission from ref. 154. Copyright American Chemical Society.)

analyzed separately. Petroleum distillate vapors may be adsorbed on carbon and then desorbed into the column.[158]

Mixed phase and multidimensional open tubular GC are valuable tools for enhancing the resolution of chemically similar compounds with isomeric or functional differences. Schwartz et al.[159] used a mixed phase (dipropyl tetrachlorophthalate/squalane) open tubular column to resolve alkylated benzenes in the aromatic fractions of petroleum. Two open tubular columns, with different stationary phases, connected in series provide a similar effect. Stuckey used columns of DC-550 (25% phenyl methylphenylsilicone) and 1,2,3-tris-2-cyanoethoxypropane in tandem for the separation of aromatic compounds in the light naphtha fractions of crude oil.[160] Multistage open tubular column GC using manually or pneumatically activated valves to implement heart-cutting, recycle, or precutting chromatographic techniques has been applied to petroleum analyses. Dean reviewed heart-cutting in GC and showed several separations of naphtha cuts of crude oil by this method.[161] Schenck and Hall[162] developed a two-stage system that separated components of a paraffinic crude, first, in sequence of boiling point on an SE-30 column, and subsequently, in order of molecular mass on a silica gel PLOT column. Heart-cutting of selected peaks was achieved by operation of a manually controlled valve. As can be seen from Figure 7.19, several additional compounds could be resolved from a single peak. A multistage GC technique for the determination of the individual paraffin, olefin, naphthene, and aromatic content of gasolines and naphthas on a single GC system has been reported by Bloch et al.[163] A precutter OV-275 open tubular column was used to retain the aromatics and elute the olefins and saturates into the inlet of a squalane SCOT column in a separate oven. By column switching, the retained aromatics were then introduced into a second open tubular column and temperature programmed for resolution of constituents that were detected with a flame ionization detector. The olefins were then separated

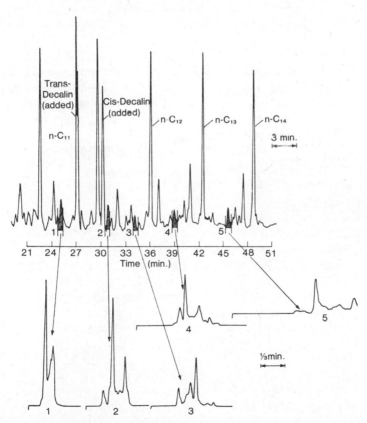

FIGURE 7.19. Chromatograms illustrating the separation of a paraffinic crude by heart-cutting. Note: SE-30 glass column. (Reproduced with permission from ref. 162. Copyright Elsevier Scientific Publishing Company.)

from the saturates (paraffins and naphthenes) by postcolumn reaction with a mercuric perchlorate–perchloric acid adsorber. A second and third FID were used to detect the separated components of these two groups of compounds. All routine GC operations, column switching, pressure flow programming, venting, and peak detection were computer controlled. Comparison of saturate, olefin, and aromatic data obtained by this GC method agreed excellently with that of standard ASTM mass spectrometry and fluorescent indicator adsorption methods.

Open tubular column GC has played an important role in the fingerprinting of petroleum crudes for the determination of sources of oil-related environmental pollution (see Section 7.2), to provide insights into the geochemical origin of crude oil,[76] and to assist in regulation enforcement in the commerce of petroleums and derived products. Clark and Jurs[164] employed computer-assisted pattern recognition methods to classify crude oils based on their open tubular column gas chromatograms. Marker compounds that have been used for fingerprinting crude oils include isoprenoids, triterpanes,[165,166] hopanic acids,[146] steranes,[165] and aromatic thiophenes.[167] Open tubular column GC has been instrumental in assessing changes due to weathering of crude oils.[167]

Open tubular column GC-MS was used to profile Alaskan crude oil at different locations along the transport route for the purpose of origin verification and embargo exemption.[168] Ratios of selected straight-chain saturates to corresponding isoprenoids were obtained from a SP-2100 open tubular column analysis and used as identification markers. Work with 25 fresh crude oils demonstrated that high-resolution chromatograms are different for oils from different continents and significant differences even exist for oils from neighboring fields.[169] Wehner and Teschner[165] used data from classical adsorption chromatography and open tubular column GC to construct three-dimensional plots of saturate, aromatic, and heterocompound content for oils from different geographical locations. From the plots, correlation of oils from different wells in the same geochemical field could be made as well as correlations of crude oils to source rocks. Open tubular column GC methods for identifying the source of oil spills have been reviewed by Bentz[170] and discussed by others.[171,172] Determinations of the fate of petroleum hydrocarbon pollutants as they are dispersed throughout the environment also rely on open tubular GC analytical techniques. Methods have been reported for the analysis of oil contaminants in marine sediments,[173,174] water,[175] marine biota,[167,175] and avian wildlife.[176]

Coal-Derived Products

Unlike their petroleum counterparts, coal-derived products contain high levels of PAC with lower concentrations of aliphatic hydrocarbons.[177–179] This poses environmental and health hazards of considerable concern since many PAC are chemical mutagens and carcinogens. Open tubular column GC has been widely used for detailed chemical analyses of coal and coal-derived products. Additionally, open tubular column GC is playing a major role in assessing the environmental implications of deploying a coal-based synfuel industry.

Coal tar has been extensively studied and chemically defined by many analytical methods. Perhaps the most detailed characterization of this coking by-product has been achieved by open tubular column GC. Lee and Wright[180] have discussed operational parameters of open tubular GC (column materials and dimensions, stationary phases, carrier gas types and velocities) as applied to the analysis of PAC in coal tar. Generally, the nonpolar to slightly polar stationary phases such as OV-1, OV-101, SP-2100, SE-52, and SE-54 coated on glass and fused silica have been universally used for coal tar analyses. Recently, however, efficient cross-linked phases with high phenyl content have shown better separations for the isomeric PAC in coal tar. Figure 7.20 shows a chromatogram of coal tar on a cross-linked 70% phenyl, 4% vinyl methylphenylpolysiloxane phase on fused silica.[181] This

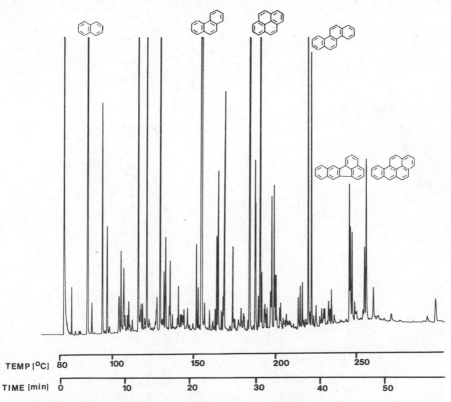

FIGURE 7.20. Chromatogram of coal tar. Note: 12 m 70% phenyl, 4% vinyl methylphenyl-polysiloxane fused-silica column, on-column injection. (Reproduced with permission from ref. 181. Copyright Friedr. Vieweg and Sohn.)

phase has shown excellent thermal stability and has been used for the separations of PAC of low volatility with molecular masses up to 400 at temperatures of 400°C. Schomburg, Borwitzky and co-workers[88,182,183] used the polar phases OV-7, OV-61, and Poly S 179 on various glass columns for coal tar PAC separations. The latter phase showed high-temperature stability (390°C) and was capable of resolving several dibenzofluoranthene/pyrene (molecular mass = 302) isomers. Kong et al.[184] demonstrated the application of a novel mesogenic polysiloxane phase, synthesized by Finkelmann et al.,[185] for the separation of hard-to-resolve PAC isomers in coal tar such as triphenylene/chrysene and the benzofluoranthenes (see Figure 7.21). Novotny et al.[186] fractionated high-temperature coal tar by a liquid–liquid partitioning scheme and determined the presence of aliphatic hydrocarbons, PAH, polycyclic aromatic nitrogen and sulfur heterocycles (PANH and PASH), and hydroxylated PAH (HPAH) by gas chromatography on SE-52,

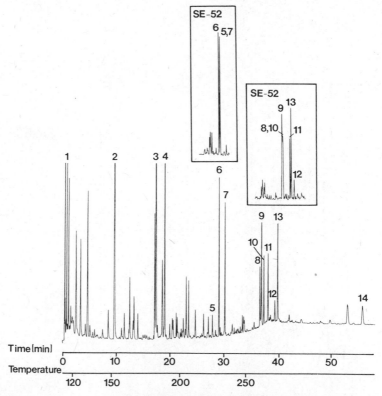

FIGURE 7.21. Chromatogram of coal tar on a novel mesogenic polysiloxane stationary phase. Note: 20 m fused-silica column, insets are on SE-52 stationary phase. Selected peak assignments: (5) triphenylene; (6) benz[*a*]anthracene; (7) chrysene; (8) benzo[*j*]fluoranthene; (9) benzo[*b*]fluoranthene; (10) benzo[*k*]fluoranthene; (11) benzo[*e*]pyrene; (12) perylene; (13) benzo[*a*]pyrene.

OV-101, and UCON 50-HB-2000 coated open tubular columns. Coal tar is often coated on pipes and storage tanks in public water supply systems to prevent corrosion. Chlorinated compounds in these materials were studied by Alben[187] using SE-54 and OV-17 open tubular columns in a GC-MS system. PAC and nitrogen-containing PAC were easily leached from the coal tar coatings, and several chlorinated naphthalenes, fluorenes, phenanthrenes, and dibenzofurans were identified. Buryan et al.[188] carried out the analysis of technical coal tar phenol products on a 50 m column coated with the selective phase tri(2,4-xylenyl)phosphate, reporting the separation of C_1– C_8 phenols as well as indanols by this method.

Synfuel materials derived from coal differ from petroleum products mainly in the concentrations of PAC, since most coal liquids are composed of 50–70% PAC. Nitrogen- and oxygen-containing PAC range from 0 to 30% (or higher) in some coal-derived liquids; PASH vary in concentration from 0 to 10%. Bertsch et al.[189] analyzed coal fluids directly without prefractionation. To protect the OV-101 coated glass open tubular column from heavy solute materials, a short plug of Tenax was used to adsorb the solvated sample. The cartridge was then placed in a modified injection port and the sample components thermally desorbed onto a liquid-nitrogen cooled U-trap at the head of the column. Over 250 compounds (aliphatics, olefins, naphthenes, and aromatics) were identified by GC-MS. Tenax purge-and-trap methods have also been used for surveying volatile components of processed coal residues[190] and vapor phase organics near a coal gasifier.[191]

Normally, because of the complex nature of synthetic fuel materials derived from coal, separation of the PAC according to chemical functionality is preferred. Traditional methods of fractionation include liquid–liquid partitioning and column chromatography on silica and alumina adsorbents,[192] and these methods have been reviewed elsewhere.[12] Schultz et al.[193] separated the polynuclear aromatic and aliphatic compounds in a solvent refined coal (SRC) recycle oil by a solvent partition procedure and identified over 146 constituents by open tubular column GC-MS. Likewise, GC-MS analysis on a methylsilicone coated fused-silica column enabled the characterization of PAC, aliphatic hydrocarbons, and phenols in fractions of an SRC I process solvent obtained by preseparation using high-performance liquid chromatography (HPLC).[194] Open tubular column GC can aid not only in the detailed separation of fractionated coal materials, but also in the development of separation methodology. Later et al.[195] described a fractionation scheme where open tubular column GC-MS was used to (a) provide detailed identification of aliphatics, PAH, PASH, N-PAC (nitrogen-containing polycyclic aromatic compounds), and HPAH, and (b) to demonstrate the precision of the chemical class separation for a heavy distillate coal liquid. Figures 7.22 and 7.23 are chromatograms on SE-52 of the PAH and 3°-PANH fractions of a coal gasification condensate separated by this method.[196] PAC retention index measurements,[16] as discussed in Section 5.7, were employed for identifications, and peak assignments were verified by open tubular column GC-

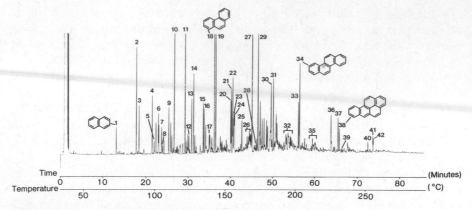

FIGURE 7.22. Chromatogram of the PAH fraction of a coal gasification condensate product. Note: 20 m SE-52 fused-silica column, H_2 carrier gas at 1.25 cm^3 min^{-1}, splitless injection.

MS. The complexity of this coal-derived product is illustrated by the presence of a wide range of parent and alkylated PAH, and analogous nitrogen heterocycles.

Of the PAC, the nitrogen-containing heterocycles are thought to be the major class of mutagens in coal liquids. Furthermore, the degree of mutagenic activity is a function of chemical class and molecular structure. Thus, the high-resolution separation of the N-PAC achievable with open tubular column GC makes it the principal analytical tool used by coal chemists for the elucidation of chemical composition and correlation with mutagenicity for the nitrogen fractions of coal-derived materials. Highly deactivated and

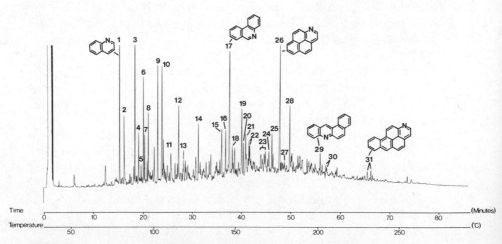

FIGURE 7.23. Chromatogram of the PANH fraction of a coal gasification condensate product. Note: same conditions as in Figure 7.22.

inert glass and fused-silica surfaces are necessary for the efficient chromatography of these compounds. Buchanan and co-workers[197,198] used an SE-52 coated fused-silica open tubular column with a novel ammonia chemical ionization mass spectrometric method for the separation and identification of nearly 100 N-PAC in coal oils. Similarly, other workers have used open tubular column separations with selective detection for the class analysis of N-PAC in various distillates of coal liquids.[199–201]

Since oxygen bridges are a predominant means by which aromatic structures are cross-linked together to form coal macromolecules, compounds with various oxygen functionalities are present in coal-derived materials. Snape et al.[202] used supercritical gas extraction to isolate aliphatic carboxylic acids from lignite and subbituminous coals. After cold diethyl ether precipitation and methylation of the extracts, open tubular column GC on an OV-1 column (Figure 7.24) enabled the determination of C_{20} to C_{34} acids. The high concentrations of phenolic compounds found in fuels derived from coal are of concern due to the toxicity of this class of compounds. SE-52 coated open tubular columns have been used for the analysis of hydroxylated aromatic compounds[195,196] and are adequate for the rapid screening of such fractions. However, resolution of all isomeric alkyl phenols has not been achieved on slightly polar phases such as SE-52, and polar stationary phases are required. Guenther et al.[203] and White and Li[204] reviewed previous open tubular GC separations of phenols and used polar polyethylene glycol phases for the resolution of isomeric phenols in SRC liquids. Guenther and co-workers used a Pluronic L64 column and reported the separation of all six

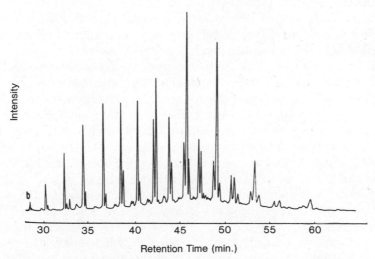

FIGURE 7.24. Chromatogram of straight-chain monocarboxylic acid methyl esters from a supercritical gas extract of Turkish lignite. Note: 20 m OV-1 glass column, temperature program from 90 to 280°C at 4°C min^{-1}, H_2 carrier gas. (Reproduced with permission from ref. 202. Copyright IPC Science and Technical Press.)

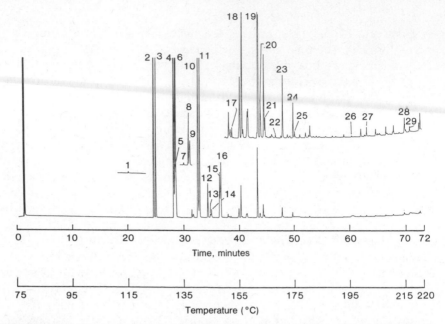

FIGURE 7.25. Chromatogram of the phenolic fraction of an SRC II middle distillate. Note: 20 m Superox 20M (0.10 μm film) fused-silica column, He carrier gas at 1.75 cm³ min⁻¹. Selected peak assignments: (2) phenol; (3) 2-methylphenol; (6) 3-methylphenol; (11) 3-ethylphenol; (18) 4-indanol; (27) 2-naphthol. (Reproduced with permission from ref. 205. Copyright American Chemical Society.)

dimethyl phenols, cresols, and phenol.[203] White and Li reported similar results using a Superox 20M coated open tubular column (Figure 7.25) and devised a retention index system using the intervals between phenol, 4-*tert*-amylphenol, and 1-naphthol to calculate indices for standard phenols which could then be used to identify phenols in coal liquids.[204,205]

Due to the high levels of aromatic and oxygen-, nitrogen-, and sulfur-containing compounds found in coal, selective detectors with open tubular column GC are extremely valuable for the determination of these compounds in coal-derived products. To differentiate aromatic from aliphatic by-products in gasifier tars and water residues, Kapila and Vogt[206] used tandem flame ionization and photoionization detectors. Classification of aromatic, alkane, or alkene hydrocarbons was based on the normalized relative molar response of the eluted peak in the two detectors. Brady described a tandem open tubular GC vapor phase infrared spectrometry system that was used to differentiate the cresol isomers in a distillate of a coal tar.[207] The components isolated by open tubular column GC were separately trapped in a vapor cell, and their infrared spectra were recorded. Similarly, matrix isolation of effluents from open tubular columns trapped on cooled disks followed by Fourier transform infrared spectrometric analysis of the PAC in SRC materials was reported by Hembree et al.[208] Perhaps the most versatile

selective detectors for the analysis of coal-derived products are the NPD
and FPD. The sulfur heterocycles from coal liquids have been analyzed by
dual trace FID/FPD; an example is shown in Figure 7.26.[209,210] Yang and
Cram[211] used the FPD in the sulfur-selective mode to analyze coal hydrogenation products, without prefractionation (Figure 5.25). The NPD has been
used for such analyses as the analysis of anilines and pyridines in SRC
distillates and aqueous extracts.[212] The N-PAC in SRC II heavy distillates
were analyzed by FID/NPD dual trace GC using a novel, efficient fused-
silica effluent splitter[213] as shown in Figure 7.27. Gangwal[214] used dual-
detectors, both FID/FPD and FID/NPD, for the analysis of gasification tars
of bituminous coal and lignite previously fractionated by solvent partitioning.
The selective detection of the highly mutagenic amino-PAH in coal liquids
can be achieved by conversion of these compounds to fluoroamides by derivatization with pentafluoropropionic anhydride and subsequent separation

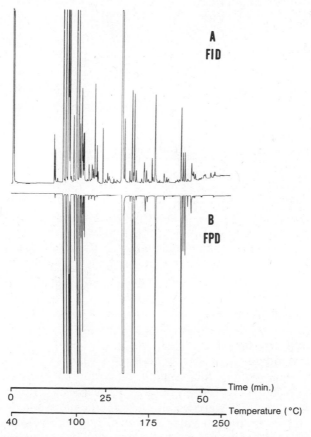

FIGURE 7.26.   FID/FPD dual trace chromatograms of the sulfur heterocycle fraction from an
SRC I product. Note: 20 m SE-52 glass column. (Reproduced with permission from ref. 209.
Copyright American Chemical Society.)

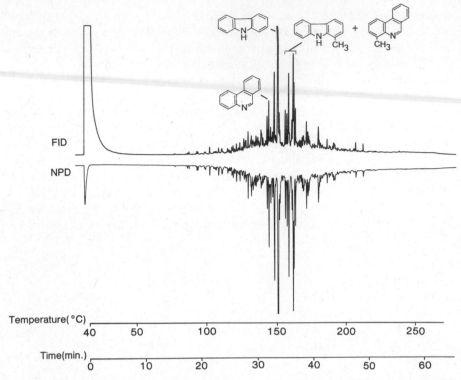

FIGURE 7.27. FID/NPD dual-trace chromatograms of an N-PAC fraction from an SRC II coal liquid. Note: 15 m SE-52 (0.25 μm film) fused-silica column, fused-silica effluent splitter. (Reproduced with permission from ref. 213. Copyright Dr. Alfred Huethig Publishers.)

by open tubular column GC and detection by electron capture and mass spectrometry.[199,200] Figure 7.28 illustrates the range of amino-containing PAC that can occur in coal conversion materials and the ease of analysis of these compounds using a selective detector. In summary, the use of element or functional-group selective detection methods increases the information obtained from high-resolution separations of coal-related materials and conversion products

As with petroleum products, open tubular column GC can be used for the determination of physical and chemical properties and fingerprinting of coal and converted products. Solash et al.[215] studied the relation between fuel properties and chemical composition of jet fuels refined from coal oils using an Apiezon L open tubular column. Fingerprint analysis of coal extracts and profiling of aliphatic and aromatic hydrocarbons on open tubular columns coated with OV-1 were used to relate chemical composition to coal rank.[216] Additionally, open tubular column GC has been used to study process treatments and reactions of coal and coal-derived materials. Some examples of studies that incorporate open tubular column gas chromatographic analysis methods include: (a) oxidative reactions that allow correlation of reaction products to coal structure;[179,217–219](b) reductive alkylation for the same purpose[220]; (c) hydrogenation of coal oil for upgrading of fuels;[221,222] and (d) process distillation for reducing concentrations of mutagenic amino functional aromatic species in coal oils.[201]

Shale Oil

Shale oil is derived from retorting the macromolecular organic kerogen content of shale rock. The bulk chemical composition of shale oils is highly aliphatic with an abundance of alkanes, alkenes, isoprenoids, and cycloparaffins, with lower concentrations of aromatics and oxygen-, nitrogen-, and sulfur-containing compounds. Open tubular column GC methods as discussed for petroleum and coal-derived products are also applicable to the analysis of shale-derived oils.

After isolation of alkane, alkene, and branched/cyclic alkane fractions, DiSanzo et al.[223] characterized these hydrocarbon classes in shale oil using a OV-101 SCOT column. Additionally, an aromatic fraction containing C_1–C_5 alkylbenzenes, parent and alkylated naphthalenes, and phenanthrene was also obtained and analyzed by open tubular column GC. The National Bureau of Standards has released a standard reference material (SRM 1580) for which certified quantitative values for several PAC have been issued.[224] Methods for the quantitative analysis of PAC in shale oil have also been reported by these same workers.[225] Several open tubular column chromatographic methods were outlined for the quantitative determination of such compounds as pyrene, fluoranthene, benzo[e]pyrene, benzo[a]pyrene,

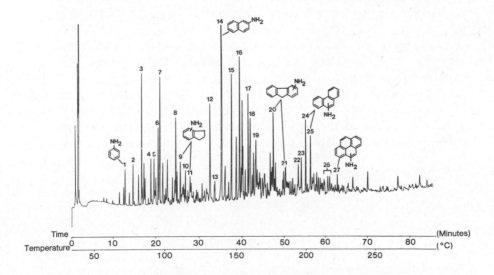

FIGURE 7.28. ECD chromatogram of the PFP-fluoroamide derivatives of the amino-PAH from a coal tar residue. Note: 20 m SE-52 (0.17 μm film) fused-silica column, H_2 carrier gas.

phenol, *o*-cresol, acridine, and 2,4,6-trimethylpyridine. SE-30 and SE-52 columns were used for the analysis of the PAH, whereas more polar phases, SP-1000 or Carbowax 20M, were used for the oxygen- and nitrogen-containing PAC.

The heterocyclic PAC found in shale oils have also been analyzed by open tubular column GC. Willey et al.[209] analyzed the thiophenic PAC isolated from a Paraho shale oil using an SE-52 coated glass open tubular column and a flame photometric detector. A nitrogen-specific detector and a Carbowax 20M glass column were employed by Crowley et al.[226] for the chromatographic analysis of polar shale oil fractions composed mainly of nitrogen heterocycles. They also analyzed the aromatic and alkane/alkene fractions of the same shale-derived oil using an SP-2100 narrow-bore fused-silica column. Alkyl pyridines and quinolines have been identified in the heterocyclic compound fractions of a retorted Green River oil shale by open tubular column GC-MS.[227] The GC column was 30 m × 0.25 mm i.d. fused-silica coated with SE-54. Phenolic compounds occur in shale-derived oils and can be separated using a polar stationary phase such as Pluronic L64.[203] The aliphatic carboxylic acids (methyl esters) from oil shale retort waters have been analyzed on OV-101 glass open tubular columns,[228] and Fish et al.[229] used an SP-2100 column to fingerprint carboxylic acids from six *in situ* oil shale retort processes. Differences in the profiles were sufficient to allow differentiation of commercial retort processes. As with petroleum fingerprinting, steranes and triterpanes in oil shale have been used as marker compounds, and Gelpi and co-workers [230] evaluated several open tubular column GC procedures for such analyses.

Two unique injection ports have been described and used for open tubular column GC analysis of oil shale and retorted products. DiSanzo et al.[231] incorporated an H_2SO_4 precolumn reactor into an injection port system. The reactor served to subtract alkenes, aromatics, and heteroatom-containing species from the complex shale oil and allowed only the alkanes to be introduced and chromatographed on the GC system. Direct analysis of shale kerogen was achieved by van de Meent et al.[232] using a Curie-point pyrolysis GC introduction system. Ideally, this GC method is used to relate detected pyrolysis products to the original structure of shale, and these workers were able to use pyrolysis data to differentiate kerogens from marine and terrestrial sources.

7.4　FOODS AND COSMETICS

Flavors and Aromas

The volatile samples that make up flavors and aromas are of such complexity that only highly efficient open tubular columns have sufficient resolution for their analysis. However, the chromatographic system that is employed must be capable not only of separating complex mixtures, but also of: (a) coping with samples containing constituents of widely differing polarities; (b) allowing analysis of head-space samples; and (c) yielding reproducible chromatograms suitable for profile analysis and in a form suitable for computer-based pattern recognition and comparison.

The availability of fused-silica columns, taken with the developments in surface deactivation, mean that most highly polar solutes can be eluted with symmetrical peaks. Schomburg et al.[233] have made a detailed study of open tubular column GC of polar compounds found in flavors and essential oils. An artificial mixture of the most demanding constituents of vine aroma extracts (with regard to tailing and irreversible adsorption) was chromatographed on nonpolar (OV-101) and polar (Carbowax 20M) fused-silica columns (Figure 7.29). Both chromatograms showed only minor peak tailing, even for free acids, although tailing on the leading edge of the peaks for these compounds was evident except when small samples were injected.

Nonetheless, problems often arise in the analysis of flavors and aromas because of the wide range of polarities of the constituents. The separation of highly polar solutes on nonpolar stationary liquids, or of nonpolar solutes on polar liquids, are both hindered because of the small k values, and the small capacities, and consequent overloading. The latter effect is especially marked if thin films are used in the GC of compounds of fairly low volatility. A compromise is the use of medium polarity stationary phases such as polypropylene glycols. But where not all the important solute pairs can be resolved on a single liquid, columns with different polarities may be coupled together. A switching unit is employed [233] which permits monitoring (by FID) of the eluate from the first column, so that appropriate parts of the chromatogram may be switched into the second column. A chromatogram of vine aroma (Figure 7.30) was obtained with monitoring by the detector at the end of an OV-101 fused-silica column. Strong leading and overlapping is observed for the free acids, but those areas can be separated by switching into a Carbowax 20M column (Figure 7.30). Even though no intermediate trapping was used, a narrow enough band was introduced into the second column for peak shapes to be narrow and symmetrical.

Although direct switching of head-space samples, obtained by milling herbs, into an open tubular column via a gas-tight valve has been reported,[234] the general restriction of small sample capacity inevitably poses severe problems in the analysis of large-volume head-space samples. A variety of methods for overcoming this limitation are available, including cryogenic focusing, precolumn concentration, and splitless injection (Section 4.4). Among these, concentration on porous polymers such as Tenax GC, Porapak Q, or the Amberlite XAD series are especially popular.[235] Stripping of volatiles is most usually done by thermal desorption, particularly if Tenax GC is used; but solvent stripping has also been advocated (see below).

An example of this commonly applied technique in flavor analysis concerns the identification of the aroma constituents of baked 'Jewel' sweet

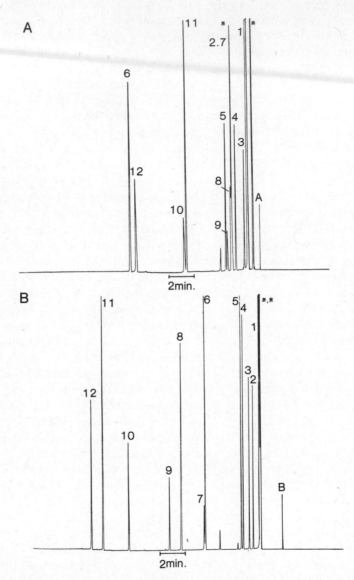

FIGURE 7.29. Chromatograms of typical polar solutes found in vine aroma extracts. Note: (A) 20 m OV-101 fused-silica column, temperature program from 35 to 100°C at 8°C min^{-1}, H$_2$ carrier gas; (B) 20 m Carbowax 20M fused-silica column, temperature program from 60 to 200°C at 8°C min^{-1}, H$_2$ carrier gas. Impurities marked *. Selected peak assignments: (2) 2-methyl-l-propanol; (5) 2-hexanol; (6) ethyl octanoate; (8) 2-methylpropanoic acid; (10) hexanoic acid; (11) phenol. (Reproduced with permission from ref. 233. Copyright Elsevier Scientific Publishing Company.)

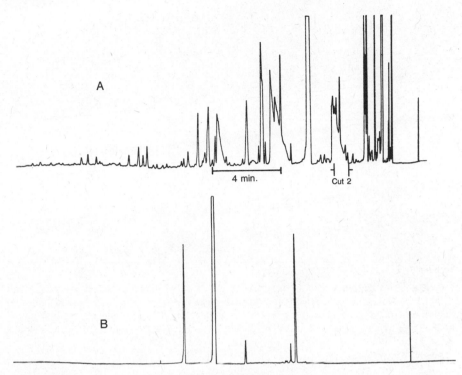

A

4 min.

Cut 2

B

FIGURE 7.30. Chromatogram illustrating multidimensional GC of an extract of vine aroma. Note: (A) 20 m OV-101 fused-silica column, temperature program from 70 to 220°C at 8°C min^{-1}. (B) 20 m OV-101 + Carbowax 20M glass column, temperature program same as for A. The cut from chromatogram A was switched into polar column B. (Reproduced with permission from ref. 233. Copyright Elsevier Scientific Publishing Company.)

potatoes.[236] Forty grams of the homogenized baked roots were heated at 90°C in a 1 dm^3 water-jacketed sampling jar and the volatiles were swept by dry helium onto 20mg of Tenax GC in a glass tube held at 25°C. This cartridge was introduced into the modified injection port and the volatiles were thermally desorbed at 250°C into an 80 m SF-96 glass column, the first 20 cm of which were formed into a trap and cooled in liquid nitrogen. The coolant was removed from the trap and the oven temperature raised to 190°C at 4°C min^{-1}. Twenty-five compounds in the resulting complex chromatogram (Figure 7.31) were identified from the retention times of standards and from spectra recorded with coupling of the column to a mass spectrometer.

Solvent extraction of the porous polymer and injection of the extract onto the column can lead to obliteration of a large part of the chromatogram by the solvent peak. However, if the cartridge is extracted with carbon dioxide under pressure, the solvent can be evaporated without loss of volatiles.[237] Figure 7.32 illustrates the chromatograms of banana aroma from the two phases obtained by carbon dioxide extraction of Porapak Q through which

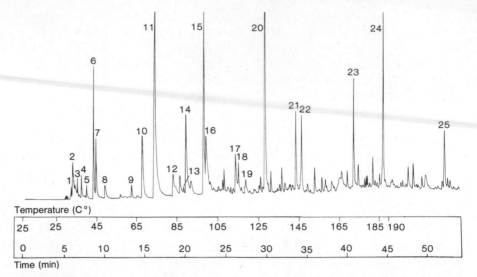

FIGURE 7.31. Chromatogram of the volatile fraction of baked 'Jewel' sweet potatoes. Note: 80 m SF-96 glass column. Components thermally desorbed from Tenax GC sampling capsule. Selected peak assignments: (6) hexane; (11) furfuraldehyde; (15) benzaldehyde; (20) acetaldehyde; (24) β-ionone. (Reproduced with permission from ref. 236. Copyright American Chemical Society.)

gas from a chamber containing ripe bananas had been passed. High-pressure carbon dioxide extraction can also be used to obtain flavor materials from foods.

Reflux trapping using Freons has also been shown[237] to be a useful method for extracting flavor volatiles from gas streams passed over foods. The stream is mixed with refluxing Freon which is condensed via a dry-ice/acetone cold finger. Freon is then evaporated to leave a flavor concentrate for injection onto the column.

Selective detectors can also often be applied with profit in the analysis of flavors and aromas, especially since nitrogen- and sulfur-containing compounds are often important constituents. The profiles of sulfur compounds obtained with a flame photometric detector for Robusta and Arabica coffees (the two major botanical varieties) are sufficiently different [238] for as little as 1% Robusta to be detected in Arabica (Figure 7.33). Similar results were obtained from the volatiles of head space of both ground dry-roasted coffee and brewed coffee.

Essential Oils

Essential oils are also extremely complex mixtures of many hundreds of constituents containing a wide range of functional groups and of varying volatilities. As noted in the discussion of the analysis of flavors and aromas,

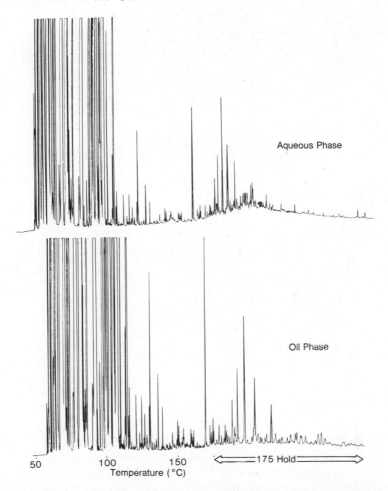

FIGURE 7.32. Chromatogram of volatiles from banana head space. Note: 60 m SP-2100 glass column, volatiles were trapped on Porapak Q and extracted with CO_2 in a high-pressure extractor. (Reproduced with permission from ref. 237. Copyright Dr. Alfred Huethig Publishers.)

the aim of the analysis of such mixtures may be in either fingerprinting or determining individual compounds with known flavor properties. Gas chromatography with open tubular columns affords significant advantages over packed-column GC because of its higher efficiency and sensitivity. The inertness of glass, and more recently fused-silica columns, has brought about significant improvements in this area, particularly in the elution of reactive compounds such as thiophenes, thiazoles, pyrazines, and furans.

Columns for essential oil analysis are generally coated with polyglycol stationary phases, particularly Carbowax 20M, although more recently the Pluronic stationary phases (polyethylene–polypropylene glycols) have been recommended[239] since they are thought to have lower bleeding rates and

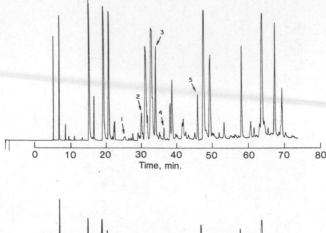

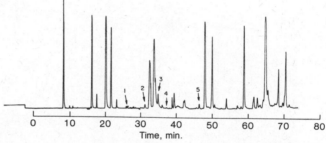

FIGURE 7.33. Chromatogram of sulfur-containing compounds in the head space of brewed coffee (upper trace, Robusta; lower trace, Arabica). Note: 100 m SF-96 glass column, FPD. (Reproduced with permission from ref. 238. Copyright Friedr. Vieweg and Sohn.)

higher maximum operating temperatures. Hydrocarbon constituents of essential oils, however, such as the sesquiterpenes, are not well separated on polar phases, and nonpolar phases may be required. Caution should be exercised since the nonpolar phases may be less satisfactory for the more volatile components. No one phase may be entirely suitable. Thus in Figure 7.34 the analysis[240] of the essential oil of California juniper is illustrated. The sesquiterpenes are not well separated on Carbowax 20M (Figure 7.34 A), while on the SE-30 column (Figure 7.34 C) the more volatile components are not resolved.

Two solutions to the above problem have been proposed: (a) double-column systems or (b) mixed stationary phases. For example, the separation of a commercial perfume oil containing compounds with a variety of functional groups was carried out[241] with an automated dual-column system with nonpolar (OV-101) and polar OS-138 (polyphenyl ether) glass open tubular columns (Figure 7.35). The components of a cut in the OV-101 chromatogram are trapped in 50 m of glass-lined 0.3 mm i.d. stainless steel capillary and injected into the OS-138 column. Identifications were made via Kovats indices. Mixed phases of Superox 20M and OV-1 can be coated homogeneously on the untreated walls of fused-silica capillaries.[242] Columns with high

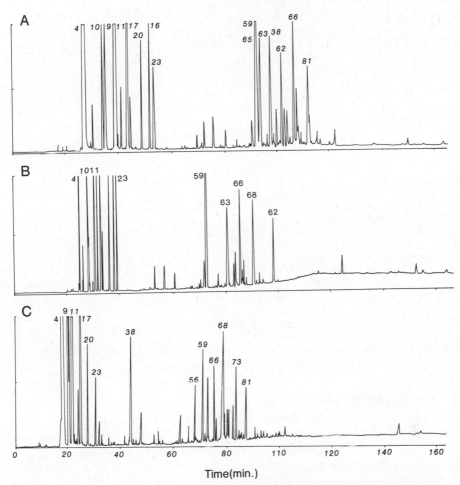

FIGURE 7.34. Chromatogram of essential oil of a California juniper. Note: (A) 100 m Carbowax 20M glass column, (B) 100 m FFAP glass column, and (C) 100 m SE-30 glass column, all columns temperature-programmed from 70 to 170°C at 1°C min^{-1}. (Reproduced with permission from ref. 240. Copyright Preston Publications, Inc.)

efficiency, inertness, and thermal stability (maximum allowable operating temperature ≥250°C) can be obtained, although some residual activity toward fatty acids and primary amines may remain. Selectivity in the analysis of essential oils can be adjusted by changing the percentage of Superox 20M in the phase (Figure 7.36). Thus, the (E)-3-hexenol (peak 2), linalool (peak 4), and carvone (peak 6) shift markedly as the proportion of polar phase is increased.

As well as novel column systems, alternative detectors can be a valuable aid in identifying crucial flavor components of essential oils. For example, simultaneous flame ionization/sulfur/nitrogen detection with the aid of three-

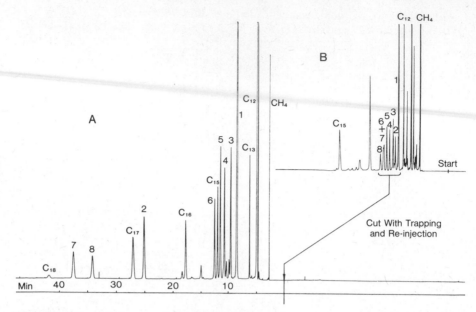

FIGURE 7.35. Chromatograms illustrating multidimensional GC for identification of selected components of a perfume oil. Note: (A) 15 m OS-138 glass column, N_2 carrier gas; (B) 20 m OV-101 glass column, N_2 carrier gas. (Reproduced with permission from ref. 241. Copyright Elsevier Scientific Publishing Company.)

way splitting of the effluent from a fused-silica column was used[243] to show the presence of: (a) alkyl alkoxy pyrazines which are responsible for the characteristic "green" aroma of galbanum oil; (b) the (lachrymatory) iso-thiocyanate compounds in horseradish essential oil (Figure 7.37); and (c) the nitrogen-containing methyl anthranilate in grape essence.

Open tubular column GC is often used in essential oil analysis as part of concerted multitechnique analyses which can also incorporate investigations of odor and composition. Two examples are discussed here—the determination of the composition of Australian tea-tree oil[244] and an investigation of oil of jasmin.[245]

Tea-tree oil is obtained by steam distillation of *Melaleuca alternifolia*, a paper-barked tree growing in swampy land on the east coast of Australia. It is used as a bactericide and in soap perfumes. Jasmin oil is one of the more important and expensive perfumery materials and is obtained in various Mediterranean countries from the flowers of *Jasminium grandiflorum L.* Both oils were first analyzed on glass open tubular columns (e.g., Figure 7.38) and, in the case of the jasmin oil, with coupling to a mass spectrometer with both electron impact and chemical ionization.

Column chromatography on silica gel, chemical group separation, and packed-column GC yielded subfractions that were also analyzed by open tubular column chromatography. Retention data or GC-MS allowed iden-

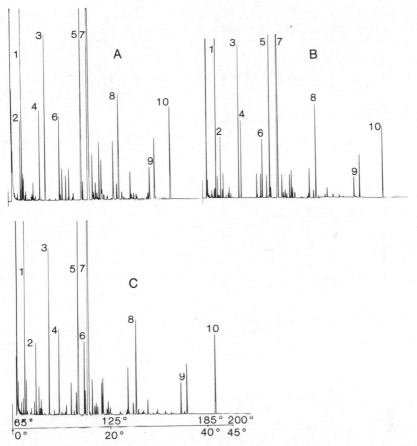

FIGURE 7.36. Chromatograms of hop oil. Note: (A) 25 m OV-1 glass column, (B) 13 m 25% Superox 20M + 75% OV-1 fused-silica column, and (C) 18 m 50% Superox 20M + 50% OV-1 fused-silica column, flow adjusted so that retention time of ethyloctanoate (peak 3) is approximately the same. Selected peak assignments: (1) myrcene; (2) (E)-3-hexenol; (4) linalool; (5) caryophyllene; (6) carvone; (7) humulene; (8) humuladienone. (Reproduced with permission from ref. 242. Copyright Friedr. Vieweg and Sohn.)

tification of many components. Where fractions were isolated in sufficient quantities and open tubular columns indicated that they were simple mixtures, or were almost pure, spectroscopic identification procedures (MS, IR, and NMR) could be applied. Viridoflorene, not previously found in nature, was hence identified in tea-tree oil.[244] A total of 48 compounds at concentrations between <0.05 and 29.4% were identified in the latter,[244] but as many as 112 were found in jasmin oil[245]—64 not previously known even after over 80 years of research into its composition.

The relationship between the composition and aroma of jasmin oil was also investigated[245] by open tubular column GC, firstly by the analysis of

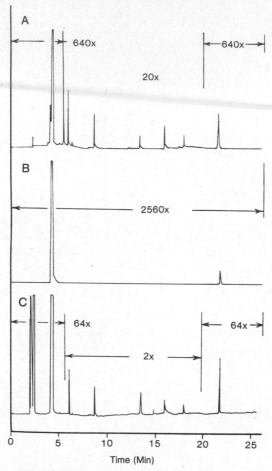

FIGURE 7.37. Chromatograms of horseradish oil. Note: 25 m OV-101 fused-silica column, (A) NPD, (B) FPD, (C) FID. (Reproduced with permission from ref. 243. Copyright Preston Publications, Inc.)

head space and low-boiling fractions of the oil. A separation was also carried out by preparative open tubular column GC on a wide-bore column (100 m × 0.7–0.9 mm i.d.) with a nondestructive microkatharometer detector. Up to 2 μL of the oil were injected, and the individual peaks were collected in traps. The amounts collected were small—often in the microgram range— but were sufficient for olfactory evaluation. This revealed that none of the compounds or fractions had a jasmin oil smell, but that "jasmin lactone" [5-(E-2-pentenyl)-5-pentanolide] is probably the most important aroma-bearing compound in the oil. Odor differences between oils from different countries, partly due to their composition, were also studied by the analysis of odor enhancing and fixing effects of odorless high-boiling components. These

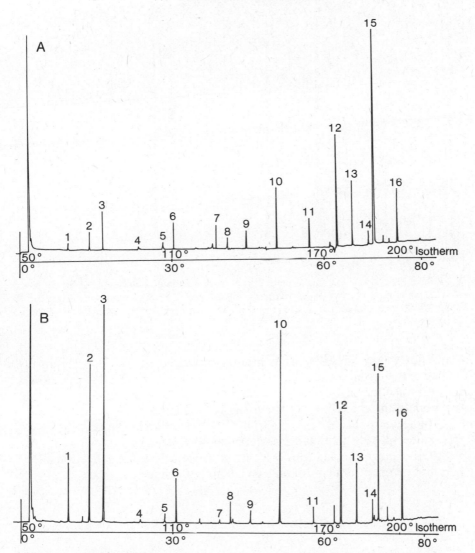

FIGURE 7.38. Chromatograms of (A) French and (B) Italian jasmin oil. Note: 30 m OV-1 glass column. Selected peak assignments: (2) linalool; (3) benzylacetate; (6) jasmone; (9) methyl jasmonate; (10) benzylbenzoate; (12) isophytol; (15) phytol. (Reproduced with permission from ref. 245. Copyright Elsevier Scientific Publishing Company.)

were carried out by open tubular column GC after evaporation of oils to which phytol, benzyl benzoate, and squalane had been added.

Simple but effective methods for odor evaluation and fraction collection with open tubular columns have been described by Sandra et al.[246] An all-glass splitter divides the column effluent, with one splitter arm connected to the FID and, for aroma sensing, the other splitter arm is led to the nose

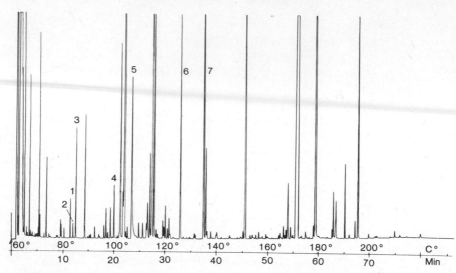

FIGURE 7.39. Chromatograms of the essential oil of *Rhododendron fragrantissimum*. Note: 35 m Superox glass column, H_2 carrier gas at 5 cm^3 min^{-1}. Peak assignments correspond to Table 7.3. (Reproduced with permission from ref. 246. Copyright Dr. Alfred Huethig Publishers.)

via Teflon tubing. Thus an aromagram is produced and the odor characterization of the essential oil of *Rhododendron fragrantissimum* flowers allowed the identification of 2-phenyl ethanol and its derivatives as the main contributors to the characteristic smell (Figure 7.39 and Table 7.3).

Off-line heart-cutting is carried out[246] without complex equipment by collecting the split effluent on a glass capillary microtrap coated with OV-101. Although nanogram quantities only are collected, repeated collection allows preparative open tubular column GC. The performance of the system was

TABLE 7.3. Odor characterization of *Rhododendron fragrantissimum* Flowers

Peak Number	Odor	Compound	Contribution toward the flower smell	Percentage in the oil
1	Green, fresh grass	*cis*-3-Hexenol	—	0.28
2	Aldehyde, sharp, unpleasant	Nonanal	—	0.11
3	Sweet almonds, oil-like	*trans*-2-Hexenol	—	0.74
4	Linalool-like	Linalool	×	0.50
5	Sharp, floral, less pleasant	Phenyl acetaldehyde	× × ×	2.74
6	Fresh, floral	2-Phenylethyl acetate	× ×	2.07
7	Pleasant, floral	2-Phenylethanol	× × × ×	4.99

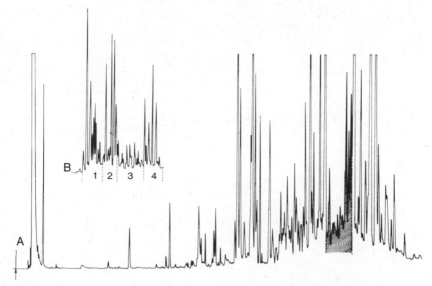

FIGURE 7.40. Chromatograms of an oxygen-containing fraction of pepper essential oil. Shaded area in A was collected and injected on column B. Note: (A) 80 m PMPE glass column, temperature program from 70 to 220°C at 2°C min^{-1}, H$_2$ carrier gas at 8 cm^3 min^{-1}; (B) 40 m FFAP glass column, temperature program from 70 to 180°C at 2°C min^{-1}, H$_2$ carrier gas at 4 cm^3 min^{-1}. (Reproduced with permission from ref. 246. Copyright Dr. Alfred Huethig Publishers.)

checked by the analysis of a mixture of monoterpenes, and then applied to a number of essential oils, such as that of pepper.[246] A small oxygen-containing fraction, separated by silica gel chromatography and preparative packed-column GC, contains a characteristic pepper-smelling cut when separated on a wide-bore (0.5 mm i.d.) polymetaphenyl ether coated column (Figure 7.40 A). This part of the chromatogram was collected and enriched in a microtrap, and then analyzed on an FFAP open tubular column (Figure 7.40 B). The outlet was again split for aroma evaluation.

Food and Food Contaminants

Since the polysaccharide and protein constituents of food are polymeric and hence of low volatility, they are beyond the scope of open tubular column GC. However, their hydrolysis products (mono- and disaccharides and amino acids) are fairly readily amenable, especially if derivatized. The analysis of cane-molasses sugars as oxime-trimethylsilyl ethers[247] and of the amino acids of fish-protein acid hydrolysate as heptafluorobutyryl isobutyl ester derivatives (Figure 7.41) by on-column injection[248] have been carried out. The analysis of monosaccharides and amino acids has been discussed fully in Sections 6.6 and 6.8.

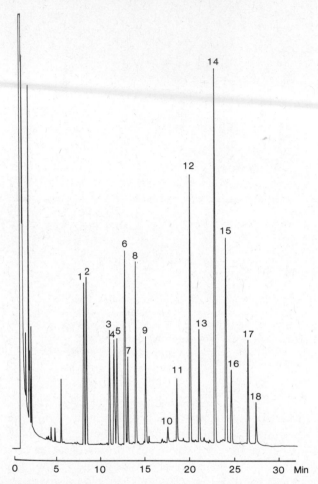

FIGURE 7.41. Chromatogram of amino acid HBB derivatives from an acid hydrolyzate of fish muscle protein. Note: 35 m glass column, temperature program from 85 to 230°C at 5°C min^{-1}. Selected peak assignments: (1) alanine; (6) leucine; (8) norleucine; (12) aspartic acid; (14) glutamic acid; (17) arginine. (Reproduced with permission from ref. 248. Copyright Dr. Alfred Huethig Publishers.)

Until recently, open tubular column GC analyses of triglycerides, the chief constituents of edible fats and fatty oils, could only be carried out after hydrolysis followed by analysis of the fatty acids, generally as methyl esters. Adipose tissue is extracted with chloroform, saponified, and the fatty acids are liberated and methylated in the presence of methanol/sulfuric acid or methanol/BF$_3$. A variety of phases have been used in this area. In early work, Apiezon L (nonpolar) and butane diolsuccinate polyester (polar) phases were applied, but presented clear disadvantages; Apiezon L gave poor separations of positional isomers, while BDS, although yielding more resolution, was less thermostable. Polyglycol phases such as Carbowax 20M have high resolution and reasonable stability, but may result in losses during analysis, possibly because of interactions with the phase by the longer chain and more unsaturated acids. Flanzy et al.[249] found that these losses could be substantially reduced if the Carbowax 20M were esterified with terephthalic acid (free fatty acid phase, FFAP). This phase was found by Jaeger et al.[250] to give good separation of *cis/trans* and double-bond isomers (Figure 7.42) suitable for the analysis of fatty acids in a wide variety of dietary oils.

More recently, there has been considerable interest in the separation of geometric isomers of fatty acids, which has biochemical importance.[251-253] Moreover, processing of dietary oils can result[252] in the presence of trans-isomers as artifacts in addition to the parent (naturally occurring) all-*cis*-acids. The cyanoalkyl polysiloxanes (Silar 10C, SP 2340, OV-275, etc.) have been preferred over polygycols and FFAP phases for these separations.[254,255] Thus, the geometric isomers of oleic acid methyl ester $(18:1\omega 9(c)$ and $18:1\omega 9(t))$† are completely separated on a 30 m glass column coated with SS-4, a cyanoethylpolysiloxane.[253] On the other hand, Grob et al.[252] have shown how the cyanoalkyl Silar phases do not resolve the group of *cis*-isomers of $C_{18:1}$ fatty acid methyl esters from the group of *trans*-isomers which are present in commercial margarine fat. On such phases (e.g., Figure 7.43) a *trans*-isomer of a given positional isomer elutes before the *cis*, whereas for FFAP the *cis* is first. For both phases, a large proportion of the *trans*-isomers are co-eluted with the *cis*, but the error in quantitation of total "*trans*" peaks is less for FFAP.

The direct analysis of triglycerides, without hydrolysis, complements the analysis of fatty acid methyl esters and has been carried out on packed columns for at least 20 years. However, the high molecular mass (> 1000) of triglycerides has limited their analysis on open tubular columns to very short columns [256] which offered only marginal improvement in resolution over that available on packed columns; very high injection temperatures were necessary. Only the availability of highly thermostable columns with cold on-column injection has allowed useful analysis of triglycerides on open tubular columns.

Grob[257] evaluated different injection techniques for triglycerides (Figure 7.44) and found that the cold on-column method was by far the most reproducible (standard deviation 1–3%). A weak discrimination against larger triglycerides was attributed to losses in the column. Combined open tubular column GC-MS of triglycerides has been carried out.[258]

On nonpolar phases (OV-1, OV-101, SE-30, and SE-52), triglycerides are separated into groups according to the number of carbon atoms[259] (Figure 7.45). Within these groups there is resolution of the structural isomers of the same relative molecular mass with different combinations of fatty acids.

---

† Shorthand notation for chain length: number of ethylenic double bonds, position of final double bond counted from terminal methyl, c is *cis* and t is *trans*.

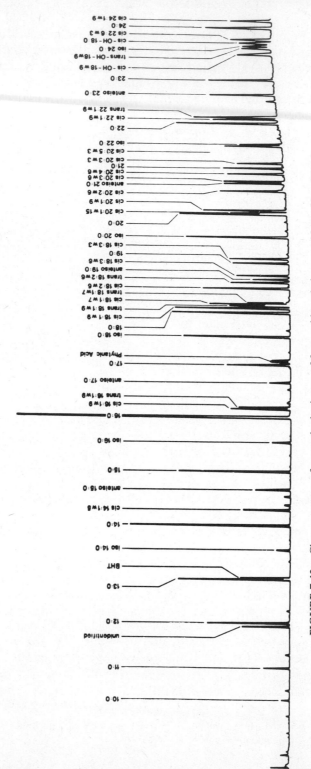

FIGURE 7.42. Chromatogram of a standard mixture of fatty acid methyl esters. Note: 50 m FFAP glass column, temperature program from 95 to 198°C at 1°C min⁻¹, H₂ carrier gas at 4 cm³ min⁻¹. (Reproduced with permission from ref. 250. Copyright Friedr. Vieweg and Sohn.)

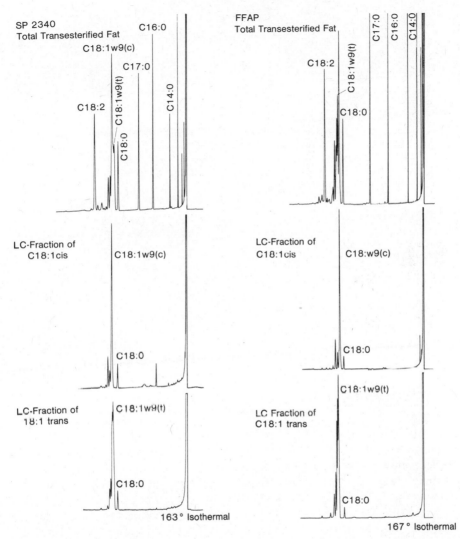

FIGURE 7.43. Chromatograms of fatty acid methyl esters of a commercial margarine fat. Note: (A) 40 m SP-2340 glass column, H_2 carrier gas; (B) 40 m FFAP glass column. (Reproduced with permission from ref. 252. Copyright Dr. Alfred Huethig Publishers.)

Thus C_{12}, C_{12}, C_{12} is the predominant C_{36} triglyceride of coconut oil, which can thus be identified in mixtures with butter for which C_4, C_{14}, C_{18} is the chief triglyceride (Figure 7.45 C). The apolar phases give enough separation of structural and stereoisomers (the latter arising from the different position of the fatty acids on the glyceride moiety) for characteristic profiles of fractionated, transesterified, or heat-treated fats to be produced on glass columns of intermediate length ($\sim$15 m) deactivated by high-temperature silylation before coating.[259] There is also some resolution[260] of triglycerides with the

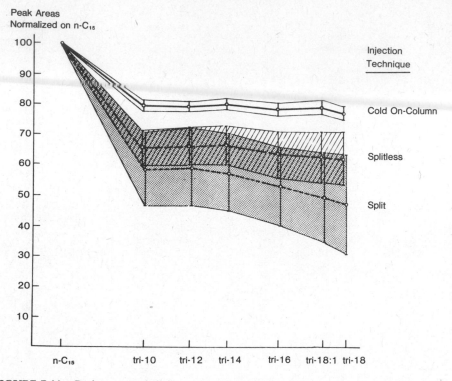

FIGURE 7.44. Peak areas and their standard deviations for triglycerides normalized on the alkane n-C_{15} obtained with different injection techniques. (Reproduced with permission from ref. 257. Copyright Elsevier Scientific Publishing Company.)

same carbon number but with different degrees of unsaturation, for example, triolein and tristearin and mixed C_{54} isomers (Figure 7.46). Better resolution of unsaturated isomers awaits the preparation of polar columns with comparable inertness, efficiency, and durability.

Naturally occurring wax esters are also constituents of oils and adipose tissue of animal origin. Short-chain wax esters (isobutyroyl, isovaleroyl, and 2-methylbutyroyl) have been chromatographed[261] on 40 m stainless steel open tubular columns coated with diethyleneglycol succinate and operated isothermally at 150°C. The new inert glass and fused-silica columns can be operated at high enough temperatures to allow direct analysis[262] of esters from C_{24} to C_{48} (Figure 7.47). Resolution of unsaturated compounds is possible even on the apolar phase (OV-1) employed; wax esters of a given carbon number are eluted in the sequence diene, monoene, and saturated (Figure 7.47). The total number of double bonds determines the elution temperature, and an ester with a double bond in the alcohol residue co-elutes with one with the double bond in the acid part.[262]

Gas chromatography with open tubular columns is at the forefront in the analysis of food contaminants of both natural and unnatural origin. Food

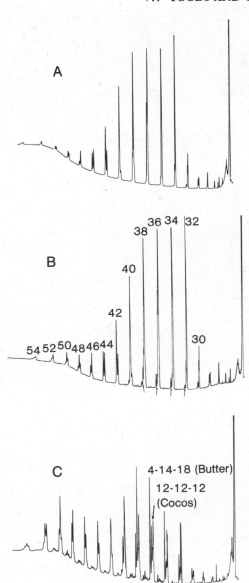

FIGURE 7.45. Chromatograms of (A) pure coconut oil, (B) 20% butter in coconut oil, and (C) 10% coconut oil in butter. Note: 14 m OV-1 (0.08 μm film) glass column, temperature program from 240 to 340°C at 4°C min^{-1}, on-column injection. Coconut oil (10%) can be easily determined in butter, but 20% of butter in coconut oil causes problems. A convenient quantitation can be achieved using an internal standard. (Reproduced with permission from ref. 259. Copyright American Oil Chemists' Society.)

may be contaminated by processes occurring naturally, or in food preparation, or by extraneous materials. The natural histidine in tuna fish may be converted into histamine during storage and hence induce scambroid poisoning. Histamine can be determined in tuna fish extracts by silylation followed by open tubular column GC on a 12 m SP-2100 fused-silica column with detection by positive-ion chemical ionization mass spectrometry using methane as reagent gas.[263]

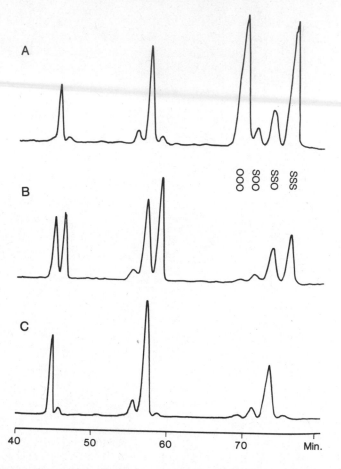

FIGURE 7.46. Chromatograms of (A) coinjection of cocoa butter and a 1:1 mixture of tri-stearin and triolein, (B) coinjection of cocoa butter and hydrogenated cocoa butter, and (C) cocoa butter original sample. Note: 8 m SE-30 (0.13 μm film) fused-silica column, temperature program from 200 to 295°C at 8°C min^{-1}, H$_2$ carrier gas, on-column injection. (Reproduced with permission from ref. 260. Copyright Dr. Alfred Huethig Publishers.)

Mass spectrometric detection was also used in the analysis[264] of roast-coffee infusion for phenols which are thought to give rise to the nausea experienced by some coffee drinkers. A roast-coffee infusion was acidified and extracted with diethyl ether. After evaporation of the solvent, the residue was silylated with N-methyl-N-trimethylsilyltrifluoroacetamide and ana-lyzed on a 25 m SE-30 glass column (Figure 7.48). The main components were identified from the electron impact and isobutane chemical ionization mass spectra—the latter being particularly useful in yielding fewer fragment ions. 1,2,4-Trihydroxybenzene, 4-ethylcatechol, and 3-methylcatechol were identified as new constituents of roast coffee with possible emetic proper-ties.[264]

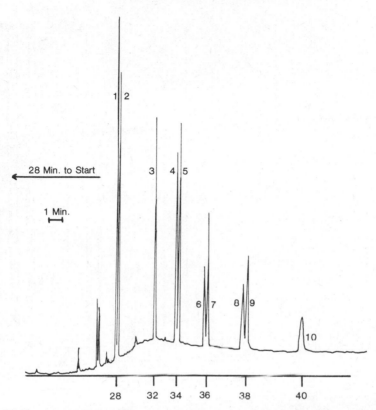

28 Min. to Start

1 Min.

FIGURE 7.47. Chromatogram of wax esters. Note: 25 m OV-1 fused-silica column, temperature program from 250 to 325°C at 5°C min⁻¹. Selected peak assignments: (1) decyl oleate (18:1–10:0); (2) decyl stearate (18:0–10:0); (4) oleoyl palmitate (16:0–18:1); (5) palmityl stearate (18:0–16:0); (8) arachidyl oleate (18:1–20:0); (9) arachidyl stearate (18:0–20:0); (10) arachidyl arachidate (20:0–20:0). (Reproduced with permission from ref. 262. Copyright Dr. Alfred Huethig Publishers.)

Food may also be contaminated with any of the numerous chemicals to which the original plants and animals may be exposed. The persistence of these compounds in food products demands separation methods with high sensitivity and resolution after appropriate clean-up procedures. Two illustrative examples are the detection and estimation of pesticide residues in vegetables by open tubular column GC with electron-capture detection,[265] and of the residues of anabolic drugs in meat by open tubular column GC-MS.[266] Vegetables are treated against mildew with pentachloronitrobenzene ("quintobenzene") which contains various chlorobenzenes as contaminants. The parent compound is converted to pentachloroaniline in the soil and on the plant. These compounds are conveniently determined at levels between 0.05 and 1 ppm by column chromatography of hexane extracts of the vegetables and splitless injection onto a 14 m glass OV-17 column with micro electron capture detector[265] (Figure 7.49).

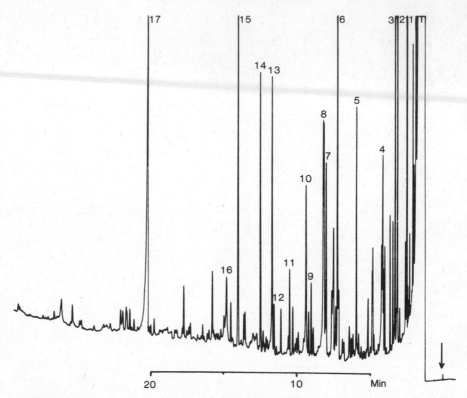

FIGURE 7.48.   Chromatogram of derivatized (MSTFA) coffee extract. Note: 25 m SE-30 glass column, temperature program from 80 to 230°C at 5°C min^{-1}, H$_2$ carrier gas, 1:25 split ratio. Selected peak assignments: (3) glycolic acid; (6) catechol; (9) 3-methyl catechol; (10) mesaconic acid; (11) 4-methyl catechol; (13) malic acid; (14) 1,2,3-trihydroxybenzene; (15) 1,2,4-trihydroxybenzene; (17) caffeine. (Reproduced with permission from ref. 264. Copyright Dr. Alfred Huethig Publishers.)

Anabolic drugs, used to increase the growth rate of animals for food production, may be determined in homogenized meat by GC on a 10 m glass open tubular column coated with SE-54, following clean-up by liquid–liquid partition and silica gel chromatography and silylation.[266] Seven estrogenic drugs used in anabolic preparations were analyzed, with a detection limit of 1–5 ppb, by electron impact MS with selective ion monitoring (usually above $m/z = 400$) (Figure 7.50).

Alcoholic Beverages

Samples containing large amounts of water and alcohol along with non volatile substances, such as alcoholic beverages, pose special problems in open tubular column GC since repeated injections can cause rapid column de-

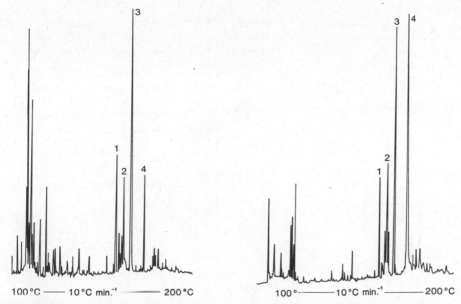

FIGURE 7.49. Chromatograms of pesticide residues in lettuce extracts. Note: 14 m OV-17 glass column. Selected peak assignments: (1) pentachlorobenzene; (2) hexachlorobenzene; (3) pentachloronitrobenzene; (4) pentachloroaniline. (Reproduced with permission from ref. 265. Copyright Friedr. Vieweg and Sohn.)

tubular column GC since repeated injections can cause rapid column deterioration. However, Grob et al.[267,268] have suggested that a 40 m glass column treated by the barium carbonate procedure and coated with Carbowax 400 will withstand repeated injections of ~1 μL (split 20:1) of aqueous solutions at 25°C, followed by programming to 150°C. The stationary phase is washed from the first 60 cm of the column to avoid leaching by water and ethanol. Beverages with a high sugar content (e.g., liqueurs) can result in pyrolysis of sugar in the injector, which is therefore equipped with a glass liner that can be replaced after 5–10 such injections.[268] After 30–50 samples, the first 50 cm of the column are removed. With these precautions, satisfactory direct analyses for higher alcohols in a variety of alcoholic drinks is possible (Figure 7.51). The broad methanol peak results from a solvent effect; artifacts also result from the injection of drinks with a high sugar content, especially furfural which arises from pyrolysis.[268]

Head-space samples of alcoholic beverages can be injected onto an open tubular column, but the large volumes required to offset the small concentrations of flavor components are not generally consistent with high resolution. Preconcentration on precolumn traps containing material with a low affinity for water and ethanol (see Section 4.4) is an attractive alternative.[269] Porapak Q has an overall higher capacity than Tenax GC for wine flavor

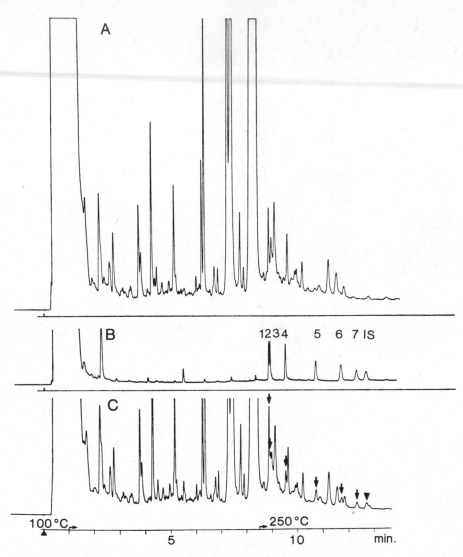

FIGURE 7.50. Chromatograms of (A) a meat sample extract free of anabolic residues, (B) a standard mixture of anabolic drugs, and (C) a meat sample extract as in A, but with the standard mixture from B added. Selected peak assignments: (1) hexestrol; (2) diethylstilbestrol; (3) dienestrol; (4) stilbestrol; (5) 17β-estradiol; (6) ethynylestradiol; (7) zeranol; (8) dodecyl gallate (internal standard). (Reproduced with permission from ref. 266. Copyright Elsevier Scientific Publishing Company.)

components (Figure 7.52), but there are also differences in the quantities of the lower molecular mass alcohols trapped. Head-space components of alcoholic beverages may also be concentrated by distillation-extraction,[269] or by a novel procedure[270] involving extraction into Freon 11; greater than 99% recovery has been claimed and the extracts do not contain impurities or

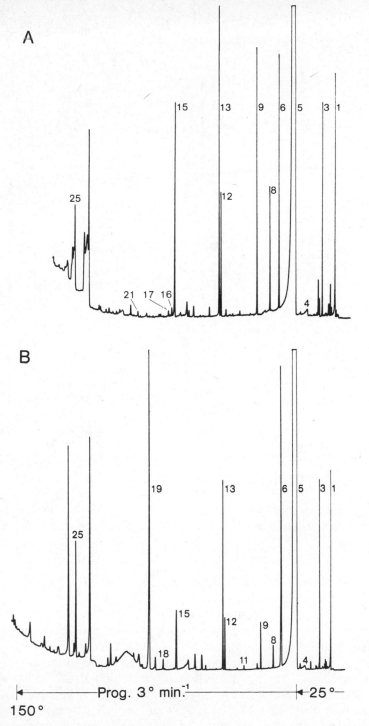

FIGURE 7.51. Chromatograms of (A) a white wine and (B) a sweet drink (Vermouth). Note: 40 m Carbowax 400 glass column, H_2 carrier gas. Selected peak assignments: (1) acetaldehyde; (4) methanol; (5) ethanol; (6) dioxan (internal standard); (9) isobutanol; (13) 3-methyl-1-butanol; (15) ethyllactate; (19) furfural; (25) furfuryl alcohol. (Reproduced with permission from ref. 268. Copyright Dr. Alfred Huethig Publishers.)

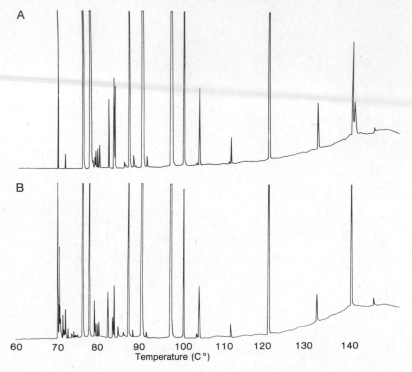

FIGURE 7.52. Chromatograms of Zinfandel wine volatiles trapped on (A) Porapak Q and (B) Tenax GC. Note: 80 m Carbowax 20M glass column, temperature program from 60 to 140°C at 1°C min⁻¹. (Reproduced with permission from ref. 269. Copyright American Chemical Society.)

artifacts , nor are special injection systems required. This method has been applied[270] to the terpenoid components of wine bouquet, with analysis on Reoplex 400.

7.5 BIOLOGICAL MATERIAL

Body Fluids and Tissues

The rapid expansion of the use of GC in clinical biology has largely arisen because of the value of metabolic profiling as an aid in the diagnosis and study of disease. The high resolution of open tubular GC columns allows either specific compounds or a range of metabolites to be separated, while the sensitivity available with sharp peaks and specific detectors has extended the method into the picogram range at which many substances are observed in body fluids and tissues.

The profiling and analysis of urine and to a lesser extent other physiological fluids such as plasma, amniotic, cerebrospinal, and seminal fluids, as well as saliva and sweat have been extensively studied. Among these, only urine is normally devoid of proteins, and procedures must normally be devised for the separation of the lower molecular weight materials in high yield, while separating from protein compounds. Because of the variety of functional groups present, the use of glass and/or fused-silica columns is mandatory, and derivatization usual for all but the analysis of volatiles.

Few areas of open tubular column analysis have attracted as much attention as the GC of steroid hormones. As long ago as 1970, a number of authors[271,272]reported how mixtures of nanogram quantities of steroids, although containing labile and involatile compounds, could be chromatographed to give symmetrical peaks as long as derivatization was carried out and glass columns were employed with splitless injection. All-glass solid injection devices are now commonly used. Much early work was directed toward the fabrication of columns with nonpolar stationary phase films which were stable in routine use at the high temperatures ($\geq 250°C$) necessary for steroid analysis.[273] Etching and silylation of the glass before coating were markedly successful in the hands of Novotny and Zlatkis,[271,273]and this procedure was also used by Bailey et al.[274] to prepare OV-101 columns for a routine procedure for analysis of neutral steroids in urine. After enzymatic hydrolysis, the compounds were separated as trimethylsilyl ethers with quantitation by reference to three internal standards added to the hydrolyzed urine (Figure 7.53).

Steroid analyses are generally performed after formation of the methoxime (all carbonyl groups but 11) trimethylsilyl ether (hydroxyl groups) derivatives, which are thermally stable, readily volatilized, and not dehydrated or adsorbed in the GC system. Other derivatization procedures for steroid hydroxyl groups involve preparation of trifluoroacetates or heptafluorobutyrates.[275] In certain instances, these may allow improved separation of some isomers, for example, the androst-5-ene/5α-androstand-3β,17β-diol pair, but for other isomers, for example, 5α-androstane-3α,17β-diol/androst-5-ene-3β,17α-diol, the resolution is decreased. The choice of derivatives depends on the desired resolution.[275]

For analysis of steroids on the more usual nonpolar stationary phases, the order of elution of the MO-TMS derivatives is well known.[276-278] However, there are often advantages in changing profiles by using selective phases. Novotny and Zlatkis investigated[273] the selectivity of stationary phases for steroids and recommended the polar but thermostable Dexsil and polyimide series. Advantages have also been claimed for columns coated with OV-225 in metabolic profiling of human urine[279] and the analysis of the estrogens of pregnant mare's urine.[280] "Whisker" columns coated with OV-17 selectively retain free steroidal ketones (e.g., 3-hydroxy-5β-androstane-11,17-dione and 3α,17α,21-trihydroxy-5β-pregnane-11,20-dione) while allowing persilylated hydroxy steroids to elute early.[281] Resolution of pregnanetriols (as MO-TMS derivatives), of clinical importance in congenital adrenal hyperplasia (21-hydroxylase defect), was achieved[282] on columns prepared by

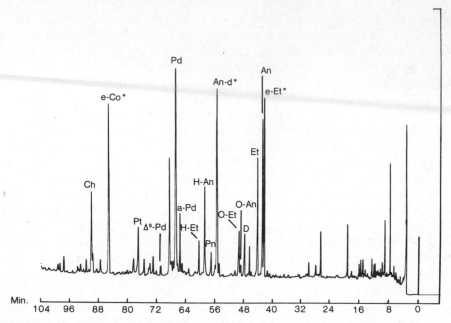

FIGURE 7.53. Chromatogram of steroid hormones (TMS ethers) from a normal female urine collected on day 20 of the menstrual cycle. Note: 50 m OV-101 glass column, temperature program from 190 to 280°C at 1°C min^{-1}. (Reproduced with permission from ref. 274. Copyright Elsevier Scientific Publishing Company.)

in situ polymerization of methylsiloxanes. The C_{87}-hydrocarbon stationary phase is also very useful in resolving the difficult 5α-stanol-$\Delta 5$-sterol TMS ether pairs.[283] Other examples of separations achievable by open tubular columns which are of clinical importance include the resolution[277] of the four tetrahydrocorticosterone isomers (SE-30); the separation[277] of the *syn* and *anti* isomers of 3-methoxime-$\Delta 4$-steroids (also on SE-30) and the four urinary acidic metabolites or cortisol on Carbowax.[284] As recently as 1977, the presence was demonstrated[285] of ten steroids previously undetected in blood by open tubular column GC on a glass column coated with SE-30.

Mass spectrometry detection is often preferred in steroid analysis; SIM at *m/z* 129 and 217 is particularly useful. Heptafluorobutyrate derivatization allows EC detection of steroids in subnanogram amounts,[286] while NP detection is four times more sensitive then FID for the MO TMS derivatives of ketonic steroids.[287]

Glass open tubular column steroid profiles can now be obtained routinely so that subtle changes in health or disease can be detected. Thus the urinary steroids of the newborn have been investigated,[288] while the urine steroid profiles of healthy adults can be divided into two groups[289] according to the different excretion rates of dehydroepiandrosterone (DHEA). Changes in DHEA excretion with time of day have also been shown,[289] the ultimate so

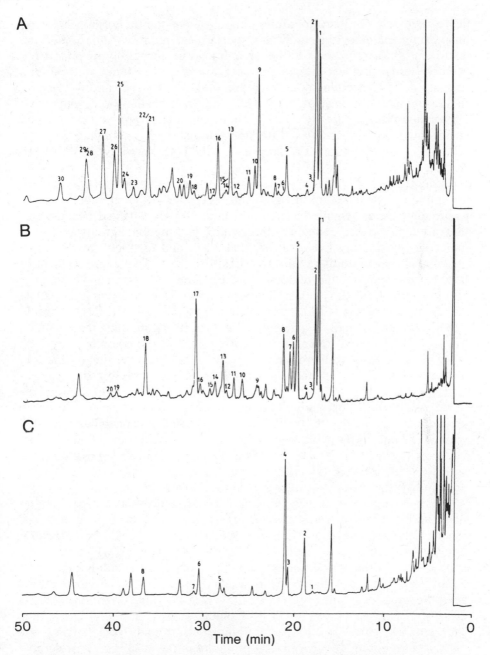

FIGURE 7.54. Chromatograms of steroids in different conjugate fractions isolated from urine of a healthy female during follicular phase. Note: 25 m SE-30 glass column, temperature program from 230 to 265°C at 1°C min⁻¹. Aliquots equivalent to 7 and 20 μL of urine were injected from the (A) glucuronide, (B) monosulfate, and (C) disulfate fractions. (Reproduced with permission from ref. 293. Copyright Elsevier Scientific Publishing Company.)

far in metabolic profiling. In addition, analyses of steroid hormones permit diagnosis of pathological conditions such as carcinomas[278] and endocrine disorders.[290] The course of a disease can also be accurately mapped from urinary steroid profiles; the gradual decrease of kidney function in uremic patients is characterized by an increase of 11-oxygenated androstanes.[291] Increasing attention is also being paid to prefractionation of steroids from plasma and urine before GC.[292,293] Column chromatography on a lipophilic strong-anion exchanger gives[293] unconjugated steroids, monoglucuronides, and mono- and disulfates for analysis as MO TMS derivatives by GC-MS (Figure 7.54).

In contrast to steroid hormones, thyroid hormones, of which T_3 and T_4 are found in human serum in significant quantities (although mostly bound to protein), are extremely difficult substances for GC analysis because: (a) they are highly polar, containing three different functional groups; (b) their molecular masses are between 900 and 1200; and (c) they are thermally unstable at temperatures required to volatilize them. Two approaches have been suggested for the use of open tubular columns. Peterson and Vouros[294] chromatographed methyl ester heptafluorobutyryl derivatives on a 20 m glass OV-101 column with SIM MS monitoring at m/z 844 $(M-213)^+$ for T_3 and RT_3 (reverse T_3), finding detection limits of 100 pg and 5 pg, respectively. They analyzed 20-cm^3 quantities of euthyroid serum containing only 2 pg cm^{-3} of RT_3, although there are doubts as to the true level since some RT_3 may arise from T_4 during dialysis. Corkill and Giese[295] preferred EC detection of the N,O-diheptafluorobutyryl methyl ester derivatives of thyroid hormone standards. Detection limits of 30 fg, with a working range of 0.4–700 pg, were achieved on short (as little as 7.5 m) fused-silica capillaries coated with SE-52 and SE-54 with on-column injection through an all-glass direct injection insert and with high carrier gas flow rate (Figure 7.55). All the glass surfaces with which the iodothyronines came into contact, including the injection syringe, were carefully cleaned and silanized.

The involvement of prostaglandins in health and disease, especially inflammation, is now a major area of investigation in biochemical analysis. Maclouf showed how prostaglandins of human plasma could be chromatographed as methyl ester methoxime trimethylsilyl ethers (E series) and methyl ester TMS ethers (F series) with the aid of a splitless solids injector.[296] All prostaglandins are derived from arachidonic acid (C_{20} polyunsaturated acid). Separation and detection of all these primary transformation products is possible[297] on glass OV-1 or OV-101 columns, again for methyl ester oxime TMS ethers (Figure 7.56), so that profiling of eicosanoids from human blood platelets is possible[298] even for concentrations as low as 300 pg. Considerably higher sensitivity is possible for F series prostaglandins as pentafluorobenzyl ester TMS ethers, and there is baseline resolution on SE-52 for the biochemically important PGH_2 metabolites as methyl ester pentafluorobenzyl oxime TMS ethers.[299] The use of a new multistep mixed derivatization approach that generates the methyl esters of n-butylboronate, pentafluoro-

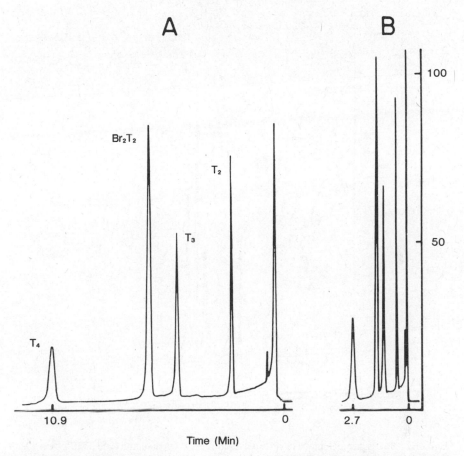

FIGURE 7.55. Chromatograms of a 0.5 μL injection of a solution containing 136, 190, 225, and 341 pg of thyroid hormone (T_2, T_3, T_4, and Br_2T_2) diheptafluorobutyryl methyl ester derivatives. Note: 7.5 m SE-52 fused-silica column, 265°C isothermal temperature, carrier flow rate at (A) 1.90 cm³ min⁻¹ and (B) 13.5 cm³ min⁻¹, electron capture detection. (Reproduced with permission from ref. 295. Copyright American Chemical Society.)

benzyloxime, and TMS ether derivatives of prostaglandins simplifies [300] the profiling of the arachidonic acid metabolites by open tubular column GC. Mass spectrometry detection with SIM at m/z 365 has been used[301] in analyses on a 10 m glass open tubular column for 7α-hydroxy-5,11-ketotetranorprostane-1,16-dioic acid, the major human urinary metabolite of prostaglandins E_1 and E_2. Mass spectrometry detection has also been used in profiling the prostaglandins of urine, plasma, and seminal fluid,[302] and CI-MS detection in the determination of 6-keto prostaglandin $F_{1\alpha}$.[303]

The organic acid fraction of physiological fluids contains intermediates and end products of many metabolic pathways and after derivatization (usually methyl esters or TMS ethers) is ideal for profiling. For example, the

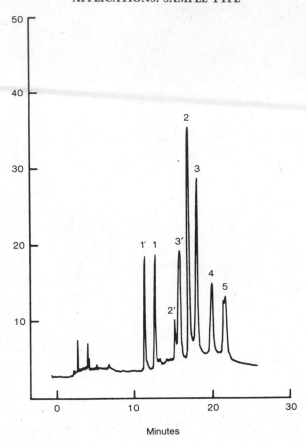

FIGURE 7.56. Chromatogram of prostaglandin methyl ester methoxime TMS ethers. Note: 40 m OV-101 glass column, 245°C isothermal temperature. Selected peak assignments: (1′,1) PGA_2 (syn and anti oxime isomers); (2′,2) PGD_2 (syn and anti); (3′,3) PGE_2 (syn and anti); (4) TxB_2; (5) 6-keto-$PGF_{1\alpha}$. (Reproduced with permission from ref. 297. Copyright American Chemical Society.)

urine of patients with maple syrup urine disease contains characteristic branched-chain 2-keto and 2-hydroxy acids which can be readily separated on a 25 m SE-30 glass column.[304] Spiteller and Spiteller[305] separated the acids of urine into eight fractions by thin-layer chromatography. Analysis of each fraction after methylation with diazomethane on a 25 m OV-101 glass column coupled to a mass spectrometer (e.g., Figure 7.57) allowed the separation of about 500 compounds of which ~200 were identified from mass spectra. Liebich et al.[306] analyzed the total profile of the organic acids of urine as methyl esters on an OV-17 column, and identified 23 dicarboxylic acids. These compounds are significant in the study of increased fatty acid metabolism; succinic and adipic acids are indicators of ketoacidosis[306] and adipic acid derivatives are elevated in uremic urine.[307]

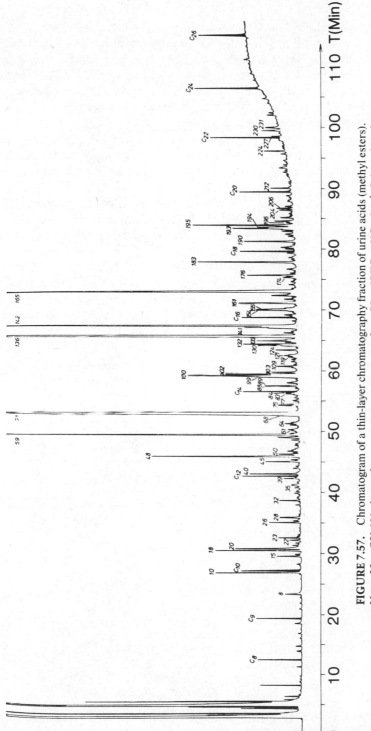

FIGURE 7.57. Chromatogram of a thin-layer chromatography fraction of urine acids (methyl esters). Note: 25 m OV-101 glass column, temperature program from 75 to 275°C at 2°C min⁻¹. Selected peak assignments: (59) $H_3COOC—C_5H_8—COOCH_3$; (71) 2,5-dimethoxycarbonylfuran; (136) methyl-3,4-dimethoxybenzoate; (142) methyl-3,4-dimethoxybenzylate. (Reproduced with permission from ref. 305. Copyright Elsevier Scientific Publishing Company.)

Gas chromatography–mass spectrometry was particularly useful in the latter studies, both with SIM and also since differences in the intensity ratios of M-31 and M-32 as well as M-59 and M-60 can be used to differentiate geometric isomers of unsaturated dicarboxylic acid methyl esters, for example, of mesaconic and citraconic acids.[306] The application of a specialized GC-MS computer system which allows accurate mass determination of the organic acids of human urine was described by Lewis et al.[308] Considerable advantages over low-resolution MS were shown in distinguishing components which produce fragments of the same nominal mass but which have different accurate masses resulting from different elemental composition[308] (Table 7.4).

Other detectors may be used in the determination of low concentrations of urine acids as appropriate derivatives. Thus, homovanillic, isohomovanillic, and vanillylmandelic acids have clinical importance in neuroblastoma, Parkinsonism, and neural chest tumors and may be analyzed as trifluoroacetylhexafluoroisopropyl esters with ECD detection.[309] However, even with high column efficiencies, it was necessary to confirm peak assignments with a second column; OV-101 and Dexsil 300 were the usual phases. Novel derivatives can be used to achieve the separation of enan-

TABLE 7.4. Accurate masses and elemental compositions of mass fragments of trimethylsilyl derivatives of urinary acid constituents which yield fragments of the same nominal masses.[a]

Accurate mass	Composition	Source
117.0372	$C_4H_9O_2Si$	Lactic acid
117.0736	$C_5H_{13}OSi$	Lactic acid
129.0736	$C_6H_{13}OSi$	β-Hydroxy-β-methylglutaric acid
129.0368	$C_5H_9O_2Si$	Unknown
205.1080	$C_8H_{21}O_2Si_2$	Gluconic acid
205.0716	$C_7H_{17}O_3Si_2$	Glycolic acid
247.1179	$C_{10}H_{23}O_3Si_2$	Erythronic acid
247.0822	$C_9H_{19}O_4Si_2$	Succinic acid
295.1037	$C_9H_{27}O_3Si_4$	Silicone
295.1550	$C_{15}H_{27}O_2Si_2$	Unknown
307.1581	$C_{12}H_{31}O_3Si_3$	Alditols, aldonic acid
307.1271	$C_{11}H_{27}O_4Si_3$	Glyceric acid
359.1166	$C_{14}H_{27}O_5Si_3$	Hexuronic acids
359.1530	$C_{15}H_{31}O_4Si_3$	Hexonic acid
435.2239	$C_{18}H_{43}O_4Si_4$	Quinic acid
435.1874	$C_{17}H_{39}O_5Si_4$	Hexonic acids

[a] The examples given are CH_4/O doublets which are 36 milli-mass-units apart and appear frequently because of the nature of the mixture.

tiomers of acids in physiological fluids by converting them to diastereroiso-
mers with a chiral reagent such as (-)-menthol.[310]

Comprehensive metabolic profiles are difficult to achieve for the organic
acids from body fluids other than urine because of the small quantities often
available (≤ 1 cm^3 for infants) and the problem of separating from protein.
Ultrafiltration through membranes can be used to remove high molecular
mass substances and this procedure has been applied in the profiling of acids,
carbohydrates, and so on, in serum from diabetics[311] and from uremic patients before and after dialysis.[312] The organic acids of uremic hemofiltrates
from the dialysis process have also been profiled on glass open tubular columns.[313] A new method, oleate replacement ultrafiltration, has been
proposed[314] to desorb organic acids from albumin before ultrafiltration. After
subsequent anion chromatography and formation of oxime trimethylsilyl
ethers, MS with SIM at m/z 73 is used to detect the plasma organic acids
eluted from an open tubular column. Profiles obtained from patients with
polyserositis and diabetes show the presence of characteristic components.[314] Organic acid profiles of human tissue biopsies have been
obtained[315] after hydrolysis and extraction with an organic solvent after acidification.

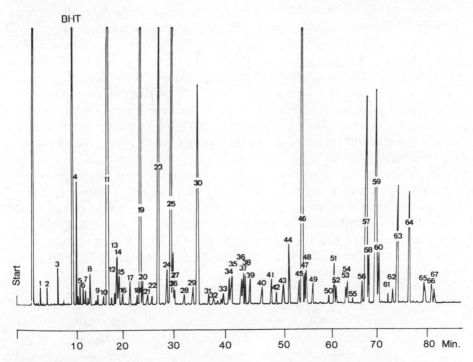

FIGURE 7.58. Chromatogram of human red cell total phospholipid fatty acid methyl esters.
Note: 30 m SP-2340 glass column, temperature program from 100 to 190°C. Selected peak
assignments: (11) C$_{16}$; (23) C$_{18}$; (46) C$_{22}$ (and others); (57) C$_{24}$. (Reproduced with permission
from ref. 317. Copyright Elsevier Scientific Publishing Company.)

The long-chain acids of human plasma (free acids)[316] and of red cell phospholipids,[317] including tocopherols and cholesterol[316] may provide important biochemical information. For example, red cell phospholipid fatty acids (Figure 7.58) show age-related compositional trends when analyzed on an SP 2340 coated glass column which separates geometric isomers. Comparisons of plasma fatty acids from stroke patients and normal subjects have been made with chromatography on a glass polyphenylether sulfone column.[316]

Amino acid profiles of physiological fluids may also reflect inborn errors of metabolism and other pathological conditions. After ion-exchange chromatography of plasma or urine, the isobutyl ester, $N(O)$-heptafluorobutyrate derivatives of the free amino acids can be effectively determined[318] by GC-MS on glass OV-101 columns (e.g., Figure 7.59). The increase in concentration of phenylalanine in the plasma of an infant with phenylketonuria was clearly shown, as was the excretion of cystathionine in a case of a cystathionase defect.[318] Similar characteristic profiles have also been obtained[319] for normal and abnormal serum and urine samples by analysis of amino acids on Carbowax 20M/Silar 5CP columns as the n-propyl, N-acetyl derivatives with nitrogen-selective detection which showed a ~10^2 improvement over flame ionization.

More specific analyses of amino acids of clinical importance have been reported. Thus the GC determination of glutamine in the presence of glutamic acid is made difficult by the rapid conversion of the former to the acid during derivatization. An ingenious solution[320] is to analyze for the methyl ester of the intermediate in the above reaction, pyrrolidone carboxylic acid. To overcome the problem of an unpredictable extent of reaction, D-glutamine is added as an internal standard and reacts to the same extent as the naturally occuring enantiomer. The enantiomers of pyrrolidone carboxylic acid ester are then separated on an open tubular column coated with the chiral stationary phase Chirasil-Val.

Mono- and polysaccharides have wide-ranging clinical significance. The presence of carbohydrates in seminal fluid may indicate a source of energy for spermatozoa, and monosaccharide profiles are important in studies of infertility.[321] The filtered fluid was passed through an ion-exchange column and the lyophilized eluate evaporated to dryness before derivatization with methoxylamine and BSTFA in pyridine. Separation was accomplished on an SE-30 glass column with MS detection.[321] Myoinositol is the major component of the urinary polyols and aldoses of infants, which may be separated[322] as acetyl and acetylaldonitrile (Wohl reaction) derivatives, respectively, on SE-30 coated columns. Smith and Novotny have shown[323] how the polyols of samples of cerebrospinal fluid may be analyzed by preliminary separation by column chromatography on Lipidex 5000, formation of trimethylsilyl ethers, and separation on a glass SE-52 column (Figure 7.60). Myoinositol is also the chief polyol of cerebrospinal fluid of infants (Figure 7.60), and its level also changes in a variety of disease states.[323] The

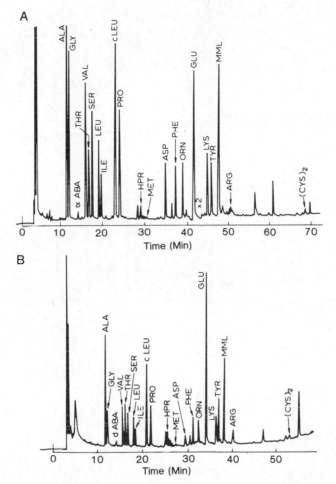

FIGURE 7.59. Chromatograms of amino acids from two normal plasma samples as $N(O)$-heptafluorobutyrate derivatives. Note: OV-101 glass column, temperature program from 90 to 270°C at (A) 2°C min^{-1} and (B) 3°C min^{-1}. (Reproduced with permission from ref. 318. Copyright Elsevier Scientific Publishing Company.)

polyol and aldose profiles of human lens tissues have been determined to assess the importance of monosaccharides in cataract formation.[324]

The analysis of body fluids for volatile compounds by collection of the head space followed by GC with open tubular columns is widely practiced. The volatiles, which represent metabolic products and intermediates, yield a clinical fingerprint for correlation with the health or disease of the subject. The technique that has received the most attention is that introduced by Zlatkis and Liebich[325] and Teranishi et al.;[326] volatiles are concentrated on a porous polymer adsorbent by passing a flow of helium over the body fluid, most often urine, and through a water condenser into sampling tubes. The

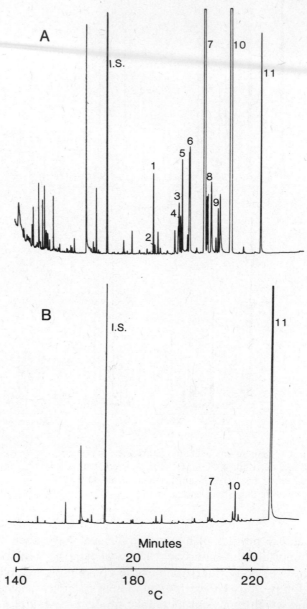

FIGURE 7.60. Chromatograms of polyols in cerebrospinal fluid from (A) normal adult male and (B) infant. Note: 50 m SE-52 glass column. Selected peak assignments: (1) arabinitol; (6) 1,5-anhydroglucitol; (7,10) glucose; (11) myo-inositol; (IS) dodecanol (internal standard). (Reproduced with permission from ref. 323. Copyright Elsevier Scientific Publishing Company.)

compounds are then desorbed thermally and condensed onto the cooled first part of an open tubular column before separation. Tenax has been shown to be a superior adsorbent over Porapak P or Carbosieve.[327]

Key components in the profiles of normal urine are acetone, 2-butanone, 2-pentanone, 3-hexanone, 3-penten-2-one, 4-heptanone, 2-heptanone, 2-nonanone, ethanol, n-propanol, dimethyl disulphide, pyrrole, N-methylpyrrole, and benzaldehyde.[328,329] The chromatograms of the urine of patients with *diabetes mellitus* show differences characteristic enough to allow screening by this method. There is an increase in the concentration of the alcohols, that is, ethanol, n-propanol, and n-butanol, and of the ketones cyclohexanone and 4-heptanone, depending on the severity of the disease.[329,330] Automated systems for recording and evaluating metabolic urine profiles have been proposed. In one definitive report, McConnell and Novotny[331] showed how repetitive sampling was possible with an apparatus comprising 2 mg of Tenax beads in a platinum gauze microbasket enclosed in an aluminum capsule. After sampling, the microbasket is cold-welded shut and carried into the injector so that the capsule base is punctured. The volatiles are then desorbed at 270°C for 45 min and condensed on a cryogenically cooled glass SF-96 column.

The reproducibility of this procedure[331] allowed pattern-recognition techniques to be applied.[332] After identification of profile components by FID and nitrogen-sensitive thermionic detection, FID data (retention times and peak areas relative to 6-undecanone and 10-nonadecanone previously added to the urine) were transferred to magnetic tape for computer processing. A method involving threshold logic units, a form of nonparametric pattern recognition, was applied and a normal versus pathological prediction rate of 93.75% was obtained for diabetics using 11 dimensions from up to 200 constituents. Increased concentrations of 4-heptanone, indole, and carvone were highly indicative of diabetic classification. In this work, a nonpolar phase was preferred even though most of the volatiles are polar, because of the long-term stability and ability of methylsilicones to elute heavy and highly polar compounds.[331] Other workers[333] have used some polar phases, however.

The Tenax adsorption procedure has also been applied to 5–10 cm^3 samples of human serum and plasma,[334,335] but the general availability of only small quantities of body fluids other than urine prompted Lee et al.[336] to propose the "transevaporator" method designed to extract the volatiles from samples in the 5–500 μL range with an average requirement of 20–50 μL. The sample is injected onto a short microextraction column packed with preconditioned adsorbent retained by glasswool plugs. The volatiles are stripped from the column with 2-chloropropane and transferred in the vapor phase to a glass bead collection column from which they are thermally desorbed into the precolumn of an open tubular column. Not only serum but breast milk, saliva, amniotic fluid, tissue homogenates, and sweat have been profiled in this way.[337]

The profiles of 50 µL serum of normal and diabetic patients have also been recorded and compared by this transevaporator technique.[338] In related work, Zlatkis recorded the chromatograms (e.g., Figure 7.61) of the volatiles from 70 µL samples of serum from normal and influenza-virus infected patients by the transevaporator method.[339] The chromatograms were interpreted by both manual and computer methods, and 85.7% of unknowns could be classified correctly using a two-peak ratio method.

As long as the polar metabolites are derivatized, GC with open tubular columns is an extremely useful method in xenobiotic chemistry. If detection is by MS, structural identification of metabolites is straightforwardly possible, in contrast with radiotracer methods in which complex degradative procedures must be carried out. Thus, Halpaap et al.[340] trimethylsilylated the rat urinary metabolites of biphenyl, both free and from conjugates after enzymatic hydrolysis. Analysis by GC-MS on an SE-30 glass column allowed identification of eight new metabolites: two hydroxybiphenyls, a trihydroxybiphenyl, a trihydroxymethoxybiphenyl, and 4,4'-dihydroxy-3-methyl-thiobiphenyl, a dihydrodiol and two hydroxydihydrodiols (Figure 7.62). Dihy-

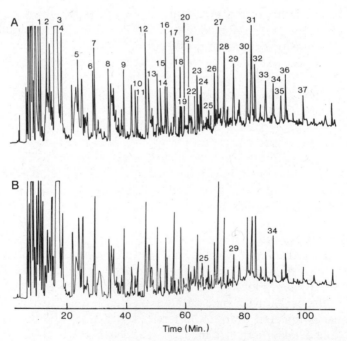

FIGURE 7.61. Chromatograms of organic volatiles from (A) virus-infected serum and (B) normal serum. Note: 100 m Witconal LA-23 stainless steel column, temperature program from 50 to 160°C at 1.5°C min⁻¹ after a 10 min isothermal period. Selected peak assignments: (7) 2-hexanone; (12) 2-heptanone; (16) 6-methyl-2-heptanone; (20) n-octanal; (27) 2-octenol; (33) acetophenone. (Reproduced with permission from ref. 339. Copyright Elsevier Scientific Publishing Company.)

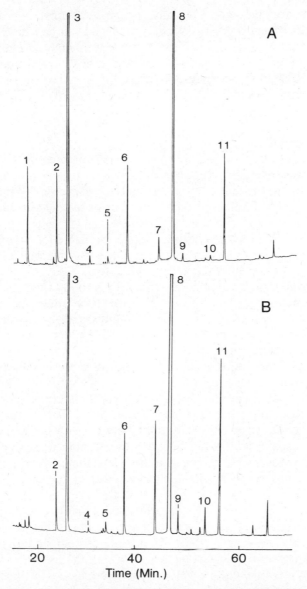

FIGURE 7.62. Chromatograms of the TMS derivatives of urinary biphenyl metabolites: (A) free metabolites and (B) metabolites extracted after hydrolysis with Glusulase. Note: 75 m SE-30 glass column, temperature program from 140°C at 2°C min^{-1}. Selected peak assignments: (2) 3-hydroxybiphenyl; (3) 4-hydroxybiphenyl; (6) 3,4-dihydroxybiphenyl; (8) 4,4'-dihydroxy-biphenyl. (Reproduced with permission from ref. 340. Copyright Elsevier Scientific Publishing Company.)

drodiols are important since they provide direct evidence for an arene epoxide-diol pathway.

The original foreign compound may also be determined in trace amounts; levels down to 10 ng cm^{-3} for 0.1 cm^3 rat plasma samples have been measured for the antioxidant 3-*tert*-butyl-4-hydroxyanisole.[341] The 2-*tert*-butyl isomer was added as internal standard, and *n*-hexane extracts were derivatized with heptafluorobutyric anhydride before injection onto an SE-52 fused-silica capillary with SIM detection at $m/z = 361$.

Drugs

The use of glass and fused-silica open tubular columns in the analysis by GC of drugs and related compounds especially in tissues and body fluids is rapidly increasing. Indeed the opinion has been expressed[342] that these columns will probably replace packed columns for all but the simplest separations.

Open tubular column GC methods are now available for a very wide range of drugs with an enormous variety of structures and pharmacological activities. Since many drugs have dangerous side effects especially if administered over long periods, their routine determination in body fluids is necessary if the pharmokinetics and possible toxicity are to be understood. Moreover, the identification and estimation of metabolites may be important since a rapidly metabolized drug may only be detectable via its metabolites, which may also be pharmacologically active.

Particular advantages, among many, in drug analysis with open tubular columns are reduced analysis time and greater specificity and, hence, simpler procedures for isolation from clinical samples.[343] In general, a straightforward clean-up step which eliminates lipids is all that is required, for example, partitioning of a plasma extract between hexane and an aqueous phase which usually leaves the lipids in the organic phase. Alternatively, some drugs may be extracted with an organic solvent from an aqueous medium if the pH of the latter is adjusted to a value at which the drug is negligibly ionized.

Although some drugs can be separated without functional-group modification, derivatization is necessary in many cases to increase volatility and to avoid thermal decomposition. The advent of cold on-column injection methods has greatly reduced complications from the latter, however, while the increasing deployment of well-deactivated glass or of fused-silica columns has made the problems of adsorption of polar drug substances less important.

Historically, drug analysis with open tubular columns began with the chromatography of steroids. Current work in the analysis of steroids of physiological origin in body fluids has already been discussed. Synthetic anabolic steroids are used as therapeutic drugs and doping agents since they stimulate the biosynthesis of proteins, and hence muscle tissue. However, because of side effects, especially the androgenic effects of these drugs, they are controlled substances and reliable analytical procedures for anabolic steroids in body fluids are being sought, particularly by GC with open tubular columns combined with mass spectrometry. For example, methandienone may be determined as a trimethylsilyl derivative in a cleaned-up extract of urine, while its metabolites have been identified[344] in this way as 7-epimethandienone and 6β-hydroxy-17-epimethandienone. The column was a well-deactivated OV-54 glass open tubular column with splitless injection.

Other drugs with important physiological properties are those used in the treatment of vascular diseases. Cyclandelate and its metabolite mandelic acid may be determined[345] in plasma, also as trimethylsilyl derivatives, on 25 m glass columns deactivated and coated with SE-30. Splitless injection allowed concentrations as low as 0.5 μg cm^{-3} plasma to be detected, and linear calibration graphs in the range 1–20 μg cm^{-3} were constructed with the aid of ethyl mandelate as internal standard; the standard deviation of repeated determinations was 6%. Analysis for the metabolites of the vasodilator isosorbide dinitrate and the 2- and 5-mononitrates, which also have physiological activity, requires[346] an open tubular column (25 m glass OV-17), since resolution from plasma components is inadequate on a packed column and concentrations are only between 1 and 20 ng cm^{-3}.

Even lower concentrations (100 pg cm^{-3}) of the chlorine-containing antihypertensive drug clonidine can be detected by electron capture if derivatization with pentafluorobenzyl bromide is carried out;[347] a cyclized derivative is, in fact, formed. OV-17 and SP-2360 are suitable phases. Glass open tubular columns have found important applications in the analysis of plasma for a number of β-andregenic blocking agents. These have been separated as TMS ethers (timolol with SIM MS detection),[348] heptafluorobutyryl derivatives (alprenolol and oxprenolol—EC detection),[349] or as trifluoroacetyl derivatives (pindolol).[350] The enantiomers of certain β blockers, for example, propranolol, were separated[351] on OV-225 open tubular columns as the diastereoisomeric N-trifluoroacetyl-1-prolyl and N-heptafluorobutyryl-1-prolyl amides after hydroxyl groups had been converted to TMS ethers.

Separations of enantiomers of optically active drugs and metabolites can also be carried out[352] on glass open tubular columns coated with the chiral phase Chirasil-Val. Such analyses are often important since enantiomers may exhibit quite different biological activities, for example, the sympathomimetic α-phenyl-β-aminoalcohols (ephedrine, phenylephrine, etc.) (Figure 7.63), D- and L-DOPA, or penicillamine. Reactive functional groups are converted to pentafluoropropionyl derivatives before analysis. Optical purity of drugs may be determined, the configuration of metabolites determined, and quantitative analyses made with the "unnatural" enantiomer as the internal standard.

Chirasil-Val was originally synthesized[353] to allow analysis of the optical purity of amino acids from peptides and has been used[354] to determine the

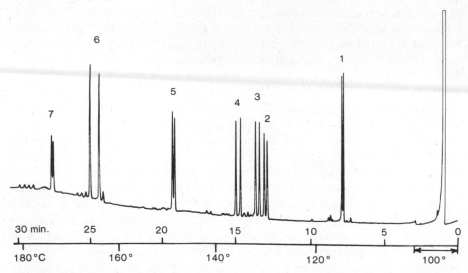

FIGURE 7.63. Chromatogram of the enantiomers of several sympathomimetic drugs and epinephrine metabolites as *N,O*-pentafluoropropionyl derivatives. Note: 20 m Chirasil-Val glass column, H₂ carrier gas. Selected peak assignments: (1) ephedrine; (2) suprifen; (5) metanephrine; (6) DOPA. (Reproduced with permission from ref. 352. Copyright Elsevier Scientific Publishing Company.)

configuration of isovaline as D-(resp.R) in a number of polypeptide antibiotics. *N*-Trifluoroacetyl amino acid *n*-propyl ester derivatives of DL-isovaline show excellent enantiomer resolution and separation from α-aminoisobutyric acid and other components.

A very large number of applications of open tubular columns in drug analysis involves psychoactive, analgesic, and narcotic drugs. The widespread abuse of such drugs also adds to the importance of this field, since overdoses are common. Rapid qualitative and quantitative analyses are often required since some drugs may be fatal at the 10 μg cm⁻³ blood level, while others may be less damaging at concentrations near 100 μg cm⁻³ and may be used medicinally at the 10 μg cm⁻³ level, for example, barbiturates.

The latter compounds are generally prescribed as water-soluble sodium salts, from which on acidification of solutions the free acids are formed and may be extracted with organic solvents. Analysis of the underivatized acids on all but the most carefully deactivated open tubular column surfaces has been complicated by adsorption. Thus Kinberger et al.[355] found very low responses for barbiturates on commercially available quartz open tubular columns, but Sandra et al.[356] were able to chromatograph free barbiturates on HCl-leached glass, silylated at 400°C and coated with OV-1 (Figure 7.64). Most open tubular column separations of barbiturates are carried out for derivatives (commonly *N,N*-dialkyl derivatives simply prepared by heating with iodoalkane in the presence of potassium carbonate—the Claisen method).[355] Detection is particularly sensitive with the NPD (Figure 7.65). A precision of ±5% has been shown with allobarbital as internal standard. Dunges et al.[357] have also recommended the Claisen method and reported the preparation of methyl, allyl, and benzyl derivatives (by "microrefluxing" of trace quantities of barbiturates with RI and K_2CO_3) followed by separation on PPG, FFAP, and SF-96 coated glass open tubular columns.

Derivatization of barbiturates can also be carried out in the heated injection port of the chromatograph by simultaneous injection of sample and reagent. For example, splitless injection of barbiturates in ethyl acetate solution with dimethylformamide dimethylacetal yields acetal derivatives for which good resolution and peak shapes on 20 m OV-1 and SE-30 glass columns have been reported.[358]

Flash-heater derivatization has also been used in the preparation of TMS derivatives of narcotic drugs; 1 μL of N,O-bis(trimethylsilylacetamide) was

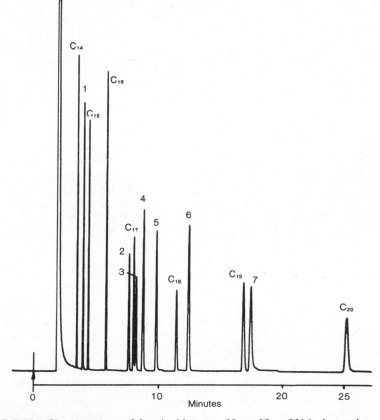

FIGURE 7.64. Chromatogram of free barbiturates. Note: 25 m OV-1 glass column, 160°C isothermal temperature, H_2 carrier gas at 3.5 mL min^{-1}. Selected peak assignments: (1) barbital; (2) amobarbital; (3) pentobarbital; (4) barbitol; (5) secobarbital; (6) hexobarbital; (7) phenobarbital. (Reproduced with permission from ref. 356. Copyright Dr. Alfred Huethig Publishers.)

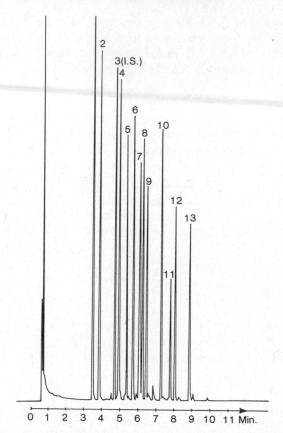

FIGURE 7.65. Chromatogram of an alkylated serum barbiturate standard obtained using a nitrogen-selective detector (NPD). Note: 25 m SP-2100 fused-silica column, temperature program from 110 to 230°C at 10°C min^{-1}, 20 μg cm^{-3} of each barbiturate. Selected peak assignments: (1) metharbital; (2) barbital; (3) allobarbital (internal standard); (6) amobarbital; (8) vinbarbital; (10) hexobarbital; (13) heptabarbital. (Reproduced with permission from ref. 355. Copyright Elsevier Scientific Publishing Company.)

heated[359] with 1 μL of an ethyl acetate solution containing 250 ng of a variety of harmful drugs in the injection port at 250°C. Excellent peak shapes and separations were achieved on a glass SE-30 column (Figure 7.66) and analyses of illicit heroin reported.[359] On the other hand, certain of the drugs may be determined on a silica column without derivatization (see below). Edlund[360] has shown how the opiates, morphine, 6-acetylmorphine, and codeine, can be determined at the nanogram per cubic centimeter level after acylation with pentafluoropropionic anhydride. The derivatives are separated on a glass column with falling glass needle injection and electron capture detection. Silyl derivatives were also used in the analysis of the hypnotics, methyprylon and pyrithildion, but here they were formed on the needle tip of an all-glass solid sample injector.[361] However, no derivatization was necessary for the open tubular column GC analysis of the intravenous anaesthetic etomidate,[362] the narcotic analgesic keto-bemidone,[363] or the anaesthetic drug of abuse phencyclidine.[364] The first of these involved nitrogen-selective detection and splitless injection,[362] as was used for the anticonvulsant hypnotic nitrazepam;[365] here detection was by electron capture.

A variety of analyses by open tubular column GC are available for psychotropic drugs. Indeed the current power of these procedures in analyses for compounds apparently unsuitable for GC is well illustrated by the determination[366] of the polar stimulant pemoline (2-imino-4-oxazolidinone) which is unstable in body fluids. The compound is hydrolyzed to oxazolidinedione, extracted, and then N-methylated. The derivative is analyzed on a short PPE/OV-1 open tubular column with splitless injection and nitrogen-selective detection. The method is five times more sensitive than with packed columns. The NPD, however, is not entirely specific in the determination of nicotine in the blood of smokers, and MS detection with SIM is recommended.[367] Kinberger et al.[368] have replaced a time-consuming and laborious procedure for the determination of a variety of stimulants in urine with a much more rapid method in which a buffered urine sample is extracted with

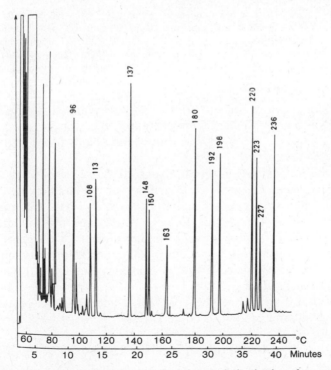

FIGURE 7.66. Chromatogram obtained by flash-heater derivatization of narcotic drugs with BSA. Note: 20 m SE-30 glass column. Selected peak assignments: (96) amphetamine; (113) phenmetrazine; (150) pethidine; (192) methadone; (223) ethylmorphine; (236) heroin. (Reproduced with permission from ref. 359. Copyright Elsevier Scientific Publishing Company.)

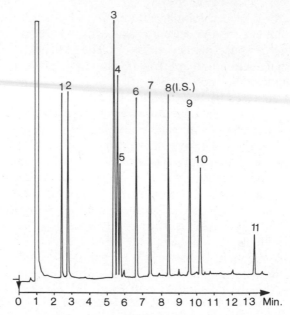

FIGURE 7.67. Chromatogram of an underivatized urine drug standard. Note: 25 m SP-2100 fused-silica column, temperature program from 110 to 220°C at 16°C min⁻¹ after a 4 min isothermal period, 5 μg cm⁻³ of each drug as free base. Selected peak assignments: (1) amphetamine; (3) nicotine; (4) chlorophentermine; (9) methylphenidate; (10) caffeine; (11) methadone. (Reproduced with permission from ref. 368. Copyright Elsevier Scientific Publishing Company.)

isopropanol/chloroform and analyzed without derivatization on a quartz SP-2100 fused-silica open tubular column (Figure 7.67). Calibration graphs were linear in the range 1–10 μg cm⁻³ with detection limits in the region of 1 μg cm⁻³.

The antidepressant psychotropic drugs comprise an important field for open tubular column GC analysis. N-trifluoroacetyl and heptafluorobutyryl derivatives, often with nitrogen-sensitive detection, have been used especially in the chromatography of doxepin[369] (*cis*- and *trans*-isomers), trancylpromine,[370] and various tricyclic antidepressants[371] on glass columns. Bailey et al.[372] showed how derivatization of the antidepressant nomfenisine to the heptafluorobutyrate followed by analysis on an OV-101 glass column with a nitrogen-selective detector allowed determinations at concentration in the region of 2 ng cm⁻³. The propyl and butyl analogs of nomfenisine were used as internal standards (Figure 7.68).

Insect Pheromones and Hormones

Pheromones are released by insects to transmit messages to individuals of the same species. There is considerable interest in these compounds in the monitoring of insect pests and in their control. The small sample sizes necessary for open tubular column GC make this technique particularly important in the analysis of insect pheromones which are often available only in minute quantities. Moreover, the compounds in question are usually unsaturated and hence can exist as a number of difficult-to-resolve geometric isomers. Thus many insect sex attractions are alkadienyl compounds, and synthesis poses problems, since often the slightest amount of geometric-isomer impurity will reduce or eliminate activity. On the other hand, the presence of an isomer may be necessary to bring about attraction synergistically.

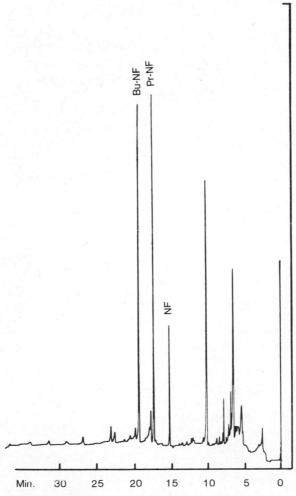

FIGURE 7.68. Chromatogram of nomfenisine (NF) in a plasma extract. Note: 50 m OV-101 glass column, 200°C isothermal temperature, 12.0 ng cm^{-3} of nomfenisine in plasma, nitrogen-selective detection (NPD). Butyl (Bu-NF) and propyl (Pr-NF) derivatives are internal standards. (Reproduced with permission from ref. 372. Copyright Elsevier Scientific Publishing Company.)

If one of the double bonds is not terminal, there are four possible isomers (E,E; Z,Z; E,Z,; and Z,E), and polar stationary phases are required for their separation. For example, the separation of alkadien-1-ol acetates including many naturally occurring sex attractants on a 15 m diethylene glycol succinate column has been reported[373] (Table 7.5). The more polarizable diene derivatives have longer retention times than corresponding saturated 1-ol acetates, while conjugated dienes are retained relative to the corresponding nonconjugated dienes. The sequence of elution of conjugated diene-1-ols and acetates is Z,E; E,Z; Z,Z; and E,E, both on this phase[373] and on FFAP-coated[374] columns, and differs from that of nonconjugated diene-1-ol acetates. Thus, Schäfer separated[374] all four 10,12-hexadecadiene-1-ols on a 50 m FFAP column and found that bombykol, the first sex-specific insect pheromone reported and thought to be an E,Z isomer, is in fact a mixture of about 84% (E,Z)-10,12-hexadecadiene-1-ol and 16% of the (E,E). Glass open tubular columns coated with cholesteryl cinnamate liquid crystal phase have also been recommended[375] in the separation of insect pheromone isomers, for example, the 11,13-hexadecadienals.

While each female *Bomby mori* contains about 200 ng bombykol,[374] much smaller quantities of sex phenomones are generally present in insects, and the small samples necessary for open tubular column GC are a particular advantage. Thus, Buser and Arn[376] analyzed cleaned-up extracts of the abdominal tips of the codling moth (*Laspeyresia pomonella*) and the European grapevine moth (*Lobesia botrana*) on a 50 m Ucon 50 HB 5100 glass column

TABLE 7.5.   Retention data for diene-1-ol acetates on 15 m DEGS column at 135°C.

Acetates of	Retention time[a] (t_R, minutes)	Relative retention time[b]
1-Tetradecanol	9.7	1.00
(Z,E)-7,9-Dodecadien-1-ol	10.6	1.10
(Z,E)-8,10-Tridecadien-1-ol	15.0	1.60
(Z,E)-9,12-Tetradecadien-1-ol	16.3	1.75
(Z,E)-9,11-Tetradecadien-1-ol	21.5	2.34
(E,Z)-9,11-Tetradecadien-1-ol	22.5	2.45
(Z,Z)-9,11-Tetradecadien-1-ol	23.0	2.51
(E,E)-9,11-Tetradecadien-1-ol	24.0	2.63
(Z,E)-7,11-Hexadecadien-1-ol	27.1	2.98
(Z,Z)-7,11-Hexadecadien-1-ol	28.0	3.08
(E,Z)-10,12-Pentadecadien-1-ol	32.3	3.57
(E,E)-10,12-Pentadecadien-1-ol	34.5	3.82

[a] Measured from the point of injection.

[b] For the calculation of relative retention, the retention times were adjusted for gas holdup measured as the retention time of methane. 1-Tetradecanol acetate was used as the standard.

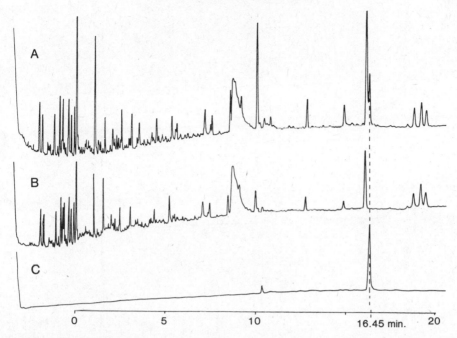

FIGURE 7.69. Chromatogram of (A) female codling moth extract, (B) male codling moth extract, and (C) synthetic (E,E)-8,10-dodecadien-1-ol. Note: 50 m UCON 50-HB-5100 glass column. (Reproduced with permission from ref. 378. Copyright Elsevier Scientific Publishing Company.)

coupled to a quadrupole mass spectrometer via a platinum capillary. Splitless injection was employed because of the very small amount of pheromone available (extracted from 250–500 moths, rather than the many thousands often employed in chemical studies). Resolution of all four isomers of 7,9-dodecadien-1-ol acetate was possible. The E,Z isomer was identified as the sex pheromone of the European grapevine moth (1.6 ng per insect) while (E,E)-8,10-dodecadien-1-ol was identified as that of the codling moth (Figures 7.69 and 7.70). Both compounds were present in high isomeric purity in the female insect and were absent in the corresponding males.

Reliable quantitative analyses for the three known insect juvenile hormones (JH-I, JH-II, and JH-III, Figure 7.71) have also been carried out[377] by open tubular column GC. SE-30 was found to be the preferred stationary phase (Figure 7.72) and the 10-heptafluorobutyryl-11-methoxy derivatives with electron capture detection the most suitable derivatives. The three hormones could be determined simultaneously in quantities as low as 0.2 ng; geometric isomers of the hormones were used as internal standards (Figure 7.72). The method was confirmed by the analysis of haemolymph from three insect species.

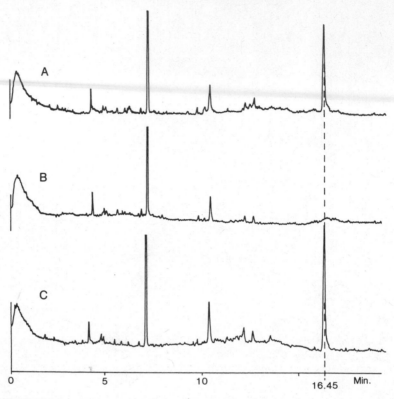

FIGURE 7.70.   Mass fragmentograms (*m/e* 182) of (A) female codling moth extract, (B) male codling moth extract, and (C) synthetic (E,E)-8,10-dodecadien-1-ol. Note: same column as in Figure 7.69. (Reproduced with permission from ref. 376. Copyright Elsevier Scientific Publishing Company.)

Microorganisms

The identification and classification of microorganisms by open tubular column GC rather than by more conventional biochemical and morphological means is readily achieved. Two approaches have been applied: (a) the analysis of metabolic products from the growth of microorganisms and the chemical composition of their cells and (b) characterization of bacteria by pyrolysis GC. The method also finds numerous applications in the analysis of specific bacterial metabolic products of biochemical interest.

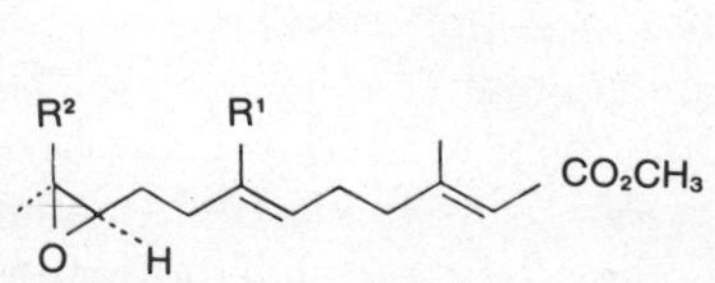

$$
\begin{array}{ll}
\text{JH-I} & R^1 = C_2H_5; \quad R^2 = C_2H_5 \\
\text{JH-II} & R^1 = CH_3; \quad R^2 = C_2H_5 \\
\text{JH-III} & R^1 = CH_3; \quad R^2 = CH_3
\end{array}
$$

FIGURE 7.71.   Structural formulas of natural juvenile hormones.

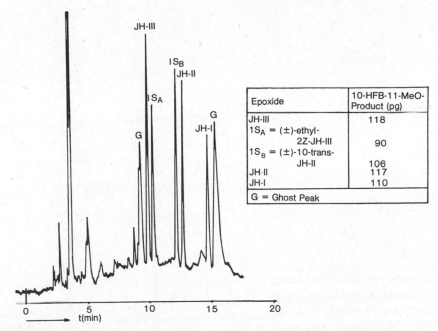

Epoxide	10-HFB-11-MeO-Product (pg)
JH-III	118
1S$_A$ = (±)-ethyl-2Z-JH-III	90
1S$_B$ = (±)-10-trans-JH-II	106
JH-II	117
JH-I	110
G = Ghost Peak	

FIGURE 7.72. Chromatogram of insect juvenile hormones. Note: 25 m SE-30 column, 216°C isothermal temperature, 10-heptafluorobutyryl-11-methoxy derivatives, electron capture detection (ECD). (Reproduced with permission from ref. 377. Copyright Elsevier Scientific Publishing Company.)

Anaerobic bacteria may be detected and identified by analysis of both acidic and neutral products of fermentation. Identification data for the fermentation products of a large number of named strains are known, and a convenient analytical procedure (Figure 7.73) involves separation on a 30 m glass or fused-silica column coated with SP 1000.[378] Nonvolatile acids and formic acid are analyzed after methylation with a temperature program of 90–160°C; volatile acids may be chromatographed isothermally at temperatures between 130 and 180°C; alcohols, aldehydes, ketones, diketones, diethyl ether, and acetoin can also be resolved isothermally at temperatures between 90°C and 170°C; volatile neutral compounds and acids from C$_4$ upwards are analyzed with a temperature program from 110 to 170°C. Significant reductions in analysis time in comparison with similar analyses on packed columns are observed.[378]

Many bacteria, however, produce only trace amounts of short-chain acids, and other profiling methods are required. The cellular fatty acids can be helpful in these circumstances and, indeed, provide an important general means of identifying and classifying bacteria. An instructive application concerns[379] the agents of Legionnaires' disease, newly discovered bacteria that differ from others in the large amounts of branched-chain acids contained in their cells. Open tubular columns are especially useful here, since

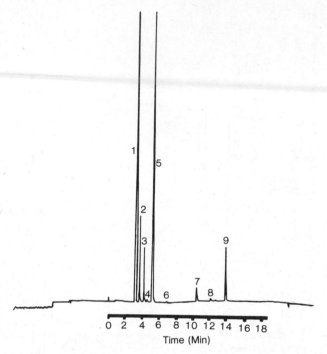

FIGURE 7.73. Chromatogram of an ether extract of acidified supernatant of *C. tetani*. Note: 30 m SP-1000 glass column, temperature program from 110 to 170°C at 10°C min^{-1} after a 7 min isothermal period. Selected peak assignments: (2) ethanol; (3) 1-propanol; (4) 2-methyl-1-propanol; (5) 1-butanol; (6) 1-pentanol; (9) *n*-butyric acid. (Reproduced with permission from ref. 378. Copyright Elsevier Scientific Publishing Company.)

on packed columns with nonpolar stationary phases (i.e., OV-101, OV-1, SE-30) positional isomers of acids with the same carbon–chain length are not resolved. However, on a 50 m SP 2100 fused-silica open tubular column, there is baseline separation of the 16-carbon methyl ester series and three 17-carbon esters.[379] This is vital since the *iso*-16:0 acid is the major component of *Legionella pneumophilia*, but comprises only 15% of the acids of a second *Legionella* species *L. bozemanii* (Figure 7.74). A third species, *L. micdadei*, differs from the other two in the presence of *anteiso*-17-carbon monoenoic acid.

Tuberculostearic acid [(R)-10-methyloctadecanoic acid] is a characteristic constituent of microorganisms of the order *Actinomycetales*.[380] Thus, pulmonary tuberculosis may be diagnosed by analysis of a patient's sputum after digestion with aqueous NaOH, neutralization, lyophilization, and extraction with chloroform/methanol. After evaporation, the organic residues were esterified and analyzed on a 25 m glass OV-17 column coupled to a mass spectrometer.[380] Both chemical ionization- and electron-impact selective-ion monitoring allowed the demonstration of the presence of tuberculostearic acid in a patient with pulmonary tuberculosis (Figure 7.75).

Less specific chromatographic profiles of cell constituents can be used to identify microorganisms. Martinez et al.[381] showed that lyophilized cells of different yeast genera could be recognized from profiles of the products

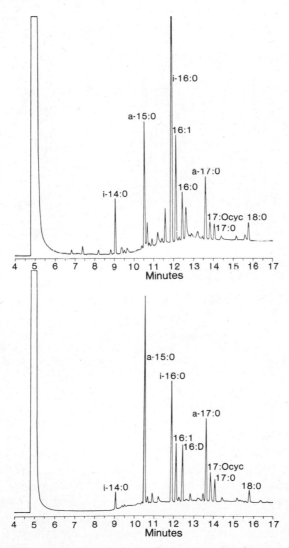

FIGURE 7.74. Chromatograms of methyl esters from two species of *Legionella*: (A) *Legionella pneumophilia* and (B) *Legionella bozemanii*. Note: 50 m SP-2100 fused-silica column. (Reproduced with permission from ref. 379. Copyright Elsevier Scientific Publishing Company.)

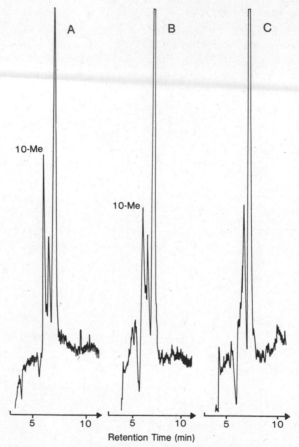

FIGURE 7.75. Mass fragmentograms of sputum specimens from patients (A) with tuberculosis, (B) without tuberculosis but with 50 pg of standard tuberculostearic acid (10-Me) added, and (C) without tuberculosis. Note: 25 m OV-17 glass column, single-ion monitoring at m/z 312. (Reproduced with permission from ref. 380. Copyright Elsevier Scientific Publishing Company.)

of silylation of the carbohydrates, sterols, fatty acids, and so on, which they contain.

Bacteria may be identified not only from the acids, and so on, contained in their cells and culture media, but from the high-resolution profiles of volatile organic compounds which they produce. Three different strains isolated from spoiled uncooked chicken were grown in a culture medium, and the volatile compounds produced were concentrated on a porous polymer precolumn.[382] Analysis was on an 80 m SF-96 glass open tubular column coupled to a mass spectrometer. Profiles unique to each of the microorganisms studied were obtained reproducibly (Figure 7.76), so that the bacterium responsible for spoilage could be identified.[382] Analysis of the "fruity" aroma from cultures of *ceratocystis moniliformis* showed, also, that these could be sources of food chemicals.[383]

Curie-point pyrolysis with open tubular columns represents a possible means of differentiating bacteria. Specific and reproducible pyrograms containing between 150 and 200 peaks are produced when bacteria are heated to 400°C and the products are separated on a 50 m glass FFAP column.[384] Early work suggested that distantly related strains such as *Neisseria meningitides* and *Neisseria sicca* could be distinguished.[385] More recently, detailed comparisons between pyrograms from eight different bacterial strains suggested[384] that they are difficult to differentiate, even if an external standard method is employed, and the chromatograms are treated by computer methods.

Finally, a good example of the use of open tubular column GC in the analysis[386] of specific products of a microorganism is that for zearalenone, a mycotoxin with oestrogenic effects produced by *Fusarium* strains and often present in grain. Analysis for zearalenone as trimethylsilylether (TMS)

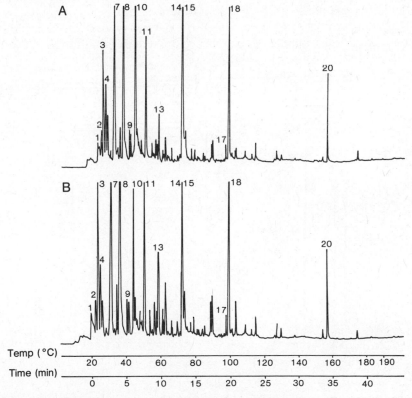

FIGURE 7.76. Chromatograms of the volatiles produced by a *Moraxella* oxidative bacteria culture illustrating the degree of reproducibility. Note: 80 m SF-96 glass column. Selected peak assignments: (7) methylisobutyrate; (8) dimethyldisulfide; (11) *n*-butylacetate; (14) benzaldehyde; (15) trimethyltrisulfide; (18) 1-undecene. (Reproduced with permission from ref. 382. Copyright American Society for Microbiology.)

and methyloxime-TMS derivatives is carried out on a 57 m SE-52 column, with solids injection, for ethyl acetate extracts of corn, cleaned up by TLC. This procedure, based on positive identification from the retention times of the TMS and MO-TMS derivatives (both *syn* and *anti* epimers) being identical with co-injected samples, provides an excellent alternative to spectroscopic detection or packed-column GC. Detection limits of 100 ppb in corn have been quoted.[386]

Nucleosides

The component units of RNA are ribose-derivatives of the purine and pyrimidine bases, adenine, cytosine, guanine, and uracil. These nucleosides are of high intrinsic interest, but analytical interest as far as open tubular column GC is concerned has centered largely on the cytokinins which are adenines and adenosines with substitutents at the 6-amino nitrogen. The presence of a variety of labile groups in such molecules makes chromatography in the gas phase a formidable problem.

However, Claeys et al.[387] separated the permethyl derivatives of *cis*- and *trans*-zeatin, and isopentenyladenine, extracted from cultures of the bac-

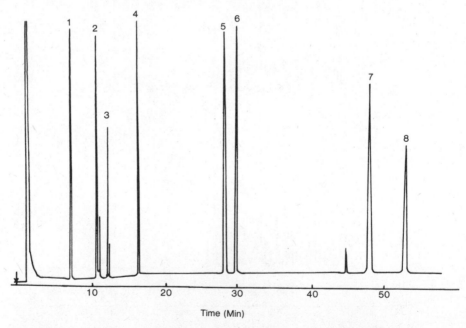

FIGURE 7.77. Chromatogram of a mixture of silylated cytokinins. Note: 30 m SE-54 fused-silica column, temperature program from 200 to 265°C at 4°C min^{-1}. Selected peak assignments: (1) isopentenyladenine; (2) dihydrozeatin; (3) *trans*-zeatin; (4) phloretin; (5) dihydroribosylzeatin; (8) *trans*-2-methylthioribosylzeatin. (Reproduced with permission from ref. 388. Copyright Elsevier Scientific Publishing Company.)

terium *Agrobacterium tumefaciens*, on a 50 m SE-30 glass column at 220°C. Detection was by selective-ion monitoring mass spectrometry with methane as chemical ionization reagent gas; derivatization was by the methylsulfinyl carbanion technique. Kemp and Andersen[388] preferred to separate the cytokinin bases and ribonucleosides as trimethylsilyl derivatives, and found the inertness of a 30 m SE-54 fused-silica column programmed from 200 to 265°C gave little degradation (Figure 7.77).

Open tubular column GC separation of low-volatile polar purinyl nucleosides prepared from riburonic acid derivatives and silylated purines has been achieved[389] on well-deactivated glass columns with on-column injection. The four linkage isomers of each nucleoside could be resolved and were identified by GC-CIMS.

7.6   GEOCHEMISTRY AND COSMOCHEMISTRY

The aim of organic geochemistry is to elucidate the origin and reconstruct the geological history of the organic matter in the geosphere. The economic importance of petroleum has made this fuel a main target of research, but there is also significant activity in the areas of coal, oil shale, and tar sands, and recent sediments are analyzed to discern the first stages of geochemical change. The main thrust of organic geochemistry is in the analysis of the lipids of ancient and recent sediments which can be both indicators of the nature of the original organic material and the process of alteration (temperature, pressure, or biodegradation) to which this material has been subjected. Geochemical markers are defined as compounds that retain the carbon skeleton of the original plant or animal biochemicals.

Gas chromatography with open tubular columns is very widely applied in geochemical studies (see also Section 7.3), most often in the separation of the extremely complex mixtures of hydrocarbons which result from the degradation of organic material over geological time. Particular attention is given to alkanes, but more recently there have also been a large number of reports of analyses of aromatic hydrocarbons; oxygenated compounds such as carboxylic acids, alcohols, and lactones may also have importance as geochemical markers.

The saturated hydrocarbon fraction separated from a sediment extract by column chromatography may contain a range of alkane types (Table 7.6). *n*-Alkanes are ubiquitous[390] with distributions generally in the range C_{10}–C_{36} (Figure 7.78), although *n*-alkanes up to C_{60} have been detected in coal extracts in low concentration and beyond C_{60} in petroleum (Figure 6.4). Analyses for *n*-alkanes are made, generally for the whole saturated fraction or an adductable fraction of this, on medium length (~20 m) glass open tubular columns with nonpolar (OV-1, SE-30, etc.) or slightly polar (SE-52) phases.[391] This allows determination of the predominance of odd-numbered over even-numbered *n*-alkanes (the carbon preference index (CPI)) and an

TABLE 7.6. Hydrocarbon types in sediments.

Type	Typical structure	Formula	Analytical conditions
Saturated			
n-Alkanes		C_nH_{2n+2}	
Monomethyl alkanes		C_nH_{2n+2}	
Acyclic isoprenoids		C_nH_{2n+2}	GC-MS with OV-1 or SE-52 open tubular columns
Steranes		C_nH_{2n-6}	
Triterpanes		C_nH_{2n-8}	
Aromatic			
Monoaromatics			GC-MS with OV-101 or Dexsil 300 or 400 open tubular columns
Diaromatics			
PAH			GC-MS with SE-52 open tubular column

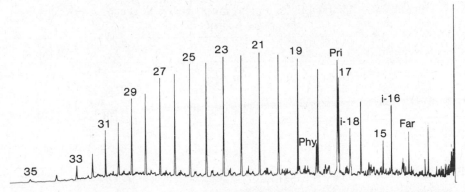

FIGURE 7.78.   Chromatogram of the saturated hydrocarbon fraction of a crude oil from Vienna Basin. Note: 20 m SE-52 glass column, temperature program from 80 to 250°C at 3°C min^{-1}. Numbered peaks represent the number of carbon atoms in the *n*-alkanes. Phy, Pri, *i*-18, *i*-16, and Far are acyclic isoprenoids (phytane, pristane, norpristane, 2,6,10-trimethyltridecane, and farnesane). (Reproduced with permission from ref. 394. Copyright Elsevier Scientific Publishing Company.)

indication of sediment maturity,[390] although injection without mass discrimination (cold on-column injection) is, of course, necessary.

The above columns allow complete separation of *n*-alkanes from singly branched alkanes[391] (which are eluted before the *n*-alkanes, on polyphenyl ether columns[392] in the sequence 7,6,5,4,2 and 3-methyl; the last two predominate) and from the more abundant acyclic isoprenoidal hydrocarbons phytane (C_{20}), pristane (C_{19}), norpristane (C_{18}), 2,6,10-trimethyltridecane (C_{16}), farnesane (C_{15}), and 2,4,6-trimethylundecane[390,391] (Figure 7.78). The latter originate from the degraded phytyl side chain of the chlorophyll of the original plant material. The ratio of pristane to phytane is an important indicator not only of sedimentation conditions,[393] but also of maturation, for example, of coals.[391] The pristane/*n*-heptadecane ratio may also be an indicator of depositional origin.[393,394] Long-chain isoprenoids (mainly C_{21}–C_{39}, but also up to C_{45}) from oil have been analyzed on a 50 m Apiezon L open tubular column.[395]

Sterols and triterpenes are common constituents of living organisms and are present in bottom muds and recent sediments. They appear in ancient geological samples as fully reduced steranes and triterpanes, or to a lesser extent as partially or fully aromatized structures. The polycyclic alkanes are of particular interest since the structure and stereochemistry is often preserved or altered in a specific and informative way.[390,391]

Full identification of steranes and triterpanes in the total alkane and branched/cyclic fractions of sediment extracts is generally achieved by open tubular column GC-MS with nonpolar columns.[391] Mass spectrometry fragmentations for cyclic alkanes usually occur at more than one bond (Table 7.7). The very intense peak characteristic of steranes at *m*/*z* 217 allows ready

TABLE 7.7.   Characteristic ions in mass spectra of geochemical marker
saturated hydrocarbons.[a]

Alkane type	Ion m/z
Acyclic isoprenoids	113, 183
Hopanes	191
Molecular ions for pentacyclic triterpanes and hopanes	370, 384, 398, 412, 426, 440, 454, 468, 482
5β steranes	217
Molecular ions for 5β steranes	260, 288, 330, 372, 386, 400
Rearranged 5β steranes	259
4-Methyl 5β steranes	231
Nuclear methylated rearranged steranes	273

[a] After I. Rubinstein, O. P. Strausz, C. Spyckerelle, R. J. Crawford, and D. W. S. Westlake, *Geochim. Cosmochim. Acta* **41**, 1341 (1978).

differentiation between steranes and triterpanes, which generally give rise to a peak at m/z 191. Typical fragments of geochemical markers are listed in Table 7.7.

The computerized GC-MS technique of multiple-ion cross-scanning[396] superimposes a mass histogram of a particular fragment ion on the total ion current chromatogram, so as to yield the distribution of a particular class of compound (Table 7.7). Several cross-scans are recorded on the same histogram, as is illustrated in Figure 7.79, which shows the parent and fragment ion maps of hopanes from crude oil. This method has had application in studies of the Alberta tar sands[396] and the crude oils from which they originate by maturation, migration, and so on, and in investigations of oil biodegradation.[397]

The hopane series of triterpanes are among the most widespread of the geolipids derived from terpene sources[398] and often comprise the major part of the branched/cyclic saturated hydrocarbon fraction of sediments, and act as universal geochemical markers. During maturation, the unstable 17βH isomer of a given hopane is converted to the more stable 17αH form.[399] These epimers are readily resolved on medium length nonpolar columns (Figure 7.80), as are the diastereoisomers of homohopane. However, high temperatures are required, and Dexsil 300 has been used as a column coating,[397] in addition to the more conventional polymethylsiloxanes.

The generation of tetra- and pentacyclic aromatics from triterpenoids during the maturation of sediments has also been the subject of detailed GC studies. As indicated during the earlier discussion of sediment pollution, a variety of hydrogenated methyl chrysenes and picenes are formed, and these may be separated on SE-52 coated open tubular columns and identified[400] by co-injection of standards and combined GC-MS. Monoaromatized steranes have[397] characteristic fragments at m/z 239 and 253. Another series of aromatic geolipids, including the substituted phenanthrenes, retene, and pimanthrene, are derived from the diterpenoids abietic and pimaric acids.[401] Retene and pimanthrene were also found in the aromatic fraction of the solvent extract[402] and vacuum distillate[179] of coal in comprehensive analyses by GC-MS (SE-52 open tubular column) in which aromatics ranging from benzene to dibenzopyrene were identified.

The oxygenated organic compounds of sediments have also received attention. For example, long-chain ketones (C_{36}–C_{39}) may be separated[403,404] on SE-52 or short OV-1 columns and identified by GC-MS. Alkanoic acids from bitumen,[405] and the alkanoic[202] and diterpenoid acids of lignite[406] have been analyzed as methyl esters on OV-1 columns.

If geochemical applications of open tubular column GC have been ex-

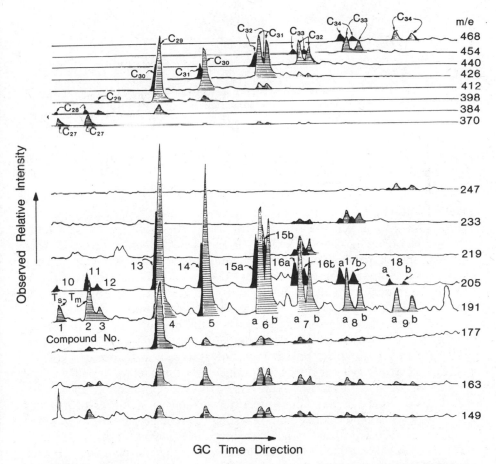

FIGURE 7.79. Mass fragmentograms of hopanes and methylhopanes in saturates of immature Jurassic oil. (Reproduced with permission from ref. 397. Copyright Pergamon Press.)

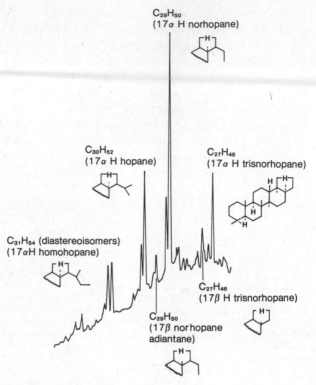

FIGURE 7.80. Section of a chromatogram of the total alkane fraction of Turkish asphaltite. Note: 50 m SE-30 (0.06 μm film) glass column.

tensive, and the area is one of rapid growth, rather few cosmochemical samples have so far become available for study. However, the possible contamination of moon samples with hydrocarbons of terrestrial origin was pointed out as a result of open tubular column analyses[407] of extracts from apparatus used to transport the samples. The organic volatiles of the Skylab 4 cabin were concentrated on Tenax and analyzed[408] on an Emulphor ON 870 coated nickel open tubular column with MS detection.

Analysis of extractable organic compounds of meteorites presents a considerable challenge for analytical methods because of the small amounts of material available and the possibility of contamination before and during analysis. Unequivocal evidence of indigenous (abiogenic) organic compounds has been adduced, however, from analyses of the amino acids present. Unlike biotic samples, these are often present as mixtures of diastereoisomers, while nonprotein amino acids may also be present. For example, the Murray meteorite was found[409] to contain 17 amino acids of which 7 were racemic mixtures, and 11 were nonprotein. Analysis on a 46 m UCON 75H 90,000 open tubular column of the N-trifluoracetyl-(+)-2-butyl esters allowed resolution of many of the diastereoisomeric pairs. Similar analyses have been made for extracts of the Murchison[410] and Orgueil[411] meteorites and, with extra precautions to avoid contaminants, the Mighei meteorite.[412]

The diastereoisomers of isovaline, *N*-methylalanine, and a number of other nonprotein acids were incompletely resolved on the above column.[413] Better resolution was observed[413] on Dexsil 400 columns and if different esterifying groups (up to *n*-octyl) and acyl groups were employed. More recent analyses[414] of extracts of a clean fragment of the Murchison meteorite by GC-CIMS on a 61 m Carbowax 20M open tubular column suggest that certain amino acids are not racemic. The *O,N*-dipentafluoropropionyl-(+)-2-butyl ester derivatives were used in this work.

7.7 FORENSIC APPLICATIONS

The potential of open tubular column GC in forensic science is very great, mainly because of the characteristic fingerprints of organic materials made available by the very high resolving power, and the small sample sizes required. These capabilities are especially valuable in analyses of flammable liquid fire accelerants, in drug identification, and in a variety of other profiling applications.[415]

Thus, the wide distribution and ready availability of liquid hydrocarbon fuels and flammable solvents make them commonly used in the acceleration of deliberate fires. Traces of these liquids often survive a fire and their identification is of vital importance in forensic work. Motor gasoline, kerosine, diesel fuels, mineral spirit, paint thinners, methylated spirit (industrial ethanol), and other industrial solvents must all be considered in this context. They are isolated by steam distillation or vacuum extraction. For volatile materials, head-space samples may be injected directly or preconcentrated on a Tenax GC trap.[416] An ingenious alternative injection technique uses a thin layer of carbon bonded to a ferromagnetic wire to adsorb vapors which are desorbed into the column via a splitless Curie-point pyrolysis inlet system.[158]

Preliminary profiling on a polar (e.g., Carbowax 20M) column reveals the presence of aromatic hydrocarbons, alcohols, and so on, since these are retained relative to the rapidly eluted alkanes under these conditions. However, fairly long (~50 m) columns coated with nonpolar phases are used in the analysis of petroleum products used as fire accelerants. These are operated isothermally at 60–70°C for samples containing alkanes up to C_{11}, or with temperature programming for less volatile materials.

About 30 major peaks have been identified in chromatograms of motor gasolines, and these are present in virtually all such samples.[415] Although the proportions vary, gasoline may be identified even if evaporation has occurred. Chromatograms of kerosine and diesel fuel are dominated by the homologous series of *n*-alkanes (C_8–C_{16} for kerosine, C_8–C_{26} for diesel). Evaporation brings about a shift to high-boiling components. Characteristic extra information can be derived from the peaks from isoprenoidal hydrocarbons (particularly pristane and phytane which partner the C_{17} and C_{18} n-alkane peaks, respectively) and the branched and cyclic alkanes eluted between the n-alkane peaks.[415]

Mineral spirit, a hydrocarbon solvent derived from petroleum with C_9, C_{10}, and C_{11} n-alkanes as major components, produces a highly characteristic chromatogram, and its identification in the head space of bedding and carpet debris from a house fire is illustrated in Figure 7.81. The more volatile components have been lost by evaporation, but there is sufficient correspondence to allow identification of this solvent as the accelerant beyond reasonable doubt. The similarity of chromatograms of debris from different materials shows that the peaks cannot originate from combustion products.

Other fire accelerants commonly used by arsonists include industrial solvents containing alcohols such as methylated spirit (ethanol plus methanol) and paint thinners (butanol and xylene isomers), and these are conveniently analyzed on a Carbowax 20M column.

Much current forensic work is directed toward drugs of abuse. Drug identification and analysis by open tubular column GC has been discussed in Section 7.5, but fingerprinting techniques are also important in dealing with chemically heterogeneous narcotics such as marijuana. The determination of whether cannabis samples confiscated at different locations have a common origin and the tracing of illicit marijuana samples to their geographical origin are both possible by a reproducible procedure.[417] The sample is ex-

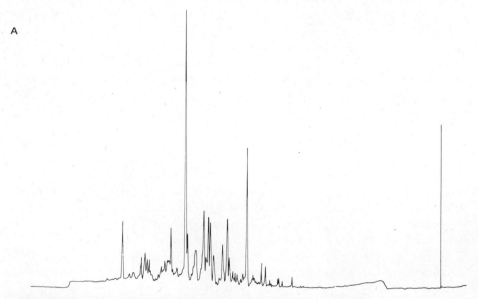

A

FIGURE 7.81. Chromatograms of head space of (A) bedding and (B) carpet from the scene of a fire. (C) is a chromatogram of "mineral spirit" under the same conditons. Note: 50 m SE-52 glass column, 60°C isothermal temperature.

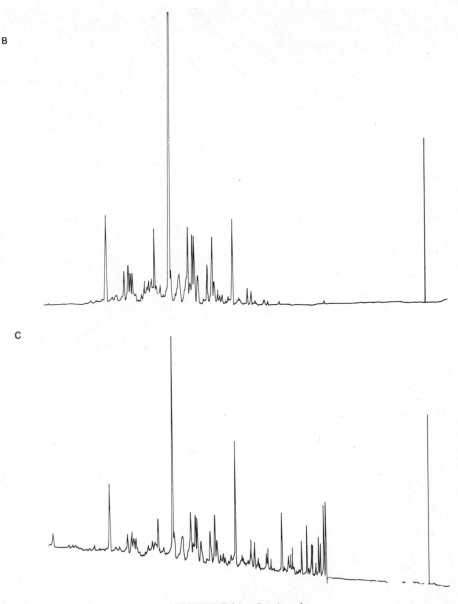

FIGURE 7.81 *Continued*

tractcd with cyclohexane and the extract is partitioned between cyclohexane and nitromethane. The cyclohexane fraction contains about 70 mainly non-polar constituents, of which 38 have been identified by GC-MS, including the cannabinoids, although most of those compounds remain in the nitro-methane. Chromatographic profiles of the cyclohexane fraction on a short

(~11 m) SE-52 coated glass column (Figure 7.82) illustrate the very clear differences exhibited by marijuana samples from different sources.

Open tubular column profiling methods may also be applied in a multitude of other areas of forensic interest, in fact, whenever high sensitivity is required in the analysis of even moderately volatile compounds, or when the mixture is complex. Examples include the analysis of mixtures of commer-

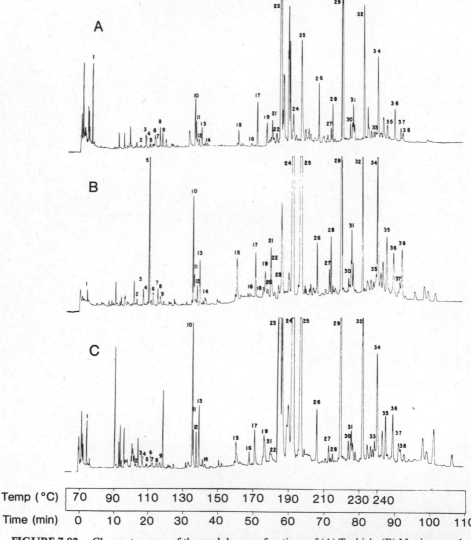

FIGURE 7.82. Chromatograms of the cyclohexane fractions of (A) Turkish, (B) Mexican, and (C) Indiana-grown Mexican marijuana. Note: 11 m SE-52 glass column. Selected peak assignments: (10) 5(9-ketodecyl)-2-furfuraldehyde; (23) cannabidiol; (24) Δ⁹-tetrahydrocannabinol; (25) cannabinol; (29) nonacosane; (32) hentriacontane; (34) α-amyrin. (Reproduced with permission from ref. 417. Copyright American Chemical Society.)

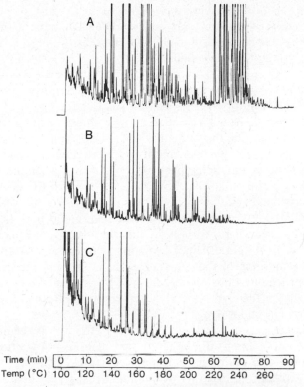

Time (min): 0 10 20 30 40 50 60 70 80 90
Temp (°C): 100 120 140 160 180 200 220 240 260

FIGURE 7.83. Chromatograms of the nitromethane extract of three different engine oils. Note: 22 m SE-52 glass column, nitrogen-selective NPD. (Reproduced with permission from ref. 148. Copyright American Chemical Society.)

cial and military explosives at the low picogram level on fused-silica OV-101 columns,[418] and the fatty acids obtained by saponifying cooking fats[419] (for comparison with acids from the stomach contents of a murder victim). The acids were converted to tetramethylammonium salts, which on pyrolysis in the injector yielded a chromatogram of methyl esters.

Element-specific detectors can also provide useful information. For example, small amounts of engine oil may be transferred to the victims of hit-and-run motor vehicle accidents or may be spilled at the scene of a crime. Polynuclear aromatic compounds (PAC) are characteristic of the combustion processes occurring in a particular engine and can provide a valuable fingerprint[148] especially if nitrogen-selective detection is used to highlight differences among nitrogen-containing PAC (Figure 7.83).

7.8 PYROLYSIS AND REACTION GAS CHROMATOGRAPHY

A variety of analyses have been reported in which a chemical reaction is first carried out in the injector of the open tubular column. The most frequently used reaction is, of course, pyrolysis, especially of polymeric materials. Thermal degradation yields very complex products, with a wide range of volatilities and polarities, and there is increasing interest in the use of open tubular column pyrolysis GC.

Various pyrolysis systems were discussed in Chapter 4, and their operation in Chapter 5, but Meuzelaar et al.[420] and de Leeuw et al.[421] have described a Curie-point pyrolysis system suitable for use with open tubular columns, which could be automated to allow continuous unattended analyses of up to 24 samples. The instrument is of general applicability, and the pyrolysis GC of insoluble organic sediments was described.[421] Splitless injection was employed here, but Schmid et al.[422] preferred a similar pyrolyzer in which gas flow is routed so that flow rates (a) through the quartz tube containing the ferromagnetic wire and then into the SE-52 column, and (b) around the quartz tube and high-frequency coil, could both be adjusted independently by means of the splitting valve of the gas chromatograph. This method still involves splitless injection of pyrolysis products onto the column but allows the purging of the remainder of the structural parts of the pyrolyzer. The procedure allowed detection of fragments of relatively high molecular mass from the pyrolysis of polystyrene, and polar fragments from the pyrolysis of pigment dyes.[422]

The low linear velocity of the carrier gas in the pyrolyzer in the splitless mode, necessary if open tubular columns of more conventional bore are used, may lead to undesirable secondary reactions. Sellier et al.[423] and Pacakova and Kozlik[424] offset this by joining a resistively heated filament or oven pyrolyzer, respectively, to 0.25 or 0.2 mm i.d. open tubular columns through a splitter; the thermal decomposition of C_6 hydrocarbons was studied.[424] Sugimura and Tsuge[425] investigated the fundamental splitting conditions for pyrolysis GC with open tubular columns using an oven pyrolyzer with provision for splitting at the coupling to the column; monodisperse polystyrene and high-density polyethylene were used as examples. Repeated injections of 30–200 µg pyrolyzed sample (compare with ~10 ng when splitless injection is employed) show that degradation product ratios for polystyrene (monomer, dimer, and trimer) were independent of splitting ratio and were highly reproducible as long as the splitting tube was heated to 250°C. For polymers such as polyethylene which produce tarry pyrolysis products, protection of the column by packing the inlet tube (see Figure 4.14) with a support that was coated with the same liquid phase (OV-101) as in the column was necessary.[425] A higher split ratio and smaller sample size also improved quantitation in such instances.

When a specific chemical reaction for the polymer degradation products is required, the inlet-tube packing can be a catalyst. For example, a 10% Pt catalyst in a tube between the pyrolyzer and inlet tube gave[425] a pyrogram exclusively composed of n-alkanes with hydrogen as carrier gas. Without hydrogenation, the pyrogram also contained each α-olefin and α,ω-diolefin (Figure 7.84). An even simpler approach to the pyrolysis/hydrogenation of

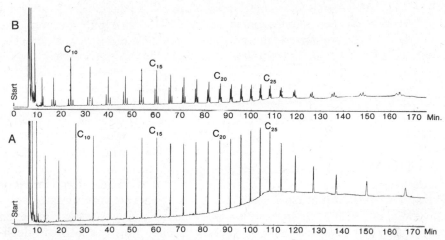

FIGURE 7.84.   Pyrograms of polyethylene (A) with hydrogenation of the degradation products and (B) without hydrogenation. Note: OV-101 glass column, split ratio of 50:1, Pt catalyst at 650°C. (Reproduced with permission from ref. 425. Copyright American Chemical Society.)

polyethylene was employed by Mlejnek[426] who inserted a tube containing 5% Pt catalyst in the injection port of the GC. A Curie-point pyrolyzer containing the sample in a loop of ferromagnetic wire was connected to the GC by penetrating the septum of the injector with a needle fixed to the pyrolyzer. Hydrogen was passed through the pyrolyzer while the sample was heated

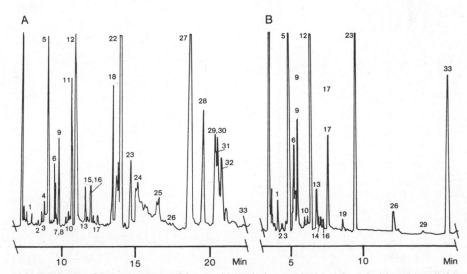

FIGURE 7.85.   Chromatograms of fatty acid methyl esters isolated from Baltic salmon lipids (A) before hydrogenation and (B) after hydrogenation. Note: 65 m OV-101 glass column, 200°C isothermal temperature. (Reproduced with permission from ref. 427. Copyright Dr. Alfred Huethig Publishers.)

for 15 s at 770°C, and the reaction products were cryogenically focused on the first loop of the column cooled in solid CO_2 or liquid nitrogen. The pyrolyzer was disconnected and the column temperature programmed. The branched-alkane products from low- and high-density polyethylenes were markedly different.

Hydrogenation after or during pyrolysis constitutes a variety of reaction chromatography. This technique can be applied to a wide variety of samples in combination with open tubular columns. Thus, injection of a fatty acid–methyl ester mixture (from lipids of Baltic salmon) onto a 1% Pt catalyst located just before the splitter and held at 220–235°C with hydrogen carrier gas allowed[427] identification of the unsaturated compounds (Figure 7.85). Similarly, carbon-skeleton GC[428] involves on-line hydrogenation on a Pd or Pt catalyst packed in a glass liner in the injection port. Mixtures of poly-chlorobiphenyls, polychloronaphthalenes, chlorinated pesticides, and even polychlorinated alkanes[429] can be analyzed, since they are all hydrodech-

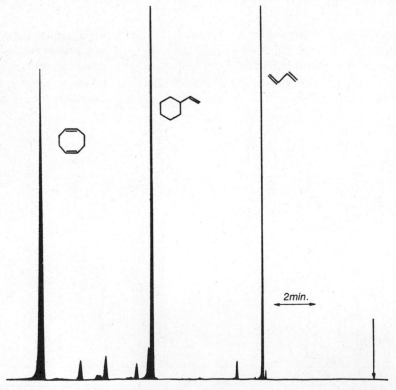

FIGURE 7.86. Chromatogram showing thermal biradical rearrangement in gold tube reactor at 405°C and 5.1 s residence time. Note: 50 m squalane stainless steel column, He carrier gas, 100°C isothermal temperature. (Reproduced with permission from ref. 430. Copyright Friedr. Vieweg and Sohn.)

lorinated to hydrocarbons. Pt catalysts are best for stripping chlorine from alkanes, but Pd is best suited to aromatic systems. The HCl inevitably generated in the reactions has little apparent effect on column performance.

A variety of thermal and catalytic conversions have been carried out in a reaction GC system proposed by Schomburg et al.[430] The products from a flow reactor are trapped and switched into the column via a pneumatically controlled valve system. The application of the system to confirmation of the proposed mechanisms of a chemical reaction is illustrated in Figure 7.86. The thermal dehydrogenation of 1,5-cyclooctadiene to styrene is thought to proceed via vinylcyclohexane, and this is confirmed by identification of this compound as a product when the reactant was passed in helium through a gold reactor tube at 405°C. Vapor phase reaction and subtraction has been applied[231] to the identification of the alkane components of light-oil fractions (up to C_{19}) of shale oil. The shale was distilled in a microfurnace and the products passed through sulfuric acid and molecular sieve 5A traps to remove all but alkanes, and then n-alkanes, respectively. The branched and cyclic alkanes were trapped and cryogenically focused on the first part of the column.

Reaction GC with an analogous apparatus can also be used, of course, in the reverse mode in that catalysts can be directly characterized by the analysis of products of thermal decomposition of single compounds. This approach has been used effectively by Kozlik and Pacakova[431] who have studied the products and degree of conversion from the thermal decomposition and hydrogenation/dehydrogenation on nickel and alumina catalysts.

REFERENCES

1. W. D. Bowers, M. L. Parsons, R. E. Clement, G. A. Eiceman, and F. W. Karasek, *J. Chromatogr.* **206**, 279 (1981).

2. K. Grob and G. Grob, *J. Chromatogr.* **62**, 1 (1971).

3. W. Bertsch, R. C. Chang, and A. Zlatkis, *J. Chromatogr. Sci.* **12**, 175 (1974).

4. A. Raymond and G. Guiochon, *Environ. Sci. Technol.* **8**, 14 (1974).

5. B. V. Joffe, V. A. Isidorov, and I. G. Zenkevich, *J. Chromatogr.* **142**, 787 (1977).

6. G. Holzer, H. Shanfield, A. Zlatkis, W. Bertsch, P. Juarez, H. Mayfield, and H. M. Liebich, *J. Chromatogr.* **142**, 755 (1977).

7. B. Versino, H. Knöppel, M. deGroot, A. Peil, J. Poelman, H. Schauenburg, H. Vissers, and F. Geiss, *J. Chromatogr.* **122**, 373 (1976).

8. M. Novotny, M. L. Lee, and K. D. Bartle, *Chromatographia* **7**, 333 (1974).

9. J. Janak, J. Ruzickova, and J. Novak, *J. Chromatogr.* **99**, 689 (1974).

10. G. Hunt and N. Pangaro, *Anal. Chem.* **54**, 369 (1982).

11. G. Becher, *J. Chromatogr.* **211**, 103 (1981).

12. M. L. Lee, M. Novotny, and K. D. Bartle, *Analytical Chemistry of Polycyclic Aromatic Compounds.* Academic Press, New York, 1981.

13. K. D. Bartle, M. L. Lee, and S. Wise, *Chem. Soc. Rev.* **10**, 113 (1981).

14. G. Lunde and A. Bjørseth, *Nature* **268**, 518 (1977).

15. M. L. Lee and B. W. Wright, *J. Chromatogr. Sci.* **18**, 345 (1980).

16. M. L. Lee, D. L. Vassilaros, C. M. White, and M. Novotny, *Anal. Chem.* **51**, 768 (1979).

17. M. L. Lee, M. Novotny, and K. D. Bartle, *Anal. Chem.* **48**, 1566 (1976).

18. M. L. Lee, D. L. Vassilaros, W. S. Pipkin, and W. L. Sorensen, in *Trace Organic Analysis,* NBS Special Publication 519. U.S. Government Printing Office, Washington, D.C., 1979, p. 731.

19. M. L. Lee, D. L. Vassilaros, and D. W. Later, *Int. J. Environ. Anal. Chem.* **11**, 251 (1982).

20. M. J. Hartigan, J. E. Purcell, M. Novotny, M. L. McConnell, and M. L. Lee, *J. Chromatogr.* **99**, 339 (1974).

21. T. Romanowski, W. Funcke, J. König, and E. Balfanz, *J. High Resoln. Chromatogr./Chromatogr. Commun.* **4**, 209 (1981).

22. P. Wauters, P. Sandra, and M. Verzele, *J. Chromatogr.* **170**, 125 (1979).

23. K. D. Bartle, M. L. Lee, and M. Novotny, *Int. J. Environ. Anal. Chem.* **3**, 349 (1974).

24. M. L. Lee, D. W. Later, D. K. Rollins, D. J. Eatough, and L. D. Hansen, *Science* **207**, 186 (1980).

25. K. Grob, *J. Chromatogr.* **84**, 255 (1973).

26. K. Grob, K. Grob, Jr., and G. Grob, *J. Chromatogr.* **106**, 299 (1975).

27. H. A. James, C. P. Steel, and I. Wilson, *J. Chromatogr.* **208**, 89 (1981).

28. W. Bertsch, E. Anderson, and G. Holzer, *J. Chromatogr.* **112**, 701 (1975).

29. D. Dowty, L. Green, and J. L. Laseter, *J. Chromatogr. Sci.* **14**, 187 (1976).

30. N. V. Brodtmann, Jr., and W. E. Koffskey, *J. Chromatogr. Sci.* **17**, 97 (1979).

31. B. G. Oliver and K. D. Bothen, *Anal. Chem.* **52**, 2066 (1980).

32. G. Eklund, B. Josefsson, and A. Bjørseth, *J. Chromatogr.* **150**, 161 (1978).

33. G. Eklund, B. Josefsson, and C. Roos, *J. High Resoln. Chromatogr./Chromatogr. Commun.* **1**, 34 (1978).

34. B. D. Quimby, M. F. Delaney, P. C. Uden, and R. M. Barnes, *Anal. Chem.* **52**, 259 (1980).

35. M. Reinhard, V. Drevenkar, and W. Giger, *J. Chromatogr.* **116**, 43 (1976).

36. R. Jeltes, E. Burghardt, T. R. Thijsse, and W. A. M. den Tonkelaar, *Chromatographia* **10**, 430 (1977).

37. D. V. Rasmussen, *Anal. Chem.* **48**, 1562 (1976).

38. E. R. Adlard, L. F. Creaser, and P. H. D. Matthews, *Anal. Chem.* **44**, 64 (1972).

39. R. Kadar, K. Nagy, and D. Fremstad, *Talanta* **27**, 227 (1980).

40. W. A. König, K. Ludwig, S. Sievers, M. Rinken, K. H. Stölting, and W. Günther, *J. High Resoln. Chromatogr./Chromatogr. Commun.* **3**, 415 (1980).

41. E. Fogelqvist, B. Josefsson, and C. Roos, *J. High Resoln. Chromatogr./Chromatogr. Commun.* **3**, 568 (1980).

42. A. R. Trussell, J. G. Moncur, F.-Y. Lieu, and L. Y. C. Leong, *J. High Resoln. Chromatogr./Chromatogr. Commun.* **4**, 156 (1981).

43. R. S. Brazell and M. P. Maskarinec, *J. High Resoln. Chromatogr./Chromatogr. Commun.* **4**, 404 (1981).

44. V. Lopez-Avila, *J. High Resoln. Chromatogr./Chromatogr. Commun.* **3**, 545 (1980).

45. E. B. Overton, J. Bracken, and J. L. Laseter, *J. Chromatogr. Sci.* **15**, 169 (1977).

46. L. S. Ramos, D. W. Brown, R. G. Jenkins, and W. D. MacLeod, Jr., in *Trace Organic Analysis,* NBS Special Publication 519. U.S. Government Printing Office, Washington, D.C., 1979, p. 713.

47. W. D. MacLeod, Jr., P. G. Prohaska, D. D. Gennero, and D. W. Brown, *Anal. Chem.* **54**, 386 (1982).

48. R. E. Laflamme and R. A. Hites, *Geochim. Cosmochim. Acta* **42**, 289 (1978).

49. S. G. Wakeham, C. Schaffner, and W. Giger, *Geochim. Cosmochim, Acta* **44**, 403 (1980).

50. Y. L. Tan and M. Heit, in *Polynuclear Aromatic Hydrocarbons*, M. Cooke and A. J. Dennis, editors. Battelle Press, Columbus, Ohio, 1981, p. 561.

51. S. G. Wakeham, C. Schaffner, and W. Giger, *Geochim. Cosmochim. Acta* **44**, 415 (1980).

52. R. L. Harless, E. O. Oswald, M. K. Wilkinson, A. E. Dupuy, Jr., D. D. McDaniel, and H. Tai, *Anal. Chem.* **52**, 1239 (1980).

53. II. R. Buser and C. Rappe, *Anal. Chem.* **52**, 2257 (1980).

54. H. R. Buser, *Anal. Chem.* **49**, 918 (1977).

55. J. J. Franken and H. L. Vader, *Chromatographia* **6**, 22 (1973).

56. H. Steinwandter, *Fres. Z. Anal. Chem.* **304**, 137 (1980).

57. M. Cooke and A. G. Ober, *J. Chromatogr.* **195**, 265 (1980).

58. J. Krupcik, P. A. Leclercq, A. Simova, P. Suchanek, M. Collak, and J. Hrivnak, *J. Chromatogr.* **119**, 271 (1976).

59. M. Zell, H. J. Neu, and K. Ballschmiter, *Fres. Z. Anal. Chem.* **292**, 97 (1978).

60. K. Ballschmiter and M. Zell, *Fres. Z. Anal. Chem.* **302**, 20 (1980).

61. H. Tausch, G. Stehlik, and H. Wihlidal, *Chromatographia* **14**, 403 (1981).

62. L. G. M. T. Tuinstra, W. A. Traag, and H. J. Keukens, *J. Assoc. Offic. Anal. Chem.* **63**, 952 (1980).

63. L. G. M. T. Tuinstra and W. A. Traag, *J. High Resoln. Chromatogr./Chromatogr. Commun.* **2**, 723 (1979).

64. J. Krupcik, P. A. Leclercq, J. Garaj, and A. Simova, *J. Chromatogr.* **191**, 207 (1980).

65. R. L. Holmstead, S. Khalifa, and J. E. Casida, *J. Agric. Food Chem.* **22**, 939 (1974).

66. J. N. Seiber, P. F. Landrum, S. C. Madden, K. D. Nugent, and W. L. Winterlin, *J. Chromatogr.* **114**, 361 (1975).

67. M. A. Salch and J. E. Casida, *J. Agric. Food Chem.* **26**, 583 (1978).

68. M. Wolf, R. Deleu, and A. Copin, *J. High Resoln. Chromatogr./Chromatogr. Commun.* **4**, 346 (1981).

69. W. Krijgsman and C. G. van de Kamp, *J. Chromatogr.* **117**, 201 (1976).

70. J. Stan, *Chromatographia* **10**, 233 (1977).

71. J. Hild, E. Schulte, and H. P. Thier, *Chromatographia* **11**, 397 (1978).

72. M. DePotter, R. Muller, and J. Willems, *Chromatographia* **11**, 220 (1978).

73. T. Wehner and J. N. Seiber, *J. High Resoln. Chromatogr./Chromatogr. Commun.* **4**, 348 (1981).

74. R. Deleu and A. Copin, *J. High Resoln. Chromatogr./Chromatogr. Commun.* **3**, 299 (1980).

75. E. Matisova, J. Krupcik, O. Liska, and N. Szentivanyi, *J. Chromatogr.* **169**, 261 (1979).

76. E. Matisova and J. Krupcik, *J. Chromatogr.* **205**, 464 (1981).

77. K. Grob, Jr., *J. Chromatogr.* **208**, 217 (1981).

78. A. deKok, I. M. Roorda, R. W. Frei, and U. A. T. Brinkman, *Chromatographia* **14**, 579 (1981).

79. T. Spitzer and G. Nickless, *J. High Resoln. Chromatogr./Chromatogr. Commun.* **4**, 151 (1981).

80. G. Nickless, J. Spitzer, and J. A. Pickard, *J. Chromatogr.* **208**, 409 (1981).

81. B. W. Wright, M. L. Lee, and G. M. Booth, *J. High Resoln. Chromatogr./Chromatogr. Commun.* **2**, 189 (1979).

82. M. L. Lee and K. D. Bartle, in *Particulate Carbon Formation During Combustion*, D. C. Siegla and G. W. Smith, editors. Plenum Press, New York, 1981, p. 91.

83. A. Bjørseth and G. Lunde, *Anal. Chim. Acta* **94,** 21 (1977); *Atmos. Environ.* **13,** 45 (1979).

84. M. L. Lee, G. P. Prado, J. B. Howard, and R. A. Hites, *Biomed. Mass Spectrom.* **4,** 182 (1977).

85. R. Jeltes, *J. Chromatogr. Sci.* **12,** 599 (1974).

86. J. M. Daisey, M. A. Leyko, and T. J. Kneip, in *Polynuclear Aromatic Hydrocarbons,* P. W. Jones and P. Leber, editors. Ann Arbor Science Publishers, Ann Arbor, Michigan, 1979, p. 201.

87. K. D. Bartle, A. I. El-Nasri, and B. Frere, *Am. Chem. Soc. Symp. Ser.,* in press.

88. H. Borwitzky and G. Schomburg, *J. Chromatogr.* **170,** 99 (1979).

89. L. Sucre, W. Jennings, G. L. Fisher, O. G. Raabe, and J. Olechno, in *Trace Organic Analysis,* NBS Special Publication 519. U.S. Government Printing Office, Washington, D.C., 1979, p. 109.

90. D. G. Nichols, S. K. Gangwal, and C. M. Sparacin, in *Polynuclear Aromatic Hydrocarbons,* M. Cooke and A. J. Dennis, editors. Battelle Press, Columbus, Ohio, 1981, p. 397.

91. L. D. Hansen, L. R. Phillips, N. F. Mangelson, and M. L. Lee, *Fuel* **59,** 323 (1980).

92. G. Prado, P. R. Westmoreland, B. M. Andon, J. A. Leary, K. Biemann, W. G. Thilly, J. P. Longwell, and J. B. Howard, in *Polynuclear Aromatic Hydrocarbons,* M. Cooke and A. J. Dennis, editors. Battelle Press, Columbus, Ohio, 1981, p. 189.

93. S. Krishnan and R. A. Hites, *Anal. Chem.* **53,** 342 (1981).

94. Y. Hirata, M. Novotny, P. A. Peaden, and M. L. Lee, *Anal. Chim. Acta* **127,** 55 (1981).

95. U. R. Stenberg and T. E. Alsberg, *Anal. Chem.* **53,** 2067 (1981).

96. M. L. Lee and R. A. Hites, *Anal. Chem.* **48,** 1890 (1976).

97. M. D. Erickson, D. L. Newton, E. D. Pellizzari, and K. B. Tomer, *J. Chromatogr. Sci.* **17,** 449 (1979).

98. Y. Hoshika and Y. Takata, *J. Chromatogr.* **120,** 379 (1976).

99. M. Yu and R. A. Hites, *Anal. Chem.* **53,** 951 (1981).

100. F. I. Onuska, A. W. Wolkoff, M. E. Comba, R. H. Larose, M. Novotny, and M. L. Lee, *Anal. Lett.* **9,** 451 (1976).

101. L. S. Ramos and P. G. Prohaska, *J. Chromatogr.* **211,** 284 (1981).

102. P. Berthou, Y. Gourmelun, Y. Dreano, and M. P. Friocourt, *J. Chromatogr.* **203,** 279 (1981).

103. S. A. Wise, S. N. Chesler, F. R. Guenther, H. S. Hertz, L. R. Hilpert, W. E. May, and R. M. Parris, *Anal. Chem.* **52,** 1828 (1980).

104. K. Grob and G. Grob, *Chromatographia* **5,** 3 (1972).

105. R. F. Severson, K. L. McDuffie, R. F. Arrendale, G. R. Gwynn, J. F. Chaplin, and A. W. Johnson, *J. Chromatogr.* **211,** 111 (1981).

106. M. Novotny, M. L. Lee, C.-E. Low, and M. P. Maskarinec, *Steroids* **27,** 665 (1976).

107. K. D. Bartle and M. Novotny, *Beitr. z. Tabakforschung* **5,** 215 (1970).

108. B. W. Good, M. E. Parrish, and D. R. Douglas, *J. High Resoln. Chromatogr./Chromatogr. Commun.* **3,** 447 (1980).

109. M. E. Parrish, B. W. Good, F. S. Hsu, F. W. Hatch, D. M. Ennis, D. R. Douglas, J. H. Shelton, D. C. Watson, and C. N. Reilley, *Anal. Chem.* **53,** 826 (1981).

110. C. R. Enzell, E. Bergstedt, T. Dalhamn, and W. H. Johnson, *Beitr. z. Tabakforschung* **6,** 41 (1971).

111. C. R. Enzell, E. Bergstedt, T. Dalhamn, and W. H. Johnson, *Beitr. z. Tabakforschung* **6,** 96 (1972).

112. G. Holzer, J. Oro, and W. Bertsch, *J. Chromatogr.* **126,** 771 (1976).

113. K. Grob, *Chemy Ind.* 248 (1973).

114. K. Grob, *Chromatographia* **7,** 94 (1974).

115. L. Blomberg, *J. Chromatogr.* **125,** 389 (1976).

116. W. Bertsch, F. Hsu, and A. Zlatkis, *Anal. Chem.* **48,** 928 (1976).

117. L. Blomberg and G. Widmark, *J. Chromatogr.* **106,** 59 (1975).

118. W. Bertsch, F. Sunbo, R. C. Chang, and A. Zlatkis, *Chromatographia* **7,** 128 (1974).

119. M. L. Lee, M. Novotny, and K. D. Bartle, *Anal. Chem.* **48,** 405 (1976).

120. M. Novotny, M. L. Lee, and K. D. Bartle, *J. Chromatogr.* **12,** 606 (1974).

121. M. Novotny, M. L. Lee, and K. D. Bartle, *Experientia* **32,** 280 (1976).

122. J. M. Schumacher, C. R. Green, F. W. Best, and M. P. Newell, *J. Agric. Food Chem.* **25,** 310 (1977).

123. M. Malaterre, J. Loheac, N. Sellier, and G. Guiochon, *Chromatographia* **8,** 624 (1975).

124. Y. Saint-Jalm and P. Moree-Testa, *J. Chromatogr.* **198,** 188 (1980).

125. R. F. Severson, R. F. Arrendale, and O. T. Chortyk, *J. High Resoln. Chromatogr./Chromatogr. Commun.* **3,** 11 (1980).

126. F. Merli, M. Novotny, and M. L. Lee, *J. Chromatogr.* **199,** 371 (1980).

127. W. N. Sanders and J. B. Maynard, *Anal. Chem.* **40,** 527 (1968).

128. J. Simekova, N. Pronayova, R. Pies, and M. Ciha, *J. Chromatogr.* **51,** 91 (1970).

129. E. R. Adlard, A. W. Bowen, and D. G. Salmon, *J. Chromatogr.* **186,** 207 (1979).

130. K. Petrovic and D. Vitorovic, *J. Chromatogr.* **65,** 155 (1972).

131. W. Jennings, *Gas Chromatography with Glass Capillary Columns,* 2nd edition. Academic Press, New York, 1980, p. 246.

132. K. Grob and G. Grob, *J. High Resoln. Chromatogr./Chromatogr. Commun.* **2,** 109 (1979).

133. K. Grob, Jr. and K. Grob, *Chromatographia* **10,** 250 (1977).

134. N. G. Johansen, *Chromatogr. Newslett.* (to be published).

135. P. Kumar, S. L. S. Sarowha, and P. L. Gupta, *Analyst* **104,** 788 (1979).

136. M. Herrera, E. Murgia, M. Dettazos, and J. Lubkowitz, *J. High Resoln. Chromatogr./Chromatogr. Commun.* **4,** 297 (1981).

137. E. J. Gallegos, I. M. Whittemore, and R. F. Klaver, *Anal. Chem.* **46,** 157 (1974).

138. J. T. Swansiger and F. E. Dickson, *Anal. Chem.* **45,** 811 (1973).

139. J. Mostecky, M. Popl, and J. Kriz, *Anal Chem.* **42,** 1132 (1970).

140. R. R. Freeman, *High Resolution Gas Chromatography,* 2nd edition. Hewlett-Packard Company, Avondale, Pennsylvania, 1981, p. 168.

141. G. Dielmann, S. Meier, and U. Rapp, *J. High Resoln. Chromatogr./Chromatogr. Commun.* **2,** 343 (1979).

142. P. C. Anderson, J. M. Sharkey, and R. P. Walsh, *J. Inst. Pet.* **58,** 83 (1972).

143. B. W. Jackson, R. W. Judges, and J. L. Powell, *J. Chromatogr. Sci.* **14,** 49 (1976).

144. J. M. Schmitter, Z. Vajta, and P. J. Arpino, in *Advances in Organic Geochemistry 1979,* A. G. Douglas and J. R. Maxwell, editors. Pergamon, Oxford, 1980, p. 67.

145. J. Shen, *Anal. Chem.* **53,** 475 (1981).

146. J. M. Schmitter, P. Arpino, and G. Guichon, *J. Chromatogr.* **167,** 149 (1978).

147. F. P. DiSanzo, *J. High Resoln. Chromatogr./Chromatogr. Commun.* **4,** 649 (1981).

148. M. L. Lee, K. D. Bartle, and M. V. Novotny, *Anal. Chem.* **47,** 540 (1975).

149. G. Grimmer, J. Jacob, K. W. Naujack, and G. Dettbarn, *Fres. Z. Anal. Chem.* **309,** 13 (1981).

150. G. Grimmer, J. Jacob, and K. W. Naujack, *Fres. Z. Anal. Chem.* **306,** 347 (1981).

151. G. Grimmer and H. Bohnke, *Chromatographia* **9**, 30 (1976).

152. E. R. Adlard, L. F. Creaser, and H. D. Matthews, *Anal. Chem.* **44**, 64 (1972).

153. L. V. McCarthy, E. B. Overton, M. A. Maberry, S. A. Antoine, and J. L. Laseter, *J. High Resoln. Chromatogr./Chromatogr. Commun.* **4**, 164 (1981).

154. M. Novotny, F. J. Schwende, M. J. Hartigan, and J. E. Purcell, *Anal. Chem.* **52**, 736 (1980).

155. F. J. Schwende, M. Novotny, and J. E. Purcell, *Chromatogr. Newslett.* **8**, 1 (1980).

156. M. Novotny, M. L. McConnell, and M. L. Lee, *J. Agric. Food Chem.* **22**, 765 (1974).

157. J. E. Picker and R. E. Sievers, *J. Chromatogr.* **217**, 275 (1981).

158. J. D. Twibell, J. M. Home, and K. W. Smalldon, *Chromatographia* **14**, 366 (1981).

159. R. D. Schwartz, R. G. Mathews, and D. J. Brasseaux, *J. Chromatogr. Sci.* **5**, 251 (1967).

160. C. L. Stuckey, *J. Chromatogr. Sci.* **9**, 575 (1971).

161. D. R. Dean, *J. Chromatogr.* **203**, 19 (1981).

162. P. A. Schenck and C. H. Hall, *Anal. Chim. Acta* **38**, 65 (1967).

163. M. G. Bloch, R. B. Callen, and J. H. Stockinger, *J. Chromatogr. Sci.* **15**, 504 (1977).

164. H. A. Clark and P. C. Jurs, *Anal. Chem.* **51**, 616 (1979).

165. H. Wehner and M. Teschner, *J. Chromatogr.* **204**, 481 (1981).

166. J. G. Pym, J. E. Ray, G. W. Smith, and E. V. Whitehead, *Anal. Chem.* **47**, 1617 (1975).

167. F. Berthou, Y. Gourmelyn, Y. Dreano, and M. P. Friocourt, *J. Chromatogr.* **203**, 279 (1981).

168. L. V. S. Hood and C. M. Erikson, *J. High Resoln. Chromatogr./Chromatogr. Commun.* **3**, 516 (1980).

169. R. D. Cole, *Nature (London)* **233**, 546 (1971).

170. A. P. Bentz, *Anal. Chem.* **48**, 454A (1976).

171. D. V. Rasmussen, *Anal. Chem.* **48**, 1562 (1976).

172. F. K. Kawahara, *J. Chromatogr. Sci.* **10**, 629 (1972).

173. E. B. Overton, J. Bracken, and J. L. Laseter, *J. Chromatogr. Sci.* **15**, 169 (1977).

174. R. Jeltes, E. Burghardt, Th. R. Thijsse, and W. A. M. den Tonkelaar, *Chromatographia* **10**, 430 (1977).

175. D. A. Murray and W. L. Lockhart, *J. Chromatogr.* **212**, 305 (1981).

176. M. L. Gay, A. A. Belisle, and J. F. Patton, *J. Chromatogr.* **187**, 153 (1980).

177. F. K. Schweighardt, in *Coal Conversion and the Environment*, D. D. Mahlum, R. H. Gray, and W. D. Felix, editors, DOE Symposium Series 54. Technical Information Center, Washington, D.C., 1981, p. 1.

178. D. W. Kopponaal and S. E. Manaham, *Environ. Sci. Technol.* **10**, 1104 (1976).

179. R. Hayatsu, R. Winans, R. G. Scott, L. P. Moore, and M. H. Studier, *Fuel* **57**, 541 (1978).

180. M. L. Lee and B. W. Wright, *J. Chromatogr. Sci.* **18**, 345 (1980).

181. P. A. Peaden, B. W. Wright, and M. L. Lee, *Chromatographia* **15**, 335 (1982).

182. G. Schomburg, R. Dielman, H. Borwitzky, and H. Husmann, *J. Chromatogr.* **167**, 337 (1978).

183. H. Borwitzky, G. Schomburg, H.-D. Sauerland, and M. Zander, *Erdol, Kohle, Erdgas Petrochemie* **31**, 371 (1978).

184. R. C. Kong, M. L. Lee, Y. Tominaga, R. Pratap, M. Iwao, and R. N. Castle, *Anal. Chem.* **54**, 1802 (1982).

185. H. Finkelmann, R. J. Laub, B. Luhmann, A. Price, W. L. Roberts, and C. A. Smith, *Macromolecules*, in preparation.

186. M. Novotny, J. W. Strand, S. L. Smith, D. Wiesler, and F. J. Schwende, *Fuel* **60**, 213 (1981).

187. K. Alben, *Anal. Chem.* **52**, 1825 (1980).

188. P. Buryan, J. Macak, and V. M. Nabivach, *J. Chromatogr.* **148**, 203 (1978).

189. W. Bertsch, E. Anderson, and G. Holzer, *J. Chromatogr.* **126**, 213 (1976).

190. R. S. Brazell and M. P. Maskarinec, *J. High Resoln. Chromatogr./Chromatogr. Commun.* **4**, 404 (1981).

191. C. E. Higgins, *Anal. Chem.* **53**, 732 (1981).

192. K. D. Bartle, *Rev. Pure Appl. Chem.* **22**, 79 (1972).

193. R. V. Schultz, J. W. Jorgenson, M. P. Maskarinec, M. Novotny, and L. J. Todd, *Fuel* **58**, 783 (1979).

194. F. P. Burke, R. A. Winschel, and T. C. Pochapsky, *Fuel* **60**, 562 (1981).

195. D. W. Later, M. L. Lee, K. D. Bartle, R. C. Kong, and D. L. Vassilaros, *Anal. Chem.* **53**, 1612 (1981).

196. D. W. Later and M. L. Lee, in *Proceedings of DOE Workshop*, Seattle, Washington, November, 1981.

197. M. V. Buchanan, C. Ho, M. R. Guerin, and B. R. Clark, in *Polycyclic Aromatic Hydrocarbons*, Vol. 5, M. Cook and A. J. Dennis, editors. Battelle Press, Columbus, Ohio, 1981, p. 133.

198. M. V. Buchanan, *Anal. Chem.* **54**, 574 (1982).

199. D. W. Later, M. L. Lee, and B. W. Wilson, *Anal. Chem.* **54**, 117 (1982).

200. D. W. Later, M. L. Lee, R. A. Pelroy, and B. W. Wilson, in *Polycyclic Aromatic Hydrocarbons*, Vol. 6, M. Cook and A. J. Dennis, editors. Battelle Press, Columbus, Ohio, 1982, p. 427.

201. B. W. Wilson, C. Willey, D. W. Later, and M. L. Lee, *Fuel* **61**, 473 (1982).

202. C. E. Snape, B. J. Stokes, and K. D. Bartle, *Fuel* **60**, 903 (1981).

203. F. R. Guenther, R. M. Parris, S. N. Chesler, and L. R. Hilpert, *J. Chromatogr.* **207**, 256 (1981).

204. C. M. White and N. C. Li, *Anal. Chem.* **54**, 1564 (1982).

205. C. M. White and N. C. Li, *Anal. Chem.* **54**, 1570 (1982).

206. S. Kapila and C. R. Vogt, *J. High Resoln. Chromatogr./Chromatogr. Commun.* **4**, 233 (1981).

207. R. F. Brady, *Anal. Chem.* **47**, 1425 (1975).

208. D. M. Hembree, A. A. Garrison, R. A. Crocombe, R. A. Yokley, E. L. Wehry, and G. Mamantov, *Anal. Chem.* **53**, 1783 (1981).

209. C. Willey, M. Iwao, R. N. Castle, and M. L. Lee, *Anal. Chem.* **53**, 400 (1981).

210. M. L. Lee, C. Willey, R. N. Castle, and C. M. White, in *Polycyclic Aromatic Hydrocarbons: Fourth International Symposium on Analysis, Chemistry, and Biology*, Vol. 4, A. Bjørseth and A. J. Dennis, editors. Battelle Press, Columbus, Ohio, 1980, p. 59.

211. F. J. Yang and S. P. Cram, *J. High Resoln. Chromatogr./Chromatogr. Commun.* **2**, 487 (1979).

212. L. J. Felice, *Anal. Chem.* **54**, 869 (1982).

213. D. W. Later, B. W. Wright, and M. L. Lee, *J. High Resoln. Chromatogr./Chromatogr. Commun.* **4**, 406 (1981).

214. S. K. Gangwal, *J. Chromatogr.* **204**, 439 (1981).

215. J. Solash, R. N. Hazlett, J. M. Hall, and C. J. Nowack, *Fuel* **57**, 521 (1978).

216. G. Alexander and I. Hazai, *J. Chromatogr.* **217**, 19 (1981).

217. R. C. Duty, R. Hayatsu, R. G. Scott, L. P. Moore, R. E. Winans, and M. H. Studier, *Fuel* **59**, 97 (1980).

218. N. C. Deno, K. W. Curry, B. A. Greigger, A. D. Jones, W. G. Rakitsky, K. A. Smith, K. Wagner, and R. D. Minard, *Fuel* **59**, 694 (1980).

219. R. Hayatsu, R. E. Winans, R. G. Scott, and R. L. McBeth, *Fuel* **60**, 158 (1981).

220. J. A. Franz and W. E. Skiens, *Fuel* **57**, 503 (1978).

221. B. W. Wilson, M. R. Peterson, R. A. Pelroy, and J. T. Cresto, *Fuel* **60**, 289 (1981).

222. R. P. Philp, N. J. Russell, and T. D. Gilbert, *Fuel* **60**, 939 (1981).

223. F. P. DiSanzo, P. C. Uden, and S. Siggia, *Anal. Chem.* **52**, 906 (1980).

224. W. E. May, J. Brown-Thomas, L. R. Hilpert, and S. A. Wise, in *Polycyclic Aromatic Hydrocarbons*, Vol. 5, M. Cook and A. J. Dennis, editors. Battelle Press, Columbus, Ohio, 1981, p. 1.

225. H. S. Hertz, J. M. Brown, S. N. Chesler, F. R. Guenther, L. R. Hilpert, W. E. May, R. M. Parris, and S. A. Wise, *Anal. Chem.* **52**, 1650 (1980).

226. R. J. Crowley, S. Siggia, and P. C. Uden, *Anal. Chem.* **52**, 1224 (1980).

227. F. F. Shue and T. F. Yen, *Anal. Chem.* **53**, 2081 (1981).

228. R. G. Riley, K. Shiosaki, R. M. Bean, and D. M. Schoengold, *Anal. Chem.* **51**, 1995 (1979).

229. R. H. Fish, A. S. Newton, and P. C. Babbit, *Fuel* **61**, 227 (1982).

230. E. Gelpi, P. C. Wszolek, E. Yang, and A. L. Burlingame, *J. Chromatogr. Sci.* **9**, 147 (1971).

231. F. P. DiSanzo, P. C. Uden, and S. Siggia, *Anal. Chem.* **51**, 1529 (1979).

232. D. van de Meent, S. C. Brown, R. P. Philp, and B. R. T. Simoneit, *Geochim. Cosmochim. Acta* **44**, 999 (1980).

233. G. Schomburg, H. Husmann, and H. Behlau, *J. Chromatogr.* **203**, 179 (1981).

234. G. Gabri and F. Chialva, *J. High Resoln. Chromatogr./Chromatogr. Commun.* **4**, 215 (1981).

235. M. Novotny, M. L. McConnell, and M. L. Lee, *J. Agric. Food Chem.* **22**, 765 (1974).

236. A. E. Purcell, D. W. Later, and M. L. Lee, *J. Agric. Food Chem.* **28**, 939 (1980).

237. W. G. Jennings, *J. High Resoln. Chromatogr./Chromatogr. Commun.* **2**, 221 (1979).

238. D. Nurok, J. W. Anderson, and A. Zlatkis, *Chromatographia* **11**, 188 (1978).

239. J. Hlavay, A. Bartha, G. Vigh, M. Gazdag, and G. Szepesi, *J. Chromatogr.* **204**, 59 (1981); *Chromatographia* **14**, 296 (1981).

240. W. Jennings, *J. Chromatogr. Sci.* **17**, 636 (1979).

241. G. Schomburg, H. Husmann, and F. Weeke, *J. Chromatogr.* **112**, 205 (1975).

242. P. Sandra and M. Van Roelenbosch, *Chromatographia* **14**, 345 (1981).

243. T. H. Parliment and M. D. Spencer, *J. Chromatogr. Sci.* **17**, 435 (1981).

244. G. Swords and G. L. K. Hunter, *J. Agric. Food Chem.* **26**, 734 (1978).

245. M. Verzele, G. Maes, A. Vuye, M. Godefroot, M. Van Alboom, J. Vervisch, and P. Sandra, *J. Chromatogr.* **205**, 367 (1981).

246. P. Sandra, T. Saeed, G. Redant, M. Godefroot, M. Verstappe, and M. Verzele, *J. High Resoln. Chromatogr./Chromatogr. Commun.* **3**, 107 (1980).

247. K. J. Chaffler and P. G. Morel Du Boil, *J. Chromatogr.* **207**, 221 (1981).

248. I. M. Moodie and J. Burger, *J. High Resoln. Chromatogr./Chromatogr. Commun.* **4**, 218 (1981).

249. J. Flanzy, M. Boudon, C. Leger, and J. Pihet, *J. Chromatogr. Sci.* **14**, 17 (1976).

250. H. Jaeger, H. U. Klör, G. Blos, and H. Ditschuneit, *Chromatographia* **8**, 507 (1975).

251. R. G. Ackman and S. N. Hooper, *J. Chromatogr. Sci.* **12**, 131 (1974).

252. K. Grob, H. P. Neukom, D. Fröhlich, and R. Battaglia, *J. High Resoln. Chromatogr./Chromatogr. Commun.* **1**, 94 (1978).

253. T. Kobayashi, *J. Chromatogr.* **194**, 404 (1980).

254. E. S. Van Vleet and J. G. Quinn, *J. Chromatogr.* **151**, 396 (1978).

255. R. G. Ackman and C. A. Eaton, *Fette, Seifen, Anstrichmittel* **80**, 21 (1978).

256. A. Monseigny, P. Y. Vigneron, M. Levacq, and F. Zwodoba, *Rev. Fr. Corps Gras* **3**, 107 (1979).

257. K. Grob, Jr., *J. Chromatogr.* **178**, 387 (1979).

258. P. P. Schmid, M. D. Müller, and W. Simon, *J. High Resoln. Chromatogr./Chromatogr. Commun.* **2**, 675 (1979).

259. K. Grob, H. P. Neukom, and R. Battaglia, *J. Am. Oil Chem. Soc.* **57**, 282 (1980).

260. H. Traitler and A. Prévot, *J. High Resoln. Chromatogr./Chromatogr. Commun.* **4**, 109 (1981).

261. R. J. Hamilton, *J. Chromatogr. Sci.* **13**, 474 (1975).

262. J. Geigert, P. Dalietos, and S. L. Neidleman, *J. High Resoln. Chromatogr./Chromatogr. Commun.* **3**, 473 (1980).

263. J. H. Henion, J. S. Nosanchuk, and B. M. Bilder, *J. Chromatogr.* **213**, 475 (1981).

264. W. Rahn and W. A. König, *J. High Resoln. Chromatogr./Chromatogr. Commun.* **1**, 69 (1978).

265. H. Brechbühler, L. Gay, and H. Jaeger, *Chromatographia* **10**, 478 (1977).

266. J. H. Stan and B. Abraham, *J. Chromatogr.* **195**, 231 (1980).

267. K. Grob, G. Grob, and K. Grob, *Chromatographia* **10**, 181 (1977).

268. K. Grob, H. P. Neukom, and H. Kaderli, *J. High Resoln. Chromatogr./Chromatogr. Commun.* **1**, 98 (1978).

269. W. G. Jennings and M. Filsoof, *J. Agric. Food Chem.* **25**, 440 (1977).

270. A. Rapp and W. Knipser, *Chromatographia* **13**, 698 (1980).

271. M. Novotny and A. Zlatkis, *J. Chromatogr. Sci.* **8**, 346 (1970); *Chromatogr. Rev.* **14**, 1 (1971).

272. A. Völlmin, *Chromatographia* **3**, 233 (1970).

273. M. Novotny and A. Zlatkis, *J. Chromatogr.* **56**, 353 (1971).

274. E. Bailey, M. Fenoughty, and J. R. Chapman, *J. Chromatogr.* **96**, 33 (1974).

275. F. L. Berthou, D. Picart, L. G. Bardou, and H. H. Floch, *J. Chromatogr. Sci.* **12**, 662 (1974).

276. A. L. German and E. C. Horning, *J. Chromatogr. Sci.* **11**, 76 (1973).

277. M. Delaforge, B. F. Maume, P. Bournot, M. Prost, and P. Padieu, *J. Chromatogr. Sci.* **12**, 545 (1974).

278. C. D. Pfaffenberger and E. C. Horning, *J. Chromatogr.* **112**, 581 (1975).

279. G. Alexander and G. A. F. M. Rutten, *Chromatographia* **6**, 231 (1973).

280. J. S. Zweig, R. Roman, W. B. Hagerman, and W. J. A. VandenHeuvel, *J. High Resoln. Chromatogr./Chromatogr. Commun.* **3**, 169 (1980).

281. P. Sandra, M. Verzele, and E. Vanluchene, *J. High Resoln. Chromatogr./Chromatogr. Commun.* **2**, 187 (1979).

282. C. Madani, E. M. Chambaz, M. Rigaud, J. Durand, and P. Chebroux, *J. Chromatogr.* **126**, 161 (1976).

283. J. A. Ballantine, K. Williams, and R. J. Morris, *J. Chromatogr.* **166**, 491 (1978).

284. C. H. L. Shackleton, E. Roitman, C. Monder, and H. L. Bradlow, *Steroids* **36**, 289 (1980).

285. H. Ludwig, J. Reiner, and G. Spiteller, *Chem. Ber.* **110**, 217 (1977).

286. H. Kern and B. Brander, *J. High Resoln. Chromatogr./Chromatogr. Commun.* **2**, 312 (1979).

287. J.-P. Thenot and A. Hung, in *Trace Organic Analysis*, NBS Special Publication 519 U.S. Government Printing Office, Washington, D.C., 1979, p. 419.

288. J. Reiner and G. Spiteller, *Monatsh.* **106**, 1415 (1975).

289. P. Pfeifer and G. Spiteller, *J. Chromatogr. Biomed. Appl.* **223**, 21 (1981).

290. W. J. J. Brunissen and J. H. H. Thijssen, *J. Chromatogr. Biomed. Appl.* **146**, 365 (1978).

291. H. Ludwig, G. Spiteller, D. Matthaei, and F. Scheler, *J. Chromatogr. Biomed. Appl.* **146**, 381 (1978).

292. M. Novotny, M. P. Maskarinec, A. T. G. Steverink, and R. Farlow, *Anal. Chem.* **48**, 468 (1976).

293. M: Axelson, B.-L. Sahlberg, and J. Sjövall, *J. Chromatogr. Biomed. Appl.* **224**, 355 (1981).

294. B. A. Peterson and P. Vouros, *Anal. Chem.* **49**, 1304 (1977).

295. J. A. Corkill and R. W. Giese, *Anal. Chem.* **53**, 1667 (1981).

296. J. Maclouf, M. Rigaud, J. Durand, and P. Chebroux, *Prostaglandins* **11**, 999 (1976).

297. F. A. Fitzpatrick, *Anal. Chem.* **50**, 47 (1978).

298. J. Maclouf, H. de la Baume, J. Caen, H. Rabinovitch, and M. Rigaud, *Anal. Biochem.* **109**, 147 (1980).

299. F. A. Fitzpatrick, D. A. Stringfellow, J. Maclouf, and M. Rigaud, *J. Chromatogr.* **177**, 51 (1979).

300. J. Rosello, E. Gelpi, M. Rigaud, and J. C. Breton, *J. High Resoln. Chromatogr./Chromatogr. Commun.* **4**, 437 (1981).

301. R. W. Walker, V. F. Gruber, J. Pile, K. Yabumoto, A. Rosegay, D. Taub, M. L. Orme, F. J. Wolf, and W. J. A. VandenHeuvel, *J. Chromatogr.* **181**, 85 (1980).

302. J. M. Halket, H. K. Albers, and B. P. Lisboa, *Fres. Z. Anal. Chem.* **290**, 124 (1978).

303. M. Suzuki, I. Morita, M. Kawamura, S. I. Murota, M. Nishizawa, T. Miyatake, H. Nagase, K. Ohno, and H. Shimizu, *J. Chromatogr. Biomed. Appl.* **221**, 361 (1980).

304. C. Jakobs, E. Solem, J. Ek, K. Halvorsen, and E. Jellum, *J. Chromatogr. Biomed. Appl.* **143**, 31 (1977).

305. M. Spiteller and G. Spiteller, *J. Chromatogr. Biomed. Appl.* **164**, 253 (1979).

306. H. M. Liebich, A. Pickert, V. Stierle, and J. Wöll, *J. Chromatogr. Biomed. Appl.* **199**, 181 (1980).

307. T. Niwa, *Clin. Chim. Acta* **94**, 71 (1979).

308. S. Lewis, C. N. Kenyon, J. Meili, and A. L. Burlingame, *Anal. Chem.* **51**, 1275 (1979).

309. J. Chauhan and A. Darbre, *J. Chromatogr. Biomed. Appl.* **183**, 391 (1980).

310. J. P. Kammerling, G. J. Gerwig, J. F. G. Vliegenhart, M. Duran, D. Ketting, and S. K. Wadman, *J. Chromatogr. Biomed. Appl.* **143**, 117 (1977).

311. T. Niwa, K. Maeda, T. Ohki, A. Saito, and I. Tsuchida, *J. Chromatogr. Biomed. Appl.* **225**, 1 (1981).

312. A. C. Schoots, F. E. P. Mikkers, and C. A. M. G. Cramers, *J. Chromatogr. Biomed. Appl.* **164**, 1 (1979).

313. D. Pinkston, G. Spiteller, H. von Henning, and D. Matthaei, *J. Chromatogr. Biomed. Appl.* **223**, 1 (1981).

314. D. Issachar, J. F. Holland, and C. C. Sweeley, *Anal. Chem.* **54**, 29 (1982).

315. S. I. Goodman, P. Helland, O. Stokke, A. Flatmark, and E. Jellum, *J. Chromatogr.* **142**, 497 (1977).

316. S. N. Lin and E. C. Horning, *J. Chromatogr.* **112**, 465 (1975).

317. H. Heckers, F. W. Melcher, and U. Schloeder, *J. Chromatogr.* **136**, 311 (1977).

318. J. Desgres, D. Boisson, and P. Padieu, *J. Chromatogr. Biomed. Appl.* **162**, 133 (1979).

319. R. F. Adams, F. L. Vandemark, and G. J. Schmidt, *J. Chromatogr. Sci.* **15**, 63 (1977).

320. H. Frank, A. Eimiller, H. H. Kornhuber, and E. Bayer, *J. Chromatogr. Biomed. Appl.* **224**, 177 (1981).

321. P. Storset, O. Stokke, and E. Jellum, *J. Chromatogr. Biomed. Appl.* **145**, 351 (1978).

322. E. C. Horning, M. G. Horning, J. Szafranek, P. Van Hout, A. L. German, J. P. Thienot, and C. D. Pfaffenberger, *J. Chromatogr.* **91**, 367 (1974).

323. S. L. Smith, M. V. Novotny, and A. Karmen, *J. Chromatogr. Biomed. Appl.* **223**, 173 (1981).

324. C. D. Pfaffenberger, J. Szafranek, and E. C. Horning, *J. Chromatogr.* **126**, 535 (1976).

325. A. Zlatkis and H. M. Licbich, *Clin. Chem.* **7**, 592 (1971).

326. R. Teranishi, T. R. Mon, P. Cary, A. B. Robinson, and L. Pauling, *Anal. Chem.* **44**, 18 (1972).

327. A. Zlatkis, A. Lichtenstein, and A. Tishbee, *Chromatographia* **6**, 67 (1973).

328. H. M. Liebich, *J. Chromatogr.* **112**, 551 (1975).

329. H. M. Liebich, O. Al-Babbili, G. Huesgen, and J. Wöll, *Fres. Z. Anal. Chem.* **279**, 148 (1976).

330. H. M. Liebich and O. Al-Babbili, *J. Chromatogr.* **112**, 539 (1975).

331. M. L. McConnell and M. Novotny, *J. Chromatogr.* **112**, 559 (1975).

332. M. L. McConnell, G. Rhodes, U. Watson, and M. Novotny, *J. Chromatogr.* **162**, 495 (1979).

333. A. Zlatkis, H. A. Lichtenstein, A. Tishbee, W. Bertsch, F. Shunbo, and H. M. Liebich, *J. Chromatogr. Sci.* **11**, 200 (1973).

334. A. Zlatkis, W. Bertsch, D. A. Bafus, and H. M. Liebich, *J. Chromatogr.* **91**, 379 (1974).

335. H. M. Liebich and J. Wöll, *J. Chromatogr.* **142**, 505 (1977).

336. K. Y. Lee, D. Nurok, and A. Zlatkis, *J. Chromatogr.* **158**, 377 (1978).

337. A. Zlatkis, C. F. Poole, R. Brazell, K. Y. Lee, and S. Singhawangcha, *J. High. Resoln. Chromatogr./Chromatogr. Commun.* **2**, 423 (1979).

338. A. Zlatkis, C. F. Poole, R. Brazell, D. A. Bafus, and P. S. Spencer, *J. Chromatogr. Biomed. Appl.* **182**, 137 (1980).

339. A. Zlatkis, K. Y. Lee, C. F. Poole, and G. Holzer, *J. Chromatogr. Biomed. Appl.* **163**, 125 (1979).

340. K. Halpaap, M. G. Horning, and E. C. Horning, *J. Chromatogr.* **166**, 479 (1978).

341. E. Bailey, L. Della Corte, P. B. Farmer, and A. J. Gray, *J. Chromatogr. Biomed. Appl.* **225**, 83 (1981).

342. W. J. A. VandenHeuvel and J. S. Zweig, *J. High Resoln. Chromatogr./Chromatogr. Commun.* **3**, 381 (1980).

343. K. H. Dudley, in *Trace Organic Analysis,* NBS Special Publication 519. U.S. Government Printing Office, Washington, D.C., 1979, p. 381.

344. H. W. Durbeck, I. Büker, B. Scheulen, and B. Telin, *J. Chromatogr.* **167**, 117 (1978).

345. G. Andermann and M. Dietz, *J. Chromatogr. Biomed. Appl.* **223**, 365 (1981).

346. M. T. Rosseel and M. G. Bogaert, *J. Pharm. Sci.* **68**, 659 (1979).

347. P. O. Edlund, *J. Chromatogr.* **187**, 161 (1980).

348. J. R. Carlin, R. W. Walker, R. O. Davies, R. F. Ferguson, and W. J. A. VandenHeuvel, *J. Pharm. Sci.* **69**, 1111 (1980).

349. D. DeBruyne, H. Kinsun, M. A. Moulin, and M. C. Bigot, *J. Pharm. Sci.* **68**, 511 (1979).

350. M. Guerret, *J. Chromatogr. Biomed. Appl.* **221**, 387 (1980).

351. S. Caccia, C. Chiabrando, P. DePonte, and R. Fanelli, *J. Chromatogr. Sci.* **16**, 543 (1978).

352. H. Frank, G. J. Nicholson, and E. Bayer, *J. Chromatogr. Biomed. Appl.* **146**, 197 (1978).

353. H. Frank, G. J. Nicholson, and E. Bayer, *J. Chromatogr. Sci.* **15**, 174 (1977); *Angew. Chem.* **90**, 396 (1978).

354. H. Brückner, G. J. Nicholson, G. Jung, K. Kruse, and W. A. König, *Chromatographia* **13**, 209 (1980).

355. B. Kinberger, A. Holmen, and P. Wahrgren, *J. Chromatogr. Biomed. Appl.* **224**, 449 (1981).

356. P. Sandra, M. Van Den Broeck and M. Verzele, *J. High Resoln. Chromatogr./Chromatogr. Commun.* **3**, 196 (1980).

357. W. Dunges, R. Langlais, and R. Schlenkermann, *J. High Resoln. Chromatogr./Chromatogr. Commun.* **2**, 361 (1979).

358. A. S. Christophersen and K. E. Rasmussen, *J. Chromatogr.* **192**, 363 (1980).

359. A. S. Christophersen and K. E. Rasmussen, *J. Chromatogr.* **174**, 454 (1979).

360. P. O. Edlund, *J. Chromatogr.* **206**, 109 (1981).

361. M. Lauwereys and A. Vercruysse, *Chromatographia* **9**, 520 (1976).

362. A. G. de Boer, J. B. Smeekens, and D. D. Breimer, *J. Chromatogr. Biomed. Appl.* **162**, 591 (1979).

363. U. Bondesson and P. Hartvig, *J. Chromatogr.* **179**, 207 (1979).

364. F. N. Pitts, L. S. Yago, O. Aniline, and A. F. Pitts, *J. Chromatogr.* **193**, 157 (1980).

365. A. G. de Boer, J. Rost-Kaiser, H. Bracht, and D. D. Breimer, *J. Chromatogr. Biomed. Appl.* **145**, 105 (1978).

366. N. P. E. Vermeulen, M. W. E. Teunissen, and D. D. Breimer, *J. Chromatogr.* **157**, 133 (1978).

367. J. Dow and K. Hall, *J. Chromatogr.* **153**, 521 (1978).

368. B. Kinberger, A. Holmen, and P. Wahrgren, *J. Chromatogr.* **207**, 148 (1981).

369. M. T. Rosseel, M. G. Bogaert, and M. Claeys, *Fres. Z. Anal. Chem.* **290**, 158 (1978).

370. E. Bailey and E. J. Barron, *J. Chromatogr. Biomed. Appl.* **183**, 25 (1980).

371. V. Romei, M. Sanjuan, and P. D. Hrdina, *J. Chromatogr. Biomed. Appl.* **182**, 349 (1980).

372. E. Bailey, M. Fenoughty, and L. Richardson, *J. Chromatogr.* **131**, 347 (1977).

373. J. D. Warthen, R. M. Waters, and D. J. Voaden, *Chromatographia* **10**, 720 (1977).

374. W. Schäfer, *J. High Resoln. Chromatogr./Chromatogr. Commun.* **1**, 71 (1978).

375. R. R. Heath, J. R. Jordan, P. E. Sonnet, and J. H. Tumlinson, *J. High Resoln. Chromatogr./Chromatogr. Commun.* **2**, 712 (1979).

376. H. R. Buser and H. Arn, *J. Chromatogr.* **106**, 83 (1975).

377. L. Huibregtse-Minderhout, A. C. Van Der Kerk-Van Hoof, P. Wijkens, H. W. A. Biessels, and C. A. Salemink, *J. Chromatogr.* **196**, 425 (1980).

378. D. B. Drucker, *J. Chromatogr.* **208**, 279 (1981).

379. C. Wayne-Moss, *J. Chromatogr.* **203**, 337 (1981).

380. L. Larsson, P. A. Mardh, G. Odham, and G. Westerdahl, *J. Chromatogr. Biomed. Appl.* **182**, 402 (1980).

381. M. Martinez, D. Nurok, A. Zlatkis, D. McQuittay, and J. Evans, *J. High Resoln. Chromatogr./Chromatogr. Commun.* **3**, 528 (1980).

382. M. L. Lee, D. L. Smith, and L. R. Freeman, *Appl. Environ. Microbiol.* **37**, 85 (1979).

383. E. Lanza, K. H. Ko, and J. K. Palmer, *J. Agric. Food Chem.* **24**, 1247 (1976).

384. K. Wasserfallen and F. Rinderknecht, *Chromatographia* **11**, 128 (1978).

385. H. L. C. Meuzelaar and R. A. in't Veld, *J. Chromatogr. Sci.* **10**, 213 (1972).

386. D. R. Thouvenot and R. F. Morfin, *J. Chromatogr.* **170**, 165 (1979).

387. M. Claeys, E. Messens, M. Van Montagu, and J. Schell, *Fres. Z. Anal. Chem.* **290**, 125 (1978).

388. T. R. Kemp and R. A. Andersen, *J. Chromatogr.* **209**, 467 (1981).

389. P. Fischer, G. R. Lösch, and D. Müller, *J. High Resoln. Chromatogr./Chromatogr. Commun.* **3**, 161 (1980).

390. J. R. Maxwell, C. T. Pillinger, and G. Eglinton, *Q. Rev. Chem. Soc.* **25**, 571 (1971).

391. K. D. Bartle, D. W. Jones, and H. Pakdel, in *Analytical Methods for Coal and Coal Products,* Vol. 2, C. Karr, Jr., editor. Academic, New York, 1978, Chap. 5.

392. H. Pichler, W. Ripperger, and G. Schwartz, *Erdöl Kohle* **23**, 91 (1970).

393. B. M. Didyk, B. R. T. Simoneit, S. C. Brassell, and G. Eglinton, *Nature* **272**, 216 (1978).

394. D. H. Welte, H. Kratochvil, J. Rullkötter, H. Ladwein, and R. G. Schaefer, *Chem. Geol.* **35**, 33 (1982).

395. J. Albaiges, in *Advances in Organic Geochemistry 1979,* A. G. Douglas and J. R. Maxwell, editors. Pergamon, Oxford, 1980, p. 19.

396. I. Rubinstein, O. P. Strausz, C. Spyckerelle, R. J. Crawford, and D. W. S. Westlake, *Geochim. Cosmochim. Acta* **41**, 1341 (1977).

397. W. K. Seifert and J. M. Moldowan, *Geochim. Cosmochim. Acta* **42**, 77 (1978).

398. A. Ensminger, A. van Dorsselaer, C. Spyckerelle, P. Albrecht, and G. Ourisson, in *Advances in Organic Geochemistry 1973,* B. Tissot and F. Bienner, editors. Technip, Paris, 1974, p. 245.

399. J. Allan, M. Bjoroy, and A. G. Douglas, in *Advances in Organic Geochemistry 1975,* R. Campos and J. Goni, editors. Enadimsa, Madrid, 1977, p. 633.

400. S. G. Wakcham, C. Schaffner, and W. Giger, *Geochim. Cosmochim. Acta* **44**, 415 (1980).

401. R. E. Laflamme and R. A. Hites, *Geochim. Cosmochim. Acta* **42**, 289 (1978).

402. C. M. White and M. L. Lee, *Geochim. Cosmochim. Acta,* **44**, 1825 (1980).

403. J. W. de Leeuw, F. W. van der Meer, W. I. C. Rijpstra, and P. A. Schenk, in *Advances in Organic Geochemistry 1979,* A. G. Douglas and J. R. Maxwell, editors. Pergamon, Oxford, 1980, p. 211.

404. J. K. Volkman, G. Eglinton, E. D. S. Corner, and J. R. Sargant, in *Advances in Organic Geochemistry 1979,* A. G. Douglas and J. R. Maxwell, editors. Pergamon, Oxford, 1980, p. 219.

405. P. I. Grantham and A. G. Douglas, in *Advances in Organic Geochemistry 1975,* R. Campos and J. Goni, editors. Enadimsa, Madrid, 1977, p. 193.

406. G. J. Shaw, R. A. Franick, G. Douglas, J. Allan, and A. G. Douglas, in *Advances in Organic Geochemistry 1979,* A. G. Douglas and J. R. Maxwell, editors. Pergamon, Oxford, 1980, p. 281.

407. J. M. Gilbert and J. Oro, *J. Chromatogr. Sci.* **8**, 295 (1970).

408. W. Bertsch, A. Zlatkis, H. M. Liebich, and H. J. Schneider, *J. Chromatogr.* **99**, 673 (1974).

409. J. G. Lawless, K. A. Kvenvolden, E. Peterson, and C. Ponnamperuma, *Science* **173**, 626 (1971).

410. K. A. Kvenvolden, J. G. Lawless, K. Pering, E. Peterson, J. Flores, C. Ponnamperuma, I. R. Kaplan, and C. Moore, *Nature (London)* **228**, 923 (1970).

411. J. G. Lawless, K. A. Kvenvolden, E. Peterson, C. Ponnamperuma, and E. Jarosewich, *Nature (London)* **236**, 66 (1972).

412. P. H. Buhl, in *Trace Organic Analysis,* NBS Special Publication 519. U.S. Government Printing Office, Washington, D.C., 1979, p. 771.

413.   G. E. Pollock, *Anal. Chem.* **44,** 2368 (1972).

414.   M. H. Engel and B. Nagy, *Nature (London)* **296,** 837 (1982).

415.   D. Willson, *Forensic Sci.* **10,** 243 (1977).

416.   L. W. Russell, *J. Forensic Sci. Soc.* **21,** 317 (1981).

417.   M. Novotny, M. L. Lee, C. E. Low, and A. Raymond, *Anal. Chem.* **48,** 24 (1976).

418.   J. F. M. Douse, *J. Chromatogr.* **208,** 83 (1981).

419.   J. B. F. Lloyd and B. R. G. Roberts, *J. Chromatogr.* **77,** 228 (1973).

420.   H. L. C. Meuzelaar, H. G. Ficke, and H. C. den Harink, *J. Chromatogr. Sci.* **13,** 12 (1975).

421.   J. W. de Leeuw, W. L. Maters, D. van de Meent, and J. J. Boon, *Anal. Chem.* **49,** 1881 (1977).

422.   J. P. Schmid, P. P. Schmid, and W. Simon, *Chromatographia* **9,** 597 (1976).

423.   N. Sellier, C. E. R. Jones, and G. Guiochon, *J. Chromatogr. Sci.* **13,** 383 (1975).

424.   V. Pacakova and V. Kozlik, *Chromatographia* **11,** 266 (1978).

425.   Y. Sugimura and S. Tsuge, *Anal. Chem.* **50,** 1968 (1978).

426.   O. Mlejnek, *J. Chromatogr.* **191,** 181 (1980).

427.   T. E. Kuzmenko, A. L. Samusenko, V. P. Uralets, and R. V. Golovnya, *J. High Resoln. Chromatogr./Chromatogr. Commun.* **2,** 43 (1979).

428.   M. Cooke and D. J. Roberts, *J. Chromatogr.* **193,** 437 (1980).

429.   D. J. Roberts, M. Cooke, and G. Nickless, *J. Chromatogr.* **213,** 73 (1981).

430.   G. Schomburg, R. Rienäcker, and R. G. Schaefer, *Chromatographia* **5,** 532 (1972).

431.   V. Kozlik and V. Pacakova, *Chromatographia* **14,** 417 (1981).

INDEX